T0295396

IET CONTROL, ROBOTICS AND SENSORS SERIES 131

Space Robotics and Autonomous Systems

Technologies, advances and applications

Other volumes in this series:

Space Robotics and Autonomous Systems

Technologies, advances and applications

Edited by
Yang Gao

The Institution of Engineering and Technology

Published by The Institution of Engineering and Technology, London, United Kingdom

The Institution of Engineering and Technology is registered as a Charity in England & Wales (no. 211014) and Scotland (no. SC038698).

The Institution of Engineering and Technology
Michael Faraday House
Six Hills Way, Stevenage
Herts, SG1 2AY, United Kingdom

www.theiet.org

British Library Cataloguing in Publication Data

A catalogue record for this product is available from the British Library

ISBN 978-1-83953-225-2 (Hardback)
ISBN 978-1-83953-226-9 (PDF)

Typeset in India by Exeter Premedia Services Private Limited
Printed in the UK by CPI Group (UK) Ltd, Croydon

Contents

PART IV System engineering

11 Verification for space robotics **377**
Rafael C. Cardoso, Marie Farrell, Georgios Kourtis, Matt Webster,
Louise A. Dennis, Clare Dixon, Michael Fisher, and Alexei Lisitsa

List of figures

List of tables

Foreword

Alistair Scott[1]

I am honoured to have been asked by Professor Yang Gao to contribute to this fascinating book. As *Space Robotics and Autonomous Systems* is primarily a textbook and, knowing of my lifelong passion for promoting wider education in astronautics, she invited me to write the Foreword. The words Robotics and Autonomy just did not exist in my childhood vocabulary or my schooling and even my early working life, but I have found this book fascinating, revealing and certainly educational.

Back in the mid-1950s, I remember hearing the unmistakeable rattle of the abacus as the local shopkeepers near our home in Bangkok totted up our bills. Our little primary school tried to teach us how to use the abacus, but I never did get the hang of it. However, technology was catching up with us fast, as I was able to watch my first TV programme there, though it was in Siamese, and our telephone system was self-dial rather than operator connected. The most advanced piece of equipment I used, apart from the transistor radio, during my school days back in Scotland, was the slide-rule, which became an essential tool at the start of my working life and at university.

In the 1960s, more rapid progress was being made on automation in aviation. The Auto-Land system was developed on the Trident airliner and came into service while I was working on the Trident production line at Hawker Siddeley Aviation. Hatfield. I think the biggest problem for the pilots was finding their way off the runway after the plane had landed in thick fog! A bit later, in 1968, when I joined the A300B Airbus wing design team, we were still using hand-drawn drawings, graph-paper and slide-rules. Though a huge mainframe computer had just been installed, I, as an apprentice, was not allowed near it and had to share two mechanical calculators, one electric-drive and the other hand-cranked, with the 100 other engineers in the department. The first Sinclair calculator actually arrived when I was there, but you needed a magnifying glass to read the LED display and the batteries only lasted 45 minutes!

What a transformation we have seen in the 50 years since. The automation I valued then has become the autonomy of the present and the future. But what I did not realise, until I moved from aircraft and missiles into Space systems in the mid-1980s and later joined the British Interplanetary Society, was where and when many of the ideas that made the use of Space feasible originated. Some of the ideas came from Science Fiction writers like Arthur C. Clarke and comic book stories like Dan Dare in the Eagle.

[1]Alistair D. Scott, TD, BSc, FBIS, MRAeS - UK Board Member, The Arthur C. Clarke Foundation, Past President, The British Interplanetary Society, alistair.scott@bis-space.com

Photo courtesy of HSA

Figure 1   The trident 3B production line, Hawker Siddeley Aviation, Hatfield

Arthur C. Clarke was one of the early members of the British Interplanetary Society (BIS) in the mid-1930s and members of the BIS were understood to have been advisors to the Eagle comic in the 1950s. In 1938 the BIS, egged on by Jules Verne's 1865 book *From the Earth to the Moon* and H.G. Wells' 1901 book *The First Men in the Moon*, set, as one of its first technical studies, the design of a Lunar Spaceship to carry a crew of three to land on the Moon and, after 14 days, return safely to Earth.

Though a few members of the BIS team had some experience in aircraft design and propulsion systems, they were starting with a blank sheet and so had to use their imaginations to create a conceptual design that would complete this challenging task using the materials and technologies available at the time. They opted for a 6-step rocket. The first 5 launch stages, each containing 168 large solid rocket motors, would be discarded after firing. The 6th stage had 45 medium motors and 1 200 small motors, which were to be controlled by the crew of the Spaceship for landing on the Moon, for lift-off 14 days later and finally as retro rockets for Earth re-entry.

It's all a far cry from the robotic and autonomous systems available today and proposed in Part I of this book, but that's how things were in the 1930s and 50s. They were actually way ahead of their time and it wasn't until the 60's that the same 'four-landing-leg' design proposed for the 1939 Lunar Lander was used on the Apollo Landers and the cooling system worn by the Apollo astronauts was similar to that proposed by the BIS for its Moon Suit in 1952.

The BIS continues to look to the future with technical study projects on interstellar travel, space colonies, nuclear propulsion and, closer to home, small launch vehicles and UK spaceports.

When I started working with communications satellites and scientific spacecraft in the 1980s and 1990s, I thought them to be the most sophisticated of robots.

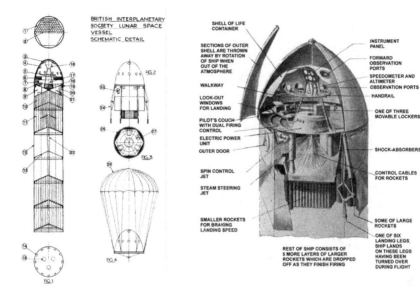

Drawings courtesy of BIS

Figure 2   The 1938 BIS lunar   Figure 3   The 1939 BIS lunar lander
spaceship

They could travel millions of miles, survive the harshest of conditions, from the high structural loadings and vibrations in launch to temperature extremes and even meteorite damage in Space, and still operate and provide essential services, from scientific exploration to communications, broadcasting, navigation and Earth observation. However, though they often had some automated sub-systems like

Photo courtesy of MMS

Figure 4   A Eurostar 2000 communications satellite

thermal control and solar array drive mechanisms, they all required some level of control or instruction from the ground for navigation or station-keeping and instrument control or transponder switching. Now, as indicated in Part II of the book, it appears that everything and anything is possible, even down to servicing and refuelling satellites in orbit to extend their lifetimes.

Space Robotics and Autonomous Systems (RAS) have advanced significantly, as revealed on almost every page of the book. This has been a bit of an eye opener for me. Where else could one find such a range of technological concepts and developments on everything from planetary rover chassis design, pneumatic muscle actuators and the use of biologically evolved mechanisms, to autonomous visual navigation for on-orbit servicing, the control challenges for the capture of orbital objects and the autonomous robotic grasping for in-orbit assembly or debris removal.

I am pleased to say there is a lot more, as in Part III of the book, one can also learn about the brain–computer interface technology for monitoring mental states and fatigue, the wearable technology for biosignal monitoring and the future of human–robotic interaction in space. We face testing times both here on Earth and in exploring our Universe. Artificial Intelligence, Autonomy and Robotics are essential tools, but we must be sure that they work as we expect them to and are safe and secure if we are to succeed in these challenging missions.

The book ends with further chapters that highlight system-level technologies critical for the future of all space activities – the verification, validation and cybersecurity of space RAS within the NewSpace era.

I would like to take this opportunity to congratulate Professor Yang Gao on consolidating such a diverse and interesting range of research topics that tackle the challenging subject of space RAS and thank the authors for their excellent and fascinating contributions. It is a good, insightful read. There is something for academia and industry, as well as anyone who is interested in RAS and space.

About the editor

Yang Gao is the Professor of Space Autonomous Systems at Surrey Space Centre of the University of Surrey, UK. Prof. Gao founded and heads the multi-awards winning Space Technology for Autonomous and Robotic systems Laboratory (STAR LAB), which specializes in robotic sensing, perception, visual GNC and biomimetic mechanisms for industrial applications in extreme environments. She has been the Principal Investigator of internationally teamed projects funded by UK Research Councils, InnovateUK, Royal Academy of Engineering, European Commission, European Space Agency (ESA), UK Space Agency, as well as industrial companies. She has also been actively involved in real-world space missions such as ESA's ExoMars, Proba3 and VMMO, UK's MoonLITE/Moonraker, and CNSA Chang'E3.

Prof. Gao is an elected Fellow of the Institute of Engineering and Technology (IET) and the Royal Aeronautical Society (RAeS). She was named by the Times Higher Education in 2008 as one of ten UK's young leading academics who are making a very significant contribution to their disciplines, and was also awarded the Mulan Award in 2019 for Contributions to Science, Technology and Engineering. Research work under her leadership and supervision has also received many international recognitions such as the IAF's 3AF Edmond Brun Silver Medal, COSPAR's Outstanding Paper Award, Top places in ESA Grand Challenges, etc.

Prof. Gao holds a B. Eng (1st Hons) and Ph.D. on electrical and control engineering from the Nanyang Technological University, Singapore.

Chapter 1

Introduction

Yang Gao[1]

The current desire to go and explore space is as strong as ever. Past space powers have been gradually joined by a flurry of new nations eager to test and demonstrate their technologies and contribute to an increasing body of knowledge. Space robotics and autonomous systems (RAS) are important to human's overall ability to explore or operate in space, by providing greater access beyond human spaceflight limitations in the harsh environment of space and operational handling that extends astronauts' capabilities. RAS can help reduce the cognitive load on humans given the abundance of information that has to be reasoned upon in a timely fashion and hence are critical for improving human and systems' safety. RAS can also enable the deployment and operation of multiple assets without the same order of magnitude increase in ground support. Given the potential reduction to the cost and risk of spaceflight both manned and robotic, space RAS are deemed relevant across all mission phases such as development, flight system production, launch, and operation [1].

This chapter introduces the book by providing the basis of space RAS, such as key technological challenges, relevant applications over the horizon as well as the recent advances to be presented in the remainder of the book.

1.1 Technologies

Modern space RAS represents a multidisciplinary field that builds on as well as contributes to the knowledge of space engineering (e.g. software and hardware harness, system engineering, and space qualifications), terrestrial robotics (e.g. sensing and perception, mobility and locomotion, and navigation), computer science (e.g. mission planning, machine learning, and soft computing) as well as many other miscellaneous subjects like advanced materials, information technologies, and bionics.

The goals of next-generation space RAS are to extend humanity's reach, exploration and exploitation of space, expand our ability to manipulate assets and resources in outer space, prepare them for human arrival, support human crews in their space operations,

[1]STAR LAB, Surrey Space Centre, University of Surrey, Guildford, United Kingdom, yang.gao@surrey.ac.uk

manage the assets they leave behind, and enhance the efficiency of mission operations across the board. Commercial endeavors already have eyes on space and actively promote the Moon and Mars as possible destinations for long-term human presence or habitation. Advances in robotic sensing and perception, mobility and manipulation, rendezvous and docking, onboard and ground-based autonomous capabilities, and human–robot interoperability will drive these goals. Furtherance of these goals can benefit the public good in a broader sense beyond the attainment of purely practical ends: Space flight has a hold on the public's imagination unlike that of any other realms of scientific endeavor; it is a barometer of national prestige as underpinned by industrial and technological sophistication; it can inspire wider participation in Science, Technology, Engineering, and Mathematics (STEM)-oriented educational programs.

The National Aeronautics and Space Administration (NASA) in its latest technology roadmap has identified several RAS areas needed by 2035. Similarly, the European Space Agency (ESA) has been developing technology roadmaps in RAS through various European Commission-funded projects such as Space Robotics Strategic Research Cluster and SpacePlan2020. Other spacefaring nations like Russia, China, India, and Japan have also announced their individual plans on future missions involving space RAS. Apart from the difference in mission timetable by different space players, there are quite a number of common technological needs or challenges in RAS that are widely acknowledged by the international space community. These technological topics typically include:

Mobility to reach and operate at sites of scientific interest on extra-terrestrial surfaces or in free-space environments (see Table 1.1 for existing successfully flown robotic mobility systems): mobility on, into, and above an extra-terrestrial surface using locomotion like flying, walking, climbing, rappelling, tunneling, swimming, and sailing; and manipulations to make intentional changes in the environment or objects using locomotion like placing, assembling, digging, trenching, drilling, sampling, grappling, and berthing.

Sensing and perception to provide situational awareness for space robotic agents, explorers, and assistants: new sensors; sensing techniques; algorithms for 3D perception, state estimation, and data fusion; onboard data processing and generic software framework; and object, event, or activity recognition.

High-level autonomy for system and subsystems to provide robust and safe autonomous navigation, provide rendezvous and docking capabilities, and enable extended-duration operations without human interventions to improve overall performance of human and robotic missions: guidance, navigation, and control (GNC) algorithms; docking and capture mechanisms and interfaces; planning, scheduling and common autonomy software framework; multi-agent coordination; reconfigurable and adjustable autonomy; and automated data analysis for decision-making.

Human–robot interaction (HRI) to enable humans to accurately and rapidly understand the state of the robot in collaboration and act effectively and efficiently toward the goal state: multi-modal interaction; remote and supervised control; proximate interaction; distributed collaboration and coordination; and common human–system interfaces.

System engineering to provide a framework for understanding and coordinating the complex interactions of robots and achieving the desired system requirements:

Table 1.1   *Successfully flown robotic mobility systems on Earth orbit, the Moon, Mars, and small bodies as of 2020, updated based on tables in [2, 3]*

Launch year	Mission name	Country	Target	Robotic mobility
1967	Surveyor 3	USA	Moon	Sampler
1970/72/76	Luna 16/20/24	USSR	Moon	Arm, drill, sampler
1970/73	Luna 17/21	USSR	Moon	Rover
1975	Viking	USA	Mars	Arm, sampler
1981/2001/08	ISS Canadarm1/2/Dextre	Canada	Earth orbit	Arm
1996	Mars Path Finder	USA	Mars	Rover
2003	Mars Express (Beagle2*)	Europe	Mars	Arm, drill, sampler
2003	Hayabusa	Japan	Asteroid	Sampler
2003	Mars Exploration Rovers	USA	Mars	Rovers, arm, sampler
2007	ISS Kibo	Japan	Earth orbit	Arm
2008	Phoenix	USA	Mars	Arm, sampler
2011	Mars Science Laboratory	USA	Mars	Rover, arm, sampler
2013	Chang'E 3	China	Moon	Rover
2004	Rosetta (Philae)	Europe	Comet	Arm, drill, sampler
2016	Aolong-1	China	Earth orbit	Arm
2018	Osiris-Rex Sample Return	USA	NEA	Arm, sampler
2018	Insight	USA	Mars	Arm, drill, sampler
2018	Chang'E 4	China	Moon-far side	Rover
2020	Mars 2020	USA	Mars	Rover, arm, helicopter
2020	Chang'E 5 Sample Return	China	Moon	Arm, drill, sampler
2020	Tianwen-1	China	Mars	Rover

*Beagle2 lander lost communication after landing. Onboard robotic mobility did not have the chance to operate.

modularity, commonality, and interfaces; verification and validation of complex adaptive systems; robot modeling and simulation; software architectures and frameworks; and safety and trust.

1.2   Applications

Space RAS covers all types of robotics for surface exploration or in orbit around the moon, planets, or other small bodies such as asteroids. They include sensors and platforms for mobility and navigation as well as for deployment of science instruments in space (see Figure 1.1). Depending on the space environments applied, *orbital robots* can be used for repairing satellites, assembling large space telescopes, capturing and returning asteroids, or deploying assets for scientific investigations, while *planetary robots* play a key role in the surveying, observation, extraction, close examination of extra-terrestrial surfaces (including natural phenomena, terrain composition, and resources), constructing infrastructures on a planetary surface for subsequent human arrival, mining planetary resources, or in situ resource utilization (ISRU) [1].

Depending on these applications (either orbital or planetary), space robots are often designed to possess *mobility* (or locomotion) to manipulate, grip, rove, drill, and/or

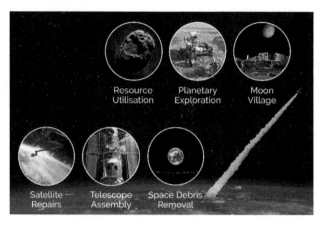

Figure 1.1   Applications and mission scenarios of space RAS, from orbital to inter-planetary [1]

sample. Driven similarly by the nature of the mission and distance from the Earth, these robots are expected to possess varying *levels of autonomy*, ranging from tele-operation by humans to fully autonomous operation by the robots themselves. Depending on the level of autonomy, space robots can act as (1) robotic agents (or human proxy) in space to perform various tasks using tele-operation up to semiautonomous operation or (2) robotic assistants that can help human astronauts to perform tasks quickly and safely, with higher quality and cost efficiency using semi- to fully autonomous operation. Some future missions will require the coworking between the robotic assistants and astronauts to achieve goals that would not be possible without such a collaboration; or (3) robotic explorers that are capable of exploring unknown territories in space using fully autonomous operation. Coordination or cooperation between autonomous robotic explorers is also envisaged within multi-robot missions to enable complex tasks such as cave exploration, construction, and resource extraction [1].

It is worth noting that the global space sector is currently moving toward the NewSpace era driven by space commercialization and resource exploitation, where RAS will play a central role and be directly responsible for meeting stringent requirements in cost, operability, reusability, and sustainability of long-lived assets in the harsh space environments. This paradigm shift is also echoed by fast emerging applications on Earth that rely on data and information from in-space assets for future smart cities, industry 4.0 (4th industrial revolution), disaster and climate change monitoring, and homeland security, etc.

1.3   Recent advances

This book presents a range of recent research outcomes beyond the state-of-the-art that covers the key technological topics described in Section 1.1 and key space applications mentioned in Section 1.2. We provide below a summary of each subsequent book chapter to help guide readers to navigate through the book in the most efficient way.

1.3.1 Part I: mobility and mechanisms

In the first part of the book, three chapters are presented to include the latest reviews, R & D work, and findings in relation to space robotic mobility and mechanism designs. These works also relate to a wide range of space applications covering both planetary and orbital environments, as well as for human spaceflight.

Chapter 2[1] entitled "Wheeled planetary rover locomotion design, scaling and analysis" offers an introduction to the novel granular scaling laws (GSL) for wheeled planetary rover design. This recently developed approach examines how to predict the performance of larger, more massive vehicles from the study of smaller vehicles. We evaluate how material properties, wheel shape, wheel position, sinkage, angular velocity, and preparation of granular media influence these predictive laws through experimental case studies. We conclude by using coupled multibody dynamics and discrete element method (MBD-DEM) simulations to examine gravity variant GSL for predicting the performance of a craft at reduced gravity.

Chapter 3[2] entitled "Compliant pneumatic muscle structures and systems for extra-vehicular and intra-vehicular activities in space environments" offers detailed discussions on how the current robotic mobility used in space environments for intra- and extra-vehicular activities can be improved to rely on more convenient structures that offer equivalent or better services. These improved structures involve one of the popular branches of soft robotics, namely pneumatic muscle actuators (PMAs). The robotic mobility systems based on PMAs have provided promising results that can potentially replace their rigid robotic counterparts. Especially it is recognized that PMAs have a higher power to weight ratio, have inherently soft properties useful for safety when interacting with humans, are compliant to colliding with objects of different shapes, are less costly in materials used to make them, and are flexible to perform motions that might be a challenge with rigid systems. The chapter presents and discusses the PMA designs on multi-fingered grippers using the self-bending contraction actuator (SBCA) and ring-shaped circular gripper that uses a circular pneumatic muscle actuator (CPMA). An extensor bending PMA (EBPMA) is used to design a power assistive glove that can be implemented for spacesuit gloves.

Chapter 4[3] entitled "Biologically-inspired mechanisms for space applications" highlights the implementation of the field of biomimetics in the development of mechanisms and mobility systems for a wide variety of space applications. This chapter collates as many of these concepts as possible into a single review study, detailing their journeys from the initial biological inspiration to the latest design and development

[1]Corresponding author: Hamid Marvi, School for Engineering of Matter, Transport & Energy, Arizona State University, 501 E. Tyler Mall, Tempe, AZ 85287, USA. Email: hmarvi@asu.edu.
[2]Corresponding author: Haitham El-Hussieny, School of Science, Engineering & Environment, The University of Salford, Salford, M54WT, UK. Email: h.e.h.a.hussien@salford.ac.uk.
[3]Corresponding author: Craig Pitcher, STAR LAB, Surrey Space Centre, University of Surrey, Guildford, GU2 7XH, UK. Email: craig.pitcher@surrey.ac.uk.

iteration, organized by which application they had been considered for in space. First to be discussed is *subsurface planetary exploration*, with drilling and mole concept designs based upon insect ovipositors and peristaltic motion. Planetary and spacecraft *surface mobility systems* have implemented gecko-inspired adhesive pads, legged and hopping locomotion mechanisms based upon spiders, scorpions, locusts, and kangaroos, as well as techniques inspired by plants including tumbleweed designs, vine-based climbing in continuum robots, and root growth processes in probes. *Object capturing mechanisms* have also used gecko-inspired adhesion techniques and a kangaroo-based vibration suppression system. Different classes of *artificial muscle actuator*, which mimic the properties of muscles, have been used in several robotic concepts, and several wing-flapping mechanisms inspired by the lift generation methods employed by insects have been designed in *aerial mobility* systems for flying on Mars. *Navigation systems* for surface traversal navigation explored methods for interfacing living organic matter into robotic systems, whereas vehicles flying around and landing on extraterrestrial bodies explored the ways insects use optic flow to solve complex problems. *Multi-agent system architectures*, including swarms of animals and the processes that govern biological cells, have been implemented into the architectures of small satellite mission concepts and scenarios. Finally, induced *hibernation for human spaceflight* has been proposed for long-term manned missions.

1.3.2   Part II: sensing, perception, and GNC

The second part of the book includes three chapters on spacecraft visual perception and GNC algorithms. These works represent the latest R & D results and demonstrations of beyond state-of-the-art solutions for robotic manipulation tasks, primarily within orbital applications such as dealing with cooperative targets (e.g. satellite servicing and space telescope assembly) and/or noncooperative targets (e.g. active debris removal). This part of the book puts focus on orbital manipulation applications but in principle the presented work can be extended or applied for manipulation on planetary surface environment too. As for GNC of other robotic platforms such as the planetary rovers, readers can refer to [3].

Chapter 5[4] entitled "Autonomous visual navigation for spacecraft on-orbit operations" presents the background on the spacecraft pose estimation and the shift in trend toward the machine learning algorithms that can help achieve 6-DOF (degree of freedom) relative orbital navigation using only monocular vision or minimum-visual processing. The two major approaches in deep learning for spacecraft pose estimation are discussed. The keypoint-based approach provides many advantages compared to the non-keypoint (or direct) approach, including better performance since the former can be easily updated to include the subcomponents offering state-of-the-art performance. One important tool for using machine learning techniques is to have realistic datasets, and hence

[4]Corresponding author: Arunkumar Rathinam, STAR LAB, Surrey Space Centre, University of Surrey, Guildford, GU2 7XH, UK. Email: a.rathinam@surrey.ac.uk.

various simulators used for orbital scene dataset generation are summarized in the chapter. Besides the digital simulators, the chapter also presents an overview of typical ground-based physical testbeds used for testing orbital GNC algorithms in close-proximity operation scenarios involving either cooperative or noncooperative targets.

Chapter 6[5] entitled "Inertial parameter identification, reactionless path planning and control for orbital robotic capturing of unknown objects" focuses on addressing the GNC challenges for spacecraft equipped with a single multiple DOF robot arm for tackling noncooperative objects in orbit. The chapter first presents the methods on inertial parameter identification of the noncooperative object through the equations of Momentum Conservation and Newton-Euler to obtain the basis of a two-step identification and then the error mechanism analysis. The chapter also includes a newly designed adaptive reactionless path planning method for the manipulator to deal with motion disturbance, a robust adaptive control strategy for arm joints, and a feedforward control strategy for spacecraft to ensure the stability of the whole system. Both computer simulations and ground-based testbeds are employed to verify parts of the overall system.

Chapter 7[6] entitled "Autonomous robotic grasping in orbital environments" provides a comprehensive review of past, present, and future scientific and engineering developments of robotic grasping for orbital applications to help assess the capabilities of existing technologies and reveal challenges for future systems. The readers will be informed of both classical and recent approaches on robotic grasping, grappling, and docking. First, an overview of experiments of human grasping in microgravity is provided, related to experiments conducted both in parabolic flights and onboard the International Space Station, with the intention of providing both inspirational and scientific background on weightless grasping and a historical overview of the importance of grasping for space activities. Then, the most important applications that require dexterous space robots are described, namely on-orbit servicing, assembly, debris removal, and astronaut–robot interaction. An extensive review of grasping methodologies for orbital targets is then given, with emphasis on coupling interface grasping, motor nozzle probing, grapple-based grasping, and usage of dexterous hands for space operations. Relevant state-of-the-art technologies in the field are analyzed, demonstrating the shift from mechanized grappling to intelligent grasping. Representative missions for the past, present, and future are also described to showcase the existing and planned technologies in robotic grasping. The chapter finishes with a summary of algorithmic, physical, and operational (verification) challenges that future space robotic grasping needs to overcome as well as help characterize relevant existing solutions.

[5]Corresponding author: Zhongyi Chu, School of Instrumentation Science & Opto-Electronics, Beihang University, No. 37, Xueyuan Road, Haidian District, Beijing, China. Email: chuzy@buaa.edu.cn.
[6]Corresponding author: Nikos Mavrakis, STAR LAB, Surrey Space Centre, University of Surrey, Guildford, GU2 7XH, UK. Email: n.mavrakis@surrey.ac.uk.

1.3.3　Part III: astronaut–robot interaction

The third part of the book includes three chapters on the latest research work and advances on HRI. These works result in new findings and technical proposals connecting wearable technologies with astronaut operations in space for the near and long terms.

Chapter 8[7] entitled "BCI for mental workload assessment and performance evaluation in space teleoperations" investigates teleoperated robotic systems used for several decades in space applications to perform complex assembly tasks in a particularly hostile environment. It explores why designing and safely teleoperating these systems require a multidisciplinary approach that relates high-fidelity simulation environments with brain–computer interfaces along with interactive design interfaces. These human-in-the-loop systems drive toward online monitoring of human mental states that relate to cognitive workload, attention, and fatigue. The chapter includes analysis of how these attributes are related to performance in human–robot-interaction systems and offers a hint of neurophysiological adaptations in space that affect brain function and the underlying signals measured with brain–computer interface (BCI) technology. These adaptations are driven by microgravity, confinement, and isolation and result in profound changes in circadian rhythm, brain waves, and autonomic system responses. The chapter also provides an overview of recent advances in machine learning and information fusion, underlying the current state of the art in BCI that could potentially be used in future space exploration.

Chapter 9[8] entitled "Physiological adaptations in space and wearable technology for biosignal monitoring" reviews the astronaut's physiological adaptations in space extreme environments that influence the cardiovascular, nervous, musculoskeletal, endocrine, and other human physiological systems. The effects on key biomarkers (e.g. for cardiovascular and musculoskeletal issues and stress) and biosignals for biomonitoring in space are further discussed. Wearable technology can facilitate our understanding of these effects, providing greater insight into human physiology in space. Wearable medical devices can also provide continuous or frequent monitoring of physiological parameters to evaluate the astronaut's status during missions and/or to help determine required robotic assistance. Thus, the rest of the chapter discusses space wearable biosignal monitoring technology with a focus on technologies for sweat analysis, cardiovascular, and endocrine responses, which can act as indices of acute stress and increased workload. Technologies for wearable thermoregulation in the inner astronaut suit are also discussed, as a means of supporting life during space missions.

Chapter 10[9] entitled "Future of human–robot interaction in space" takes a view that as opposed to becoming obsolete in the face of autonomous systems, HRI

[7]Corresponding author: Fani Deligianni, School of Computing Science, Glasgow University, Glasgow G12 8RZ, UK. Email: fani.deligianni@glasgow.ac.uk.
[8]Corresponding author: Panagiotis Kassanos, Hamlyn Centre, Institute of Global Health Innovation, Imperial College London, SW7 2AZ, UK. Email: p.kassanos@imperial.ac.uk.
[9]Corresponding author: Stephanie Sze Ting Pau, Hamlyn Centre of Robotics Surgery, Bessemer Building, Imperial College London, SW7 2AZ, UK. Email: stephanie.pau@network.rca.ac.uk.

research becomes more important. New tools and approaches are necessary for the successful adoption and integration of autonomous space robots, shifting from human–robot systems in teleoperations to human–robot teams in shared autonomy. The chapter presents "new" definitions of HRI by bringing in the Data-Information-Knowledge-Wisdom (DIKW) pyramid. This includes descriptions of robotic agent capability and limitation in different space environments and the HRI design in relation to Don Norman's work from the 1990s, which has been deeply embedded into human–computer interaction theory. The chapter further presents a projected future of HRI based on technological trends in the space industry. A case study on CIMON, a future crew assistance robot using ground-based artificial intelligence (AI) (IBM Watson), is used to help demonstrate the long-term vision. The case study shows the feasibility of a novel conversational interface and the use of AI in space, representing the potential benefits that AI can bring to HRI in space.

1.3.4   Part IV: system engineering

The fourth part of the book provides two further chapters that address the verification, validation, and security design issues for future space RAS. These system engineering topics are equally important to sustainable, long-term development of RAS for space, particularly under the NewSpace era, where there will be fast-growing intelligent capabilities and potential new threats to manage for robotic machines operating in space.

Chapter 11[10] entitled "Verification for space robotics" discusses verification and validation for autonomous space robotics that help to assess the safety, reliability, functional correctness, and trustworthiness of systems. The chapter covers both formal verification, a mathematical analysis of systems, and nonformal techniques such as simulation. Following an overview of verification for RAS, the chapter discusses a range of tools and techniques including logical specification, model checking, temporal theorem proving, runtime verification, and simulation as applied to space robot architectures, the Mars Curiosity rover, robot astronaut teamwork, and multiple object systems such as satellites.

Chapter 12[11] entitled "Cyber security of new space systems" presents the work on examining the security of emerging space RAS. A novel reference architecture for space systems (RASA) is used for the identification of the attack surface of RAS in the NewSpace era. An autonomous debris collection use case is analyzed using the proposed RASA and developing attack trees of the specific threats. Then the existing threat modeling approaches are evaluated based on the identified requirements of the space systems. Threat modeling aims to provide a broader view of the system, which allows establishing opportunities for risk management. At this point, space systems have distinctive security vulnerabilities. A risk management strategy, which is based on the practices drawn from other sectors, is introduced. The common practice is that

---

[10]Corresponding author: Clare Dixon, Department of Computer Science, University of Manchester. Email: clare.dixon@manchester.ac.uk.
[11]Corresponding author: Carsten Maple, WMG, University of Warwick, Coventry, CV4 7AL, UK. Email: cm@warwick.ac.uk.

of conducting threat modeling to analyze and manage threats and conducting formal verification independent from the threat modeling. However, it is ineffective since each analysis can independently lead to significant changes and it may be challenging to converge. Thus, a security-minded formal verification methodology, which is inspired by techniques in agile software development, is proposed to integrate threat modeling and formal verification analysis. An autonomous docking use case is analyzed, to illustrate it can be applied in these settings.

1.4   Acknowledgements

The book editor and coauthors would like to thank IET for publishing this work and the editorial team for their support. Some parts of the book are based on scientific results, knowledge, and experience gained from funded R & D activities. The following funded projects would be acknowledged: "Future AI and Robotics for Space (FAIR-SPACE)" funded by UK Research and Innovation and UK Space Agency for Robotics and AI Hubs in Extreme and Hazardous Environments under grant number EP/R026092; "Chair Professorship in Emerging Technologies" for Michael Fisher supported by the Royal Academy of Engineering; "dual reciprocating drill technique for use in horizontal drilling" funded by British Telecom; "in-orbit robotic assembly for large space science telescope" funded by CAS-CIOMP; "Inertial Parameter Identification, Reactionless Path Planning and Control for Orbital Robotic Capturing of Unknown Object" funded by National Natural Science Foundation of China under grant number 51975021; "Crew Interactive MObile companion (CIMON)" commissioned by the DLR Space Administration, with funding from the Federal Ministry for Economic Affairs and Energy of Germany.

The editor and coauthors also thank the following organizations or colleagues for their support to the work presented in this book: Aleksander Maslowski and Richard Gillham-Darnley at Surrey Space Centre for their lab technical support in relation to Chapters 5 and 7; Matthew Bradbury, Sara Cannizzaro, Marie Farrell, Clare Dixon, Michael Fisher, and Chronis Kapalidis for coauthorship of the research papers which Chapter 12 is based upon; Saurav Sthapit for proof-reading of Chapter 12; and major space agencies like NASA, ESA, and DLR for their public or authorized access to images used by several chapters.

References

[1]   Gao Y., Jones D., Ward R., Allouis E., Kisdi A. *UK-RAS network White Paper on space robotics and autonomous systems: widening the horizon of space exploration* [online]. 2018. Available from www.ukras.org/wp-content/uploads/2018/10/UK_RAS_wp_Space_080518.pdf [Accessed 28 April 2021].

[2]   Gao Y., Chien S. 'Review on space robotics: toward top-level science through space exploration'. *Science Robotics*. 2017;**2**(7):eaan5074.

[3]   Gao Y. (ed.). *Contemporary Planetary Robotics – An Approach to Autonomous Systems*. Berlin: Wiley-VCH; 2016. pp. 1–450. Available from eu.wiley.com/WileyCDA/WileyTitle/productCd-3527413251.html.

Part I

Mobility and mechanisms

Chapter 2

Wheeled planetary rover locomotion design, scaling, and analysis

Andrew Thoesen[1] and Hamidreza Marvi[1]

Rover locomotion on extraterrestrial surfaces is of general interest to the space community. Understanding and characterizing the surface processes that contribute to locomotion can increase efficiency, safety, and mission duration. This chapter presents an explanation of recent methods developed for modeling rover locomotion in granular media at a fundamental level. We begin by examining a brief progression of granular locomotion modeling, important regolith characteristics, and how these inform the choice to use a scaling approach in the remainder of the chapter. We then address recent experiments that reveal limitations of such theories, and how one can develop simple criteria and test methods that may allow better design of roving vehicles. Finally, we close by examining current and future directions that can possibly lead to better modeling.

2.1   Background: modeling the granular environment

Vehicles traverse deformable and granular media through complex interactions. These mobility mechanisms are the interest of the robotics, terramechanics, and physics communities [1–12]. Granular mechanics as a field tends to favor empirical or semiempirical approaches. This was the precedent set by Bekker [13, 14] including for extraplanetary mobility [15]. He popularized many of the theories at the time, and we will briefly dive into the meaning behind these formulas so that it is better understood why modern methods are important to incorporate. The first of the two foundational relationships is that between pressure and sinkage, seen here:

$$p = kz^n \tag{2.1}$$

The equation states that the average pressure of a plate p pushed into homogeneous terrain is dependent on an exponential function of sinkage z. Both the sinkage exponent, n, and the proportionality constant, k, are properties dependent on the soil and empirically derived fitted parameters for each particular terrain. Bekker

---

[1]Arizona State University, School for Engineering of Matter, Transport & Energy

approximates this for wheels by assuming large diameter wheels experiencing only small amounts of sinkage, such that the contact patch (contact area) of the wheel is invariate. This is sometimes acceptable as this scenario would produce contact in a relatively flat and rectangular plane, but often does not apply well for smaller rover wheels. Later, Bekker expanded the proportionality constant:

$$p = (\tfrac{k_c}{b} + k_\phi)z^n \tag{2.2}$$

where k_c is a cohesive constant, k_ϕ is a frictional constant, and b is the smaller dimension of contact, typically wheel width. For dry sand dominated by frictional forces, k_c is typically negligible. Likewise, for a wet clay dominated by cohesion (tension) between particles, k_ϕ is typically negligible. These constants are again soil sinkage parameters, empirically derived. As they are flat plate approximations, they ought to be used with caution, especially at the sizes of small-wheeled craft.

Meirion-Griffith examined such issues in his work *Advances in Vehicle-Terrain Interaction Modeling for Small, Rigid-wheeled Vehicles Operating on Deformable Terrain* [16] and proposed several corrections. For example, using a diameter-dependent pressure model, the estimations for a pressure-sinkage relationship were found to be far more accurate with a diameter term:

$$p = kz^n D^m \tag{2.3}$$

He also notes, in results for dry quartz sand, the convergence of the diameter-dependent and Bekker flat-plate models for large wheel diameters. While the wheel diameter at which the model converges will vary with soil type, it was found to be 2 meters in that work. While there are gradations of accuracy, it is clear the flat plate model should be examined as wheel size decreases.

The second foundational relationship in terramechanics is that between the shear and normal stresses in soil. The shear stress in soil is related to the normal pressure by the Mohr-Coulomb criteria:

$$\tau = c + \sigma \tan \phi \tag{2.4}$$

where τ is the shear stress of the soil, c is the apparent cohesion, σ is the normal pressure, and ϕ is the internal friction angle. The cohesion and friction angle are, again, measured quantities. However, to label them as "material properties" would be rather dubious. In experimental science, an important mantra to bear in mind is "the observer changes the measurement." This reality is a recurring issue when discussing classic terramechanics because the method of measurement can often have significant impact on the value. Moreland noted in his 2012 work *Traction Processes of Wheels in Loose, Granular Soil* that "There are three methods that are commonly used to determine the two soil strength values (c,ϕ), each yielding different results with respect to wheel mechanics. The measured soil properties are strongly dependent on the geometry of the measurement unit plate compared to that of the wheel of interest for modeling." The derivations for wheels that follow also assume that the pressure $p \approx \sigma$ from (2.2) and (2.4), an assumption that may not hold for a variety of reasons. Examining a formulation for drawbar pull will show the reliance on empirical parameters:

$$DP = H - R \tag{2.5}$$

where DP is drawbar pull, H is thrust, and R is the sum of motion resistance [14]. Thrust, H, is a measure of the positive tractive force a wheel can generate on a given soil. It is a direct function of the maximum shear stress the terrain at the wheel–soil interface can accommodate before complete shear failure occurs. Bekker calculated ideal thrust directly from Mohr-Coulomb theory by applying the contact area A of the wheel–soil interface to both sides of (2.4):

$$H = Ac + W \tan \phi \tag{2.6}$$

where W is the normal force caused by the pressure over contact area. Given the previously discussed issues with measuring (c,ϕ), this is a cause for concern. There is an important context that also needs to be given to R. The expression is given [17] as follows:

$$R = \int_0^{z_0} pbdz \tag{2.7}$$

where p is rim pressure described by the pressure-sinkage equation, b is wheel width, z is sinkage along the rim, and z_0 is total sinkage. The physical meaning of this equation describes a net motion resistance that is entirely composed of the horizontal component of the normal soil pressure on the wheel rim, created by some level of sinkage, resisting forward motion. This ignores the complexities of soil transportation and flow during wheeled locomotion. In fact, this equation was originally formed for predicting the motion resistance of *towed* wheels [18], which is a more physically accurate scenario for such a formulation. Additionally, the estimation for both motion resistance and thrust relies on the pressure-sinkage equation. For the topic of this chapter, wheeled planetary rovers, this poses particular problems because these vehicles often have smaller wheels than these relationships were intended to estimate (given the flat plate assumptions).

Bekker acknowledged concerns about the pressure-sinkage equation fairly early: "Predictions for wheels smaller than 20 inches in diameter become less accurate as wheel diameter decreases, because the sharp curvature of the loading area was neither considered in its entirety nor is it reflected in bevameter tests" [14]. Instrumentation assumptions are not the only ones that need to be accounted for. Wong [17], another pioneer in the field, made advancements in examining many different soil-geometry models, including those with gravity variation. One important discovery was the insight that weight-offset testing can have erroneous or opposite results compared to identical experiments in gravity-varied parabolic flight testing [19] due to the gravitational compaction of grains. This does not mean such classic terramechanics cannot be useful or adapted in a modern context to some extent. However, many of these empirical equations require a priori knowledge of various granular characteristics, derived from time-consuming and possibly inaccurate geotechnical tests, sometimes requiring additional complimentary testing for fit parameters. Combined with the previously discussed limitations, it is clear that additional techniques might be useful.

One possible solution is to approach capturing the complex locomotive behavior holistically. If planetary vehicle roboticists are mostly concerned with the behavior of certain rover outputs or performance numbers, perhaps the best strategy is to adapt some frameworks from other fields. For example, recent efforts to understand granular dynamics and craft motion from a more theoretical point of view have produced new semiempirical models [20]. Granular resistive force theory (RFT), an examination of granular material reactions against geometries [3, 21–28], is driven by a similar theory used in fluid dynamics. RFT utilizes superposition and discretization of intruders into smaller geometries to sum the resultant forces for analysis. This typically takes the form of flat plates or slender cylinders. RFT has shown impressive predictive results for certain intruders with a few characterizing tests of the media [25], including reasonable approximations obtained with a single plate test. RFT has been reconciled with other theoretical granular physics by assuming the target environment to be a continuum obeying a frictional yield criterion without cohesion [10, 29–31]. What's more, both RFT and continuum approaches can be explained by "frictional plasticity" theories [31]. A comparison of dimensional analysis of both approaches resulted in the same set of scaling parameters [32], leading to a general set of predictive granular scaling laws (GSLs) for wheeled locomotion in granular media.

The power of this framework is the ability to predict certain dependent wheel outputs, such as power and velocity, based upon a careful selection of wheel parameter inputs such as diameter and mass. This can lead to the prediction of larger vehicle performance from smaller vehicles with no a priori knowledge of soil characteristics, as long as the soil is identical. This dimensionless approach is similar to wind tunnel testing in aerodynamics, and the gravity-variant laws provide a unique opportunity for planetary vehicle design. In the remainder of the chapter, we examine the GSLs for the boundary case of lightweight space vehicles. These are often used in practice for space robotics and in labs for prototype and development. Sojourner [33], Moonraker [34], and PUFFER [35, 36] were designs of 11, 10, and sub-1 kg mass, respectively. Use cases on Earth also include laboratory developments of new grouser approaches [37] or angled granular mobility [38].

The study by Slonaker *et al.* encompassed a range of the masses between 13.4 and 45.7 kg for all sets [32] and mobilizing at angular velocities below 5 RPM. The experiments were run with a single wheel on a gantry with the direction of travel constrained in a planar fashion. The reported error for these experiments was below 3 percent for set trend lines. The studies of this chapter explore an unconstrained two-wheeled vehicle at masses of 1.5–5.2 kg, running at 13–75 RPM nominally. This different set of lighter, faster parameters was targeted to explore the design space occupied by the laboratory robots and small prototypes previously mentioned in hopes of advancing planetary rover experimentation. To understand how lightweight rovers may deviate from established GSLs, we examine the derivation of the theory.

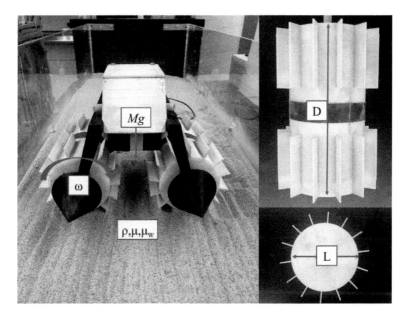

Figure 2.1   Granular scaling parameters labeled for craft and straight grousered wheel

2.2   Methods

2.2.1   Theory: wheel parameter scaling for output parameters of mechanical power and translational velocity in granular media

This section will serve as an expansion and addendum to the scaling laws presented by Slonaker *et al.* [32] to make them more accessible to a general audience. The end result of these laws is the ability to predict power and velocity of a larger vehicle by the performance of a smaller model. By discussing the derivation of and the assumptions behind the scaling laws, the utility of using these laws for planetary rover investigation will be more readily seen. The steps behind the dimensional analysis will be presented to further inform terramechanicists and rover designers. A functional relationship of mechanical power (referred to hereafter as simply "power") and translational velocity for a wheel traversing granular media is presented in (2.8) with parameters labeled on their physical counterparts in Figure 2.1:

$$(P, V) = \psi(d, l, m, \omega, t, f, g, \rho, \mu, \mu_w) \tag{2.8}$$

The output parameters of interest are power P and translational velocity V. The wheel is described by its characteristic length dimension (typically radius) l, its thickness (depth into the page) d, its mass m, a driving rotational velocity ω, and a consistent shape outline f. This shape outline f is a dimensionless set of points that sets the condition that although the wheel shape may be arbitrary, it must be consistent when scaling. We cannot, for example, examine a grousered wheel and

Table 2.1   Locomotion parameters and dimensionless expressions

Parameters	Dimensions	Dimensionless parameters	Variable expression
$[P]$	$[\frac{ML^2}{T^3}]$	$\bar{P}$	$[\frac{P}{mg\sqrt{lg}}]$
$[V]$	$[\frac{L}{T}]$	$\bar{V}$	$[\frac{V}{\sqrt{lg}}]$
$[d]$	$[L]$	$\bar{d}$	$[\frac{d}{l}]$
$[l]$	$[L]$	$\bar{l}$	1
$[m]$	$[M]$	$\bar{m}$	1
$[\omega]$	$[\frac{1}{T}]$	$\bar{\omega}$	1
$[f]$	$[none]$	$\bar{f}$	f
$[g]$	$[\frac{L}{T^2}]$	$\bar{g}$	$[\frac{g}{l\omega^2}]$
$[r]$	$[\frac{M}{L^3}]$	$\bar{\rho}$	$[\frac{\rho l^3}{m}]$
$[\mu]$	$[none]$	$\bar{\mu}$	$[\mu]$
$[\mu_w]$	$[none]$	$\bar{\mu}_w$	$[\mu_w]$
$[t]$	$[T]$	$\bar{t}$	$t\sqrt{\frac{g}{l}}$

predict results for a non-grousered wheel, even of the same size. The environment is described by time t, gravity g, and the granular characteristics ρ, μ, and μ_w; these are density, internal friction, and wheel-grain friction, respectively. They occur as a property of the granular environment and its interaction with the geometry. They are also assumed consistent between scaled comparisons. One could not, for example, examine results in silica sand and predict results for planetary regolith due to differing granular mechanics.

The first step toward transforming this function begins by using dimensional analysis and expressing each variable by its base quantities. In our case all variables are defined by the base dimension quantities of length (L), mass (M), and time (T). These dimensional expressions are seen in the second column of Table 2.1. While this expresses our variables in terms of dimensions, it is necessary to also express our dimensions in terms of our variables for the analysis. To do so, we choose a subset of variables to define these dimensions as follows:

$$L = l \quad M = m \quad T = \frac{1}{\omega}, \quad \sqrt{\frac{l}{g}} \tag{2.9}$$

This relationship allows us to generate dimensionless numbers for our variables. We create the dimensionless numbers by "canceling" the base quantities; the results are seen in the rightmost column of the table. An example of the procedure is given using gravity:

$$\bar{g} = g * \frac{T^2}{L} = \frac{g}{l\omega^2} \tag{2.10}$$

Gravity is an acceleration and has units of length over time squared. To create a dimensionless number, we multiply it by our variable choices in such a way that these units cancel out. The new function expressed using these bar variables is now:

$$(\bar{P}, \bar{V}) = \Psi(\bar{d}, \bar{t}, f, \bar{g}, \bar{\rho}, \mu, \mu_w) \tag{2.11}$$

The bar variables are a composition of the original variables as given in Table 2.1. We note here that *l*, *m*, and *ω* were used as the expressions for their respective dimensions, and divide by themselves; their fractions reduce to 1. The shape outline and friction were already dimensionless expressions; they do not change. It is useful to see (2.11) in terms of the original variables before applying our assumptions:

$$[\frac{P}{mg\sqrt{lg}}, \frac{V}{\sqrt{lg}}] = \Psi(\frac{d}{l}, t\sqrt{\frac{g}{l}}, f, \frac{g}{l\omega^2}, \frac{\rho l^3}{m}, \mu, \mu_w) \tag{2.12}$$

Equation 2.12 is (2.11), simply expressed in a more accessible way using the original variables. We can now address the first three assumptions for the problem, which will reduce its complexity:

1. We assume a homogeneous granular environment between experiments with different wheels. This means identical granular media between the mass/size pairings, but also importantly means a sufficient depth/sinkage is reached in both cases such that neither has surface effects or other deviating behavior. As such, the friction coefficient of grain-screw interaction, the internal friction of the granular media, and the expression for density are assumed consistent; therefore, we eliminate these from our function.
2. We assume constant gravity and will use a gravity-invariant version of the laws for now. This assumption will be relaxed toward the end of the chapter. Gravity is therefore eliminated as a functional variable.
3. We constrain the wheel thickness *d* and mass *m* to be scaled dependently, both by some arbitrary integer *n*, and assume that granular motion under the wheel is trivial in the depth dimension so it does not affect power output. If this is the case, then the resulting power change will be *nP*. This constraint results in the equivalent of running *n* copies of the wheel next to each other. For example, under this constraint, let us say that the mass of the wheel has been doubled and necessarily so has the thickness. This is the same as if we simply ran the original values for two wheels. The power of this constraint is that it necessitates that *d* and *m* are *not* independent inside Ψ. To create this dependence, we constrain *dρ* by their product so we are dependent on the $\frac{d}{m}$ ratio.

Using dimensional analysis and the above set of assumptions, the result is as follows:

$$\left[\frac{P}{mg\sqrt{lg}}, \frac{V}{\sqrt{lg}}\right] = \widetilde{\Psi}\left(t\sqrt{\frac{1}{l}}, f, \frac{1}{l\omega^2}, \frac{dl^2}{m}\right) \tag{2.13}$$

We have derived these equations in such a way that they match the original expressions given by Sloanker. However, we employ two additional adjustments not explicitly stated in the original use. By doing so, we will set the functions in their formal formulas with correct dimensions. It will also make the relationship between variables more intuitively clear for rover engineers.

4. We have already discussed that while the wheel shape *f* may be arbitrary (while remaining thickness invariant), the relationships of the outline are constraints to scale with the fundamental length *l*. Since this shape remains consistent, it is eliminated from our function for simplicity.
5. We observe that while power and velocity are a function of time, engineers are often concerned only with the steady-state performance of rovers and not with the time-variant characteristics of such. If we assume time-averaged steady-state values, then we eliminate time as a functional concern. Note this means eliminating t and not simply the variable t; the difference is that the entire dimensionless parameter is gone from our final expression.

Rearranging (2.13) to two separate formulas expressing the dependent variables with proper dimensions, we have the following:

$$
\begin{aligned}
P &= mg\sqrt{lg} * \left[\widetilde{\Psi}(\tfrac{1}{l\omega^2}, \tfrac{dl^2}{m})\right] \\
V &= \sqrt{lg} * \left[\widetilde{\Psi}(\tfrac{1}{l\omega^2}, \tfrac{dl^2}{m})\right]
\end{aligned}
\tag{2.14}
$$

Note that $\widetilde{\Psi}$ is a dimensionless function dependent only on vehicle parameters. After examining m, g, and l we see that power and velocity both have correct dimensions. We observe now that we have four variables remaining inside the dimensionless expression. We also observe that both mass and length are outside of the dimensionless expression (and our gravity value will not change). If we change our characteristic length l by some scalar r, change mass m by s, and constrain our remaining variables angular velocity ω and thickness d by scalars (b, a), such that the fractions inside $\widetilde{\Psi}$ always remain consistent, then we know the value of $\widetilde{\Psi}$ *itself* will always be consistent. This is the meaning of the scalar variables seen in (2.15). Moreover, it implies a change in vehicle mass and wheel diameter will always lead to a predictable change in power and velocity if the conditions during this derivation are followed. Using this second set of variables we see how this is possible.

$$\acute{l} = rl, \quad \acute{m} = sm, \quad \acute{d} = ad, \quad \acute{\omega} = b\omega \tag{2.15}$$

$$
\begin{aligned}
\widetilde{\Psi}_1 &= \tfrac{1}{l\omega^2} \;\rightarrow\; \acute{\Psi}_1 = \tfrac{1}{rlb^2\omega^2}, \\
\widetilde{\Psi}_2 &= \tfrac{dl^2}{m} \;\rightarrow\; \acute{\Psi}_2 = \tfrac{adl^2r^2}{sm}
\end{aligned}
\tag{2.16}
$$

Our Ψ terms are now constrained to be equivalent. By solving for the (d, ω) scaling factors (a, b) from (2.15) in terms of (r, s), it becomes possible to develop design parameters that allow prediction.

$$\acute{\psi}_1 = \widetilde{\Psi}_1, \quad \acute{\psi}_2 = \widetilde{\Psi}_2 \tag{2.17}$$

$$\frac{1}{rlb^2\omega^2} = \frac{1}{l\omega^2}, \quad \frac{adl^2r^2}{sm} = \frac{dl^2}{m} \tag{2.18}$$

$$a = \frac{s}{r^2}, \quad b = \frac{1}{\sqrt{r}} \tag{2.19}$$

Note that because we have set the constraint that our function of unknown shape is always the same value, and we have chosen l, m as our variables to modify, we have also produced a new set of dependent variables for power and velocity, $\acute{P}, \acute{V}$, in (2.20) and (2.21), which will scale predictably. If the values from (2.15) are substituted into (2.14), keeping in mind the equivalency from (2.17), then this becomes apparent:

$$\acute{P} = (s\sqrt{r})(mg\sqrt{lg} * \widetilde{\Psi}) = s\sqrt{r}P \tag{2.20}$$

$$\acute{V} = (\sqrt{r})(\sqrt{lg} * \widetilde{\Psi}) = \sqrt{r}V \tag{2.21}$$

Simply put, if a wheel with the inputs of (l, m, d, ω) is modified by positive scalars r and s and the above constraints, then the relationship with a second wheel is $(\acute{l}, \acute{m}, \acute{d}, \acute{\omega})$. By adhering to these constraints, the time-averaged, steady-state mechanical power and the translational velocity of the second wheel are predictable from the first by way of (2.20) and (2.21): $(\acute{P}, \acute{V}) = (sr^{1/2}P, r^{1/2}V)$. The final equivalent relationships for the new variables can be easily expressed in terms of original variables and the two scalars:

$$\acute{l} = rl, \quad \acute{m} = sm, \quad \acute{d} = sr^{-2}d, \quad \acute{\omega} = r^{-1/2}\omega, \quad \acute{P} = sr^{1/2}P, \quad \acute{V} = r^{1/2}V \tag{2.22}$$

The above expression shows that if the characteristics of a larger, heavier wheel are related to a smaller, lighter wheel by certain constraints, then the power and velocity outputs are similarly constrained and thus predictable. This allows for a method of rover testing, which should capture the subtleties of the rover's movement in a particular granular media without relying on externally measured granular qualities or empirical fit data.

2.2.2 Experimentation: planetary regolith simulants and testing environments

The focus and scope of this chapter is centered on rover testing methods and not the geology or granular characteristics of rover environments; however, since the two are closely related we will discuss the utility of using a lunar regolith simulant to approximate rover behavior in the lunar environment. While the studies in this chapter only use a lunar simulant, the techniques and principles apply to experimentation with all planetary simulants. A planetary regolith simulant is a material meant to approximate the chemical, mechanical, or other engineering properties of planetary regolith. This can require various characteristics such as similar mineralogy and particle size distributions within the material. The mechanics of wheel–grain interface can vary significantly by environment; preliminary Earth testing of Mars Curiosity Rover traversability showed the variability in wheel performance and interaction that will occur in different types of granular media [39]. Therefore evaluating the performance of GSLs in a lunar simulant is important to using GSL for the testing and design of planetary rovers.

The physical lunar simulant used in experimental studies for this chapter is known as Black Point 1 (BP-1). BP-1 is derived from crushed mining tailings found in the Black Point basalt flow of the San Francisco Volcanic Field. This simulant is used for testing lunar robotics and in the robotic mining competition [40, 41] at NASA Kennedy Space Center, where it was obtained. A full geotechnical assessment of BP-1 is available in the literature [42, 43]. There are two distinct characteristics that separate this from many other testing materials. The first is a D60 value of 0.11 mm. This means 60 percent of the tested particle sizes were found finer (smaller) than 110 microns. The second characteristics is the particle shape: It is classified as angular to sub-angular. The combination of these two characteristics results in a granular media that is predisposed to resist flowing, shows more discrete deformation, and is sensitive to plastic deformation memory and high trenching [44] due to high particle interlock. Most importantly, it is [45, 46] geotechnically similar to lunar regolith and reacts in a similar physical manner. The wide range of material properties in deformable and granular media means that experimental or theoretical techniques ought to be verified for planetary regolith simulants before they are used in planetary rover design and testing.

As a control, a Quikrete medium sand used in the Slonaker *et al.* was selected for direct comparison in experiments. It is silica sand that is kilned, sieved, and washed before packing. Particles are primarily 0.3–0.8 mm diameter in size and far less angular in contrast with the BP-1. To complete these comparative studies of BP-1 lunar simulant and silica sand, two separate granular testing environments were constructed. BP-1 presents a respiratory hazard when dusted in a confined space due to the small and angular nature of the particles. Testing of planetary simulants is often performed with some level of containment or mask. Owing to this, a portable, custom-made simulant containment unit on wheels was manufactured for the purposes of the studies in this chapter. For protection, the chamber contains a top-opening door with a deformable rubber seal to allow cable pass-through. On the sides, sealed glove holes are added to access tools and the testing vehicle. LED lights were added to control conditions for visual tracking of craft in videos. Bed dimensions are 37.5×67.5 cm. It is advised that similar devices are used when experimenting with small-grained, angular simulants.

A much simpler testing environment for the silica sand was also constructed. Because the smaller particles of the commercial sand had already been sieved and washed away ahead of time by manufacturer, there was no general dusting hazard and the sandbox could remain open at all times. It also did not require additional illumination for visual tracking or any specialized setup for tool access. This 80×250 cm test bed can be seen alongside the BP-1 testing chamber in Figure 2.2. These environments were used for three experimental studies to examine the scaling laws in the context of lightweight, planetary rovers.

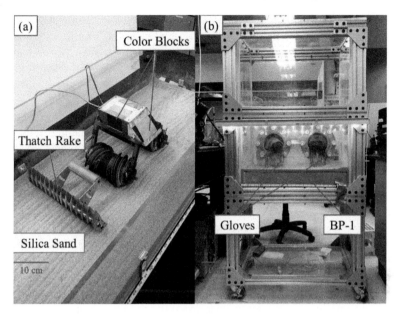

*Figure 2.2   (a) Craft with curved grousered wheels attached in silica sand bed
and (b) simulant containment unit with tools displayed*

In the first study, the effects of environmental media, angular velocity, and wheel
shape on these scaling laws are explored with a lightweight (1.5–3.0 kg) rover. A set
of experiments are performed in which power draw and velocity prediction of larger
wheels from smaller ones are analyzed. Wheels with both straight and curved grousers
are evaluated. The second study examines the performance of a heavier (1.5–5.2 kg)
lightweight rover equipped with similar straight grouser wheels and 80-grit sandpaper
wheels. These experiments are examined for variation of prediction by mass, angular
velocity, and motor placement. In the third study, the sinkage of the craft and the magnitude of the sand preparation patterns from the second study are examined for influence on scaling law accuracy. The final study examines the gravity-variant scaling laws
entirely with simulations and discusses the utility and accuracy of this subset.

2.3   Studies giving context to scaling theory

2.3.1   Study one: mechanical power and translational velocity
prediction variance by granular material and wheel shape

To form a comparative basis for testing the law, three different sets of parameters
were chosen to be tested identically in both BP-1 and silica sand and for two grouser
shapes. The wheel dimensions, craft mass, and angular velocities for experiments
were selected using the relationships from (2.22). The straight grousered wheels are
seen in Figure 2.1 and the curved grouser wheels are seen attached in Figure 2.2. A

Table 2.2   Wheel and experiment design parameters

Design	Diameter (cm)	Mass (kg)	Length (cm)	Target ω (RPM)
GSL1-G	7.5	1.459	14	15, 30, 45, 60, 75
GSL2-G	10.0	2.594	14	13, 26, 39, 52, 65
GSL3-G	9.375	2.918	18	14, 27, 40, 54, 67
GSL1-B	7.5	1.477	14	15, 30, 45, 60, 75
GSL2-B	10.0	2.626	14	13, 26, 39, 52, 65
GSL3-B	9.375	2.954	18	14, 27, 40, 54, 67

full set of input parameters are shown in Table 2.2. Naming designation was created using the theory (GSL), the size and mass (1, 2, and 3), and the shape of the wheels ("B" for curved and "G" for straight grouser). Neither straight nor curved grousers had been previously examined for these laws. Both grousers have equal thickness and length. The grouser lengths are 1.25, 1.67, and 1.56 cm to maintain the characteristic radial scaling.

A characteristic of using the scaling laws is the ability to predict power requirements and translational velocity of a larger craft from a smaller one. In the context of planetary vehicles, which often have very limited margins and strict requirements, this can confer a design advantage. For example, using the test parameters shown along with the equations in (2. 22), the predicted power of P_2 (the power required in GSL2 sizing) should be $2.05P_1$, regardless of the P_1 (the power required in GSL1 sizing) value. P_3 is predicted at $2.24P_1$. By the same logic, V_2 should be equal $1.15V_1$ and V_3 should be $1.12V_1$, regardless of the V_1 velocity. This is, again, respective of both wheel shape and environment; we cannot use the results of the straight grouser in sand to predict results in either BP-1 or with curved grousers. These relationships also require that the larger sets are evaluated at specific speeds as listed in the table. Using a Proportional-Integral-Derivative controller, we ensured speeds were very close to those targeted (within 3 percent error) and extrapolated to those targets via linear regression.

Experimental procedure for all trials was identical. Soil was tilled with a thatch rake to eliminate any large soil stress concentrations. The vehicle was placed in one of the chambers, on top of the material, at a far end. When the power supply was engaged, electric power draw was recorded via serial monitor signals from the in-line hall effect current sensors. Velocity was extracted from camera recordings. Printed colors attached to the side of the craft enabled a MATLAB® tracking algorithm to compare position as a function of time. Mechanical power was evaluated using sensor readings of individual motor currents, comparing to the no-load current drawn from benchtop testing, and converting to torque via the motor's torque constant. The time-averaged torque and rotational speed during steady state were multiplied to produce a time-averaged power. A total of 12 trials were performed for each set of 5 speeds, using 2 wheel shapes at 3 different masses/dimensions and in both BP-1 and silica sand, for a total of 720 trials.

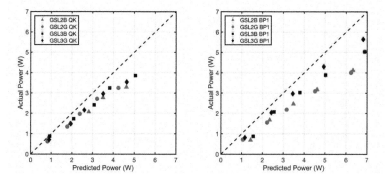

Figure 2.3   Predicted power versus actual power consumption with black line indicating perfect prediction. Quikrete on the left and BP-1 on the right

The results of the study [56] were analyzed for both power and velocity predic-tive deviation. In general, the power of the GSL2 and GSL3 sets was overpredicted; that is, the heavier sets drew less power than scaling indicted they ought to. The translational velocity of both GSL2 and GSL3 were also underpredicted in gen-eral; that is, the heavier sets moved faster than scaling indicated they ought to. The sources for these general deviations will be explored in a later study. However, there are several differentiating factors in the power and velocity predictions that ought to be discussed, primarily the granular material as seen in Figure 2.3. For a set to be correctly predicted, all data ought to fall on the black line.

Higher error is present in all four pairings for BP-1 than Quikrete. Errors for all but the lowest speeds were 29–35 percent, 19–27 percent, 30–36 percent, and 13–17 percent for GSL2B, GSL3B, GSL2G, and GSL3G, respectively. Errors for the slowest speeds of those sets were 52 percent, 44 percent, 33 percent, and 30 percent. It is apparent that the heavier of the two sets, "3," resulted in lower error for both shapes in BP-1. The silica sand remained relatively consistent in error across sets, with lower general error. Errors for all speeds were 9–29 percent, 5–24 per-cent, 16–25 percent, and 16–24 percent for GSL2B, GSL3B, GSL2G, and GSL3G, respectively. After examining the comparisons, one concludes that the shape of the grouser made little difference predicting power, that the larger mass difference in the "3" case made them marginally more accurate in BP-1 and, in general, that BP-1 was marginally worse than Quikrete in predicting power results. In all cases, the data deviation from the power prediction line increased with speed. The context of the power prediction results is further explored in the next study.

Differentiating the velocity prediction by material yields a much starker contrast than the power predictions (Figure 2.4). In BP-1, the error for GSL2B, GSL3B, GSL2G, and GSL3G was 7–24 percent, 0–21 percent, 17–27 percent, and 3–15 per-cent. This trend generally increased with angular velocity. For the Quikrete, results were much closer in both accuracy and precision. GSL2B, GSL3B, GSL2G, and GSL3G resulted in errors of 0.1–3.8 percent (except 17 percent error at 65 RPM),

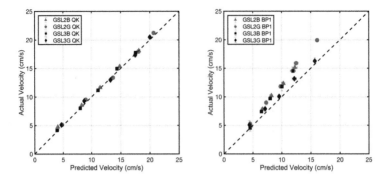

Figure 2.4   *Predicted velocity versus actual velocity achieved with black line indicating perfect prediction. Quikrete on the left and BP-1 on the right*

0.1–4.9 percent, 0.6–6.8 percent, and 0.8–9.1 percent with lower errors found at higher speeds. A likely explanation for the translational velocity error was significantly lower than the power error is because angular velocity was driven by the craft controls. More important to examine is the velocity prediction deviating significantly less in Quikrete. Recall that both results mean the rover is considered to have "overperformed" in BP-1; the mechanical power drawn is less and the velocity is greater than predicted for both larger vehicles. This seems to contrast with conventional wisdom about regolith (and its simulants) and the difficulty of achieving low slip mobility. To investigate the cause of the power deviations, we control for environment and begin another study into the effects of mass and velocity for power-scaling predictions in silica sand.

2.3.2   Study two: mechanical power prediction variance by mass, velocity, and motor placement

The second study [48] is designed to examine specifically the failure of the scaling laws to predict power in Quikrete in the previous study. Since the granular media was identical to the successful trials in the literature, it was determined wheel shape and mass would be primarily investigated. This study was performed with the same craft as the first study, entirely in the Quikrete sand bed shown in Figure 2.2, and with the parameters selected in Table 2.3. The GSL1G and GSL2G wheels from the first study were retained, and sandpaper wheels GSL1SP and GSL2SP were introduced. The sandpaper wheels have an identical diameter to the two previously featured wheels but 80-grit sandpaper is adhered around the entire surface (the same grit as the original studies) instead of grousers.

A light, medium, and heavy pair of masses for grousered wheels and a light and heavy pair of masses for the sandpaper wheels were selected based on vehicle performance capabilities. The diameters for the sandpaper wheels were made identical to the grousered wheels. We also note the grouser length of 1.25 cm for GSL1G

Table 2.3   Scaling parameters tested

Name	L (cm)	M (kg)	D (cm)	ω (RPM)
GSL1G	7.5	1.46, 2.19, 2.92	14	15, 30, 45, 60, 75
GSL2G	10	2.59, 3.84, 5.19	14	13, 26, 39, 52, 65
GSL1SP	7.5	1.46, 2.92	14	15, 30, 45, 60, 75
GSL2SP	10	2.59, 5.19	14	13, 26, 39, 52, 65

and 1.67 cm for GSL2G as proportional with diameter length scaling. For additional simplification, the ratio of mass/length scalars was chosen such that the thickness of the wheel remained constant while diameter increased. This can be done if the scalar relationship in (2.22) is $s = r^2$. Angular velocity was selected for the GSL1 sets to span a wide range of speeds while maintaining consistency, and the necessary angular velocities for the GSL2 sets were found from the scaling parameter relationships (2.22) and maintained to predict power. According to this equation, and given the scalars of $(r, s) = (1.333, 1.776)$, this means that for every set of wheel parameters paired together, the power draw of the second set should be slightly larger than twice the first set. For example, taking the first set $(l, m, d, \omega)_1 = (7.5, 1.46, 14, 75)$ and second set $(l, m, d, \omega)_2 = (10, 2.59, 14, 65)$ produces a predicted $\frac{P_2}{P_1}$ of 2.05 based on scalars. This should occur in all pairings for this study by design. Experimental procedure occurred in an identical fashion to the Quikrete procedure in the first study, with the notable exception that velocity was not tracked as this study was focused on power comparisons.

A consistent power ratio of 2.05 was not reached for any of the experiments as predicted. Instead, several dependencies were noted. A mass dependency observed in the results is shown in Figure 2.5. The data is graphed using the heavier mass of the pair on the horizontal axis. The errors across all speeds for the light, medium, and heavy grousered pairs averaged 23.5 percent, 18.3 percent, and 12.1 percent.

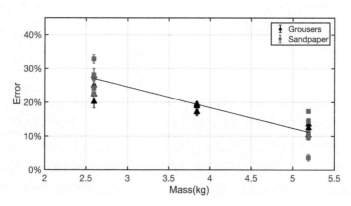

Figure 2.5   Prediction error percentage as a function of mass is displayed

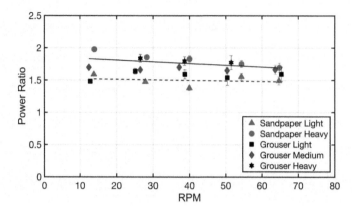

Figure 2.6   The power ratio trend of all data points shows a decrease with wheel rotational speed, with significantly more decline in heavier sets

For the sandpaper wheels, observed error was 27.0 percent for light and 11.2 percent for heavy. Addressing wheel shape first, we note that wheel shape was not observed to have any impact on outcome, similar to the previous study. The results indicate that for both wheels, a strong error dependency on mass emerges. This dependency is consistent such that error decreases for *every* individual data point when compared to the same angular velocity data point of a heavier pair. Accounting for all data, the overall linear regression approximated that this error disappears when the heavier mass of the pair is at 9 kg. Two important pieces of information give context to this error intercept. The first is from the original experiments; the heaviest wheels of similar dimensions in the original experiments approached 45 kg without reported error. It seems unlikely that any quantity of "excessive" mass would somehow induce error. The second item of information is from this study: the 13 RPM data for the heaviest sandpaper pair only deviated by 3 percent. This is partially due to the second dependency observed, that of velocity.

In Figure 2.6, power ratios are shown compared to wheel rotational speed. Addressing the grouser wheels first, we see that the lightest pair shows similar error to the first study; prediction deviation was 20.1–27.7 percent compared to 16–25 percent in study one (§ 2.3.1). The medium mass pair had both less error and a narrower band, ranging 17.1–19.5 percent. The heaviest pair was tested at the three middle speeds, and error was 10.3–13.5 percent. For the light pair of sandpaper wheels, error ranged from 22.6 to 32.8 percent, slightly worse than grousered wheels. For the heaviest pair, drawing a direct comparison to grousered wheels, error ranged 9.6–14.5 percent in the three middle speeds. The heavy sandpaper pair is quite illustrative when examining all five speeds; with errors of 3.6 percent, 9.6 percent, 10.9 percent, 14.5 percent, and 17.3 percent the velocity dependence is quite clear. To quantify this particular set, the slope of linear regression is −0.0052; the increase from 13 to 65 RPM was predicted to reduce power ratio by 0.27. For context, the original scaling law experiments presented in [32]

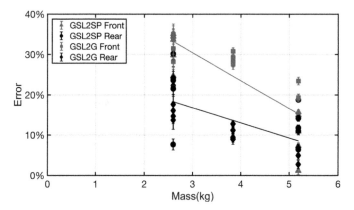

Figure 2.7   *Power prediction error is differentiated by front or rear motor; rear motors show significantly less deviation from predicted values*

had speeds between 2.33 and 4.76 RPM and no relationship between power prediction and either velocity or mass was reported.

Limitations at higher speeds or lower masses are important to bear in mind when choosing to apply these laws. However, the final observed deviation from prediction may be more difficult to avoid. A predictive difference between leading and lagging motors in the wheels is observed in Figure 2.7.

When analyzing the two wheels independently and not as a whole system, the power ratios of the leading and lagging wheels showed significant difference. For the grouser wheels, the rear wheel recorded a higher power ratio and therefore lower error in every case. Notably, although the sandpaper wheels did not show equally high differentiation in error between leading and lagging wheels, this trend was still present. This particular deviation generated a hypothesis for the error found in both studies: The effect of a lagging wheel having significantly decreased error could be explained by the pre-compaction or smoothing of the sand grains by the first wheel before the second wheel rolls over it. This might counteract a lack of uniformity in the grains and lead to a more consistent outcome. This hypothesis is explored in the next study.

2.3.3   Study three: context of deviations and examination of scaling law application sinkage threshold

Returning to the Quikrete velocity data in study one (§ 2.3.1), all predictive pairs but one have errors below 10 percent. Generally close agreement of velocity predictions occurs across shape and mass. This is interesting in light of the range of angular velocities here (13–75 RPM) compared to the original studies (2.33–4.76 RPM). Conversely, power prediction significantly deviated (marginally more in BP-1) with a much lower mass here (1.5–2.9 kg) than in the original studies (13.4–45.7 kg). One might infer from this that the scaling laws deviations are much more dependent on mass variation than velocity range. Interestingly, linear regression of power prediction error as a function of craft mass for all speeds in silica sand with straight grouser wheels was very similar for the first two studies in this chapter. Analysis of error in the first study predicts elimination of error at approximately 8 kg and at 9 kg during the second study using linear regression. While the craft could not be configured to handle additional mass, the effects on the grains could be more closely examined.

An important assumption for the scaling predictions is a consistent effective friction. This generally requires a depth of 10–20 grain diameters to assure no surface effects. Given that the particle size of the Quikrete sand is generally between 300 and 800 microns, a depth of 10 grain diameters would be between 3 and 8 mm. In the third study, we investigated the errors by repeating a subset of experiments from study two (§ 2.3.2) while measuring the depth of sinkage using pre- and post-run caliper measurements. The depth of our tilling preparation was also examined. Both wheel shapes were tested, using the largest sizing, at lightest and heaviest masses, and at the lowest speed to control for inertial effects. Each set was performed for ten trials apiece. This was done with the tilling preparation from the previous 2 studies, and then repeated with a leveled and smoothed surface for a total of 80 trials. A sliding rigid bar was used with an attached caliper to measure the difference between the characteristic peaks and valleys of the raking pattern before beginning each of tilled trials. The sinkage changes were measured in the same manner after both sets.

An average depth difference in the characteristic pretrial peak/valley pattern between high and low points for all 40 trials was found to be 1.58 ± 0.18 mm. Sandpaper wheels, with measurements made from the high point of the raked sand, showed sinkage of 2.69 ± 0.29 mm and 2.82 ± 0.18 mm for the light and heavy cases. The grousered wheels showed 2.02 ± 0.19 mm and 2.51 ± 0.28 mm for the light and heavy cases. Given the depth of raking pattern and the measurements from peaks, one could approximate the sinkage with respect to an "averaged" prepared plane by subtracting half of the peak/valley amplitude (0.79 mm). A total of 40 trials of unraked sand were also examined. The smallest and largest masses of the sandpaper wheels had 1.00 ± 0.24 mm and 1.09 ± 0.24 mm sinkage, respectively. The grousered wheels had 0.69 ± 0.12 mm and 0.92 ± 0.12 mm of sinkage for light and heavy wheels. There are several implications from this data set.

One immediately notes that the sinkage for all raked cases was greater, even accounting for the amplitude of the peak. The sinkage for all cases increases with mass (though within the standard error for some) and the grousered wheels sunk less than sandpaper in all cases. Most importantly, the sinkage is not sufficient to surpass even the most generous estimation of minimum depth to avoid surface effects. Neither the level sand sinkage nor the raked sand sinkage was observed to exceed 3 mm on average for these tests. Additionally, by comparison, we see the characteristic pattern from preparation by raking is of similar depth magnitude to the sinkage.

Wheel sinkage of approximately ten grain diameters or smaller, defined as shallow, likely subjected the wheel to grain-size effects. This effect on sand response caused a deviation from the frictional plasticity and consistent granular environment assumption, which underpin the scaling theory. In particular, shallow sinkage alone and/or combined with a characteristic raking pattern of similar magnitude are the probable causes of scaling error. This supports the hypothesis why the lagging wheels displayed lower error. The front wheels partially counteracted the lack of uniformity in the friction coefficient caused by packing fraction issues from raking and low sinkage, and, therefore, reducing scaling discrepancies.

This leads to three general recommendations for using the GSLs for vehicle testing:

1. Utilize a level of sinkage adequate to avoid surface effects and disruption of the internal friction of the grains. This could be qualitative in nature; larger craft may observably far exceed a 10–20 grain diameter depth during experimental conditions. For lighter, smaller craft this may require some level of baseline testing and measurement. This response will depend on multiple factors such as granular media, wheel shape, and mass.
2. Ensure testing environment contains a minimal level of surface depth variation (either from preparation or from other conditions). This ought to be of a sufficiently smaller magnitude than the average sinkage to avoid similar interference effects.
3. Examine scaling predictions for wheel power draw independently on a multi-wheeled vehicle. There is not enough data to draw a conclusion about under what conditions a multi-wheeled vehicle, with at least one following wheel, will obey the scaling predictions given that the dynamics necessarily will change the granular environment between front and back wheel. The front wheel may be disruptive to back wheel predictions, or the differences may be trivial with a heavier, higher sinkage vehicle.

To summarize the three experimental studies: results in the first study show general power prediction deviation between 20 and 35 percent for crushed basalt and 15–25 percent for silica sand. Velocity prediction deviation showed high dependence on material, with silica sand generally between 4 and 10 percent and crushed basalt varied between 0 and 27 percent. Analysis of the second study affirms similar error for grousered and sandpaper wheels, a mass (pressure) dependency for all five pairs and a velocity dependency for both of the heaviest sets. The overall experimental deviations from prediction were investigated in a third study and shown to be likely caused by the scaling dependency on sinkage thresholds. Shallow sinkage during trials (~1 mm, less than 10 grain diameters) resulted in violation of a critical assumption of constant frictional plasticity, subjecting the wheels to surface effects. This is thought to be the source of the mass dependant errors. An alternating peak-and-valley pattern of 1.5 mm retained in the sand after raking during trial preparation was implicated as the source of the leading/lagging wheel differences. This is thought to also cause general error. The findings create insights for using the laws with lightweight robots in granular media and generalizing GSLs. Preliminary sinkage tests ought to be performed at the extremes of planned experimental parameters (heaviest and lightest masses, fastest and slowest speeds) in the material of question to determine if sinkage is adequate. One cannot extrapolate the criterion of adequate sinkage in one media from sinkage results in other granular media.

Speaking to BP-1 specifically and generalizing to planetary simulants, the current hypothesis is that poor sinkage is the most likely cause of power prediction error in this material as well. While both silica sand and BP-1 overpredicted power, BP-1 significantly underpredicted speed. The larger and heavier craft was able to move faster than predicted, pointing to an increased tractive ability not seen in the Quikrete results, possibly from very low flowability due to lack of shear stresses. Quikrete shows small velocity prediction error and power prediction error could likely be reduced with adequate sinkage to ensure a uniform frictional constant, as was reported in the original findings. While we cannot state conclusively if the granular mechanics of BP-1 would exhibit similar behavior for a wheeled vehicle, it is highly likely to occur given an adequate amount of sinkage. Preliminary works [49, 57] indicate BP-1 will obey a dimensionless scaling law under sufficient sinkage engagement with non-wheeled mobility. Exploration of a wheeled GSL in planetary regolith simulant is important future work. To conclude this chapter, we examine the applicability of these laws for wheeled planetary vehicles in the fourth and final study: testing the gravity-variant formula in simulations.

2.3.4   Study four: investigating gravity-variant scaling using MBD-DEM simulations

Recall (2.12), the form of scaling law function before assumptions were applied. We now lift assumption 2, constant gravity, and reduce the equation in a similar manner as before to the following:

$$[\frac{P}{mg\sqrt{lg}}, \frac{V}{\sqrt{lg}}] = \Psi_g(t\sqrt{\frac{g}{l}}, f, \frac{g}{l\omega^2}, \frac{dl^2}{m}). \tag{2.23}$$

If we assume that gravity is changed by some scalar q as mass and length are changed by (s, r), apply the remaining assumptions 4 and 5, and use similar transformation reductions as before, the result is a new set of independent and dependent variables for predicting performance of a wheel of different mass and size in a different gravity:

$$\acute{g} = qg, \quad \acute{l} = rl, \quad \acute{m} = sm, \quad \acute{d} = sr^{-2}d, \quad \acute{\omega} = q^{-1/2}r^{-1/2}\omega$$
$$\acute{P} = q^{3/2}sr^{1/2}P, \quad \acute{V} = q^{1/2}r^{1/2}V \tag{2.24}$$

Note that these terms have identical meaning to those in (2.22), with the addition of the gravity and its scalar. The gravity of the Moon is approximately 1/6 of Earth's, a suitable order of magnitude for evaluating gravity-variant GSLs. The use of computational tools should complement results that may be difficult to obtain experimentally, and this is the case for examining reduced gravity granular–wheel interfaces. Recall that studies [19, 50] mentioned in Section 2.1 found

Figure 2.8   *MBD-DEM simulation displays the vehicle with larger wheels in lunar gravity; color warmth corresponds to sinkage with blue indicating deeper impressions*

that for granular locomotion, results from weight-offset rover experiments in earth gravity did not match identical experiments performed on parabolic flights with direct gravity variation and, in some cases, showed inverted trends due to the physics of granular compaction. If granular locomotion tests at Earth gravity could be theoretically extrapolated to predict performance in lower-gravity environments, it would be of great benefit to planetary rover roboticists. For an initial examination of this, a multi-body dynamics (MBD) and discrete element method (DEM)-simulated environment was created as seen in Figure 2.8.

DEM simulations model individual particle collisions via established and selected physics-driven contact models. Each particle-to-particle collision is modeled. Similar wheel-terrain deformation simulations have been performed before [51]. MBD models the kinematic linkages of the vehicle itself and its reaction forces, motions, and other dynamic details. The MBD-DEM class of simulations is an important emerging tool for insight into how a vehicle system may behave in reduced gravity looking at wheel–terrain interfaces [52, 53]. Earth and lunar gravity velocity scaling comparisons have been previously performed with simulated glass beads with success [4].

All simulation properties were chosen to match that of BP-1 or basalt as best found in the literature in Table 2.4. Rolling and static friction of BP-1 on ABS plastic were determined experimentally by spraying spheres and a plate with adhesive, dusting with BP-1, and running trial experiments. Bulk density measurements of BP-1 were taken; unconsolidated (experimental conditions) bulk density was found to be 1 561 kg/m^3 while consolidated density was 1 633 kg/m^3. Both of these are well within the range previously noted [42]. Young's modulus was reduced, particle size increased, and particle size normally distributed to make simulations computationally feasible. This technique has been employed before [4] for MBD-DEM vehicle

Table 2.4   Properties of simulated BP-1 and craft

Material properties	BP-1	ABS
Poisson's ratio	0.25	0.35
Density (kg/m^3)	3 150	1 070
Young's modulus (Pa)	73E7	1.8E9
Interactive properties	**BP1-BP1**	**BP1-ABS**
Coefficient of restitution	0.8	0.8
Coefficient of static friction	0.56	0.57
Coefficient of rolling friction	0.07	0.17
Other properties	**Value**	
Size of bisphere clump	3 mm	
Size of tetrasphere clump	3.75 mm	
Simulation timestep	9.6E-6 s	
Saved data timestep	0.05 s	

dynamics and is the recommended technique for simulation acceleration [54]. These simulations would otherwise take prohibitively long to complete and, thus, this is a common practice when using DEM simulations. Particles in DEM are also created as combinations of spheres due to the particular physics models, creating less angular particles than their real-world counterparts.

One wheel, the GSL-B set, was chosen to perform three sets of simulations. The craft body, the mass, dimensions, and speeds were identical to experiments in study one (§ 2.3.1). GSL1B was run in a simulated BP-1 environment using Earth gravity. GSL2B/3B were run using identical starting environmental conditions with lunar gravity. The particles were given adequate time to settle to steady granular compaction. In a manner similar to the experimental trials, the craft rests on one side of the simulated bed and the wheels are commanded to rotate at a set angular velocity for the duration of the simulation. A variety of results can be examined, and notably the power and velocity can be easily measured directly. The power and velocity for GSL1B in Earth gravity is used to predict the outcome for GSL2B/3B in lunar gravity and the results are shown in Figure 2.9.

A close match with the gravity-variant scaling laws for predicting both power and velocity was found. GSL2B power deviated 5–8 percent from predicted output for most speeds whereas GSL3B power deviated 1–5 percent from predicted output for most speeds. The fastest speeds had 19 and 12 percent error in GSL2B and 3B, respectively. This error is lower than experiments in either material in study one (§ 2.3.1). In all cases but slowest, the craft drew more power than predicted in the lunar gravity as well. This is a deviation from experimental results, as they showed quite the opposite trend for BP-1 in study 1. It should be noted that this material, though as accurate as possible, cannot be directly compared to BP-1 as it is an approximation composed of spherical composites. These results are nevertheless interesting. A second important difference is the perfect repeatably of the environmental conditions in a simulation and lack of characteristic pattern from terrain preparation. This likely had an influence on the results,

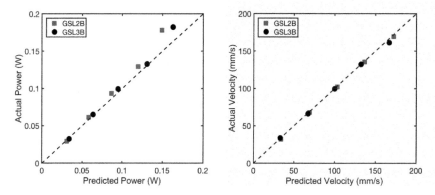

Figure 2.9   *Simulation results paired with their respective predictions with solid black line indicating perfect prediction*

especially the precision of them. Similar to the experiments, we also see less error with increased mass for all data, adding support to the importance of examining sinkage.

Velocity scaling predictions were both more precise and accurate than power. Most had less than 2 percent prediction error; GSL2B error was below 2 percent for all but slowest speed (5 percent) and GSL3B ranged from 0.2 to 3.3 percent. We again acknowledge that the simulation environment was able to provide a much more ideal preparation of material than is typically achieved with experiments. The prediction error was lower than either experimental set in study one (§ 2.3.1), similar to power prediction. This is attributed to the same simulation reasons as power.

Additional insight was gained with sample simulations meant to investigate the error occurrences in study two (§ 2.3.2). The light grouser pair (GSL1G and 2G) from this study were simulated at 30 and 60 RPM and 26 and 52 RPM, respectively, in accordance with scaling law factors. These sample simulations were run with silica sand properties [55]. The scaling law predictions showed 6.5 percent error for DEM simulation results at the same light masses that showed 25 percent error in experiments. All of these results highlight the power of computational tools but also require a note of caution about the ideal conditions of simulations. Identical parameters that performed with significant deviation in experiments in two materials performed with little significant error in the simulated environment. This is not a flaw of the laws themselves, but rather the reality of experimentation. Notably, the surface in a simulation was perfectly leveled (no imperfect surface height variations) and also unraked, which has been discussed as a source of error. This is the most likely explanation for lower DEM error. We conclude that GSLs, with gravity variance, can be used as a design tool for evaluating small, lightweight craft in lunar gravity provided that adequate sinkage threshold is achieved.

2.4   Recommendations and future work

The results of the four studies give new context to GSLs for planetary rover roboticists in several ways:

1. Evaluation of applicability to a lunar simulant, characterized by a weakly cohesive, highly angular, and interlocked media. Identical tests performed with Quikrete silica sand show differences in predictive accuracy, particularly for predicting velocity.
2. Evaluation of grousered wheels, a typical aspect of planetary wheel design in granular media to avoid high slip. Wheel shape does not appear to impact prediction accuracy, even for smaller lightweight rovers.
3. Evaluation of a two-wheeled vehicle with free dynamic movement. Differentiation of front and back wheels showed significant differences in prediction accuracy of power for lighter, smaller vehicles.
4. Evaluation of lighter rovers at faster angular and translational speeds; the laws were originally evaluated at 2.33–4.76 RPM with 13.4–45.7 kg of mass. The set in this chapter spans 13.1–75.0 RPM and approximately 1.5–5.2 kg mass for a vehicle with similar wheel sizing to the original tests (although with weight distributed to two wheels instead of one). During study two (§ 2.3.2), it was shown that mass (sinkage) had a significant effect on prediction capabilities. Importantly, this study also showed that even at suboptimal sinkage levels, accuracy decreases with speed at higher mass levels. This implies possible limitations when tests are performed even at adequate sinkage levels.
5. Evaluation of effects of characteristic preparation patterns, both as solitary factors and compared to sinkage magnitudes. The results of all studies indicate that these preparation patterns can disrupt predictions, and that caution should be undertaken with the prepared environment.
6. Evaluation using three-dimensional, two-wheeled, MBD-DEM co-simulations at lunar gravity. The results of the fourth study support the utility of these computational tools and the accuracy of the scaling laws to predict power and velocity at one gravity from another.

These results guide us to several recommendations for proposed future avenues to expand this work:

1. To better understand the inaccuracies, a study in a well-characterized sand could evaluate a range of masses from 2.6 to 29.3 kg. This would span the masses used in these studies to the smallest mass used for a large wheel in Slonaker *et al.,* which reported no errors. This would evaluate the relationship between sinkage and scaling error, ideally through direct measurement of sinkage and pressure.
2. Additional work with BP-1 or similar media ought to be performed to examine whether a granular environment with comparable angularity and particle size will obey the GSLs for wheels and at what sinkage thresholds.

3. If BP-1 (or another simulant) is shown to obey wheeled scaling laws with sufficient sinkage, an important advancement would be examining the accuracy while subjecting the environment to vacuum. If this step was proven accurate, it is likely that wheeled planetary vehicles could be designed with accurate estimates of performance on the Moon and many other bodies of interest.
4. Examination of the mass scalar s magnitude is also important. In the first study of the chapter, 1.778 and 2.0 are used. In the second study, 1.778 was used for all sets. In the Slonaker *et al.* study, mass scalars were 2.18–2.40. In plainer terms, confirming that a value of $s = 3$ leads to accurate results would mean predicting large vehicle performance from a 1/3 weight vehicle. Is there a maximum limit to this, or is there an acceptably small error for larger scalars? The Curiosity Rover is 900 kg; if certain performance characteristics could be predicted easily by a 90 kg miniaturized version, how might that change the ability to explore the rover design space?
5. Developing other parameters of interest, such as non-dimensionalized drawbar pull force, was mentioned as a potential avenue in the original GSL paper; it would be beneficial to predict drawbar pull forces using a scaling approach.
6. MBD-DEM simulations similar to study four (§ 2.3.4) should continue to be explored. This computational tool has shown high accuracy with regard to the laws, and attention and improvement to the simulated granular environment will make such simulations all the more relevant. In particular, examining the accuracy of laws with regard to the gravity of smaller bodies such as the Martian, Saturnian, and Jovian moons may be of interest.

The data presented in this chapter highlight important context of feasibility for using GSL. The power predictions for a lightweight (1.5–5.2 kg), two-wheeled, unrestrained craft with three wheel shapes significantly deviated across two sets of studies. Significant deviations occurred in both silica sand and crushed basalt. The root cause of these power prediction deviations was identified as inadequate sinkage in the third study. This violated the assumption of a constant friction coefficient in the environment and trivial particle effects. A first-order experiment evaluating the sinkage of a target craft, aiming for sinkage levels of ten times the average particle diameter is suggested. This is arguably the most important influence on law accuracy for power, although we also recommend a preparation method that avoids any characteristic patterning on the granular surface. We see greater deviations in velocity predictions in BP-1 than silica sand. The underprediction of velocity for all wheel types in this case is likely caused by a lack of granular flow, driven by the small, angular, interlocking particles and might be eliminated at higher sinkages. Finally, MBD-DEM simulations show a promising avenue of evaluating gravity-variant rover prediction with GSL and as a design evaluation tool for planetary vehicles.

References

[1] Marvi H., Gong C., Gravish N., *et al.* 'Sidewinding with minimal slip: snake and robot ascent of sandy slopes'. *Science*. 2014;**346**(6206):224–9.

[2] Thoesen A., McBryan T., Mick D., Green M., Martia J., Marvi H. 'Comparative performance of granular scaling laws for lightweight grouser wheels in sand and lunar simulant'. *Powder Technology*. 2020;**373**(2):336–46.

[3] Thoesen A., Ramirez S., Marvi H. 'Screw-generated forces in granular media: experimental, computational, and analytical comparison'. *AIChE Journal*. 2019;**65**(3):894–903.

[4] Thoesen A., McBryan T., Marvi H. 'Helically-driven granular mobility and gravity-variant scaling relations'. *RSC Advances*. 2019;**9**(22):12572–9.

[5] Thoesen A., Ramirez S., Marvi H. 'Screw-powered propulsion in granular media: an experimental and computational study'. 2018 IEEE International Conference on Robotics and Automation (ICRA). IEEE; 2018. pp. 1–6.

[6] Grujicic M., Marvi H., Arakere G., Bell W.C., Haque I. 'The effect of up-armoring of the high-mobility multi-purpose wheeled vehicle (HMMWV) on the off-road vehicle performance'. *Multidiscipline Modeling in Materials and Structures*. 2010;**6**(2):229–56.

[7] Grujicic M., Marvi H., Arakere G., Haque I. 'A finite element analysis of pneumatic-tire/sand interactions during off-road vehicle travel'. *Multidiscipline Modeling in Materials and Structures*. 2010;**6**(2):284–308.

[8] Bagheri H., Taduru V., Panchal S. 'Animal and robotic locomotion on wet granular media'. *Conference on Biomimetic and Biohybrid Systems*. 2017:13–24.

[9] Kamath S., Kunte A., Doshi P., Orpe A.V. 'Flow of granular matter in a silo with multiple exit orifices: jamming to mixing'. *Physical Review E*. 2014;**90**(6):062206.

[10] Kamrin K. 'Nonlinear elasto-plastic model for dense granular flow'. *International Journal of Plasticity*. 2010;**26**(2):167–88.

[11] Lee S., Marghitu D.B. 'Multiple impacts of a planar kinematic chain with a granular matter'. *International Journal of Mechanical Sciences*. 2009;**51**(11–12):881–7.

[12] Omori H., Murakami T., Nagai H. 'Validation of the measuring condition for a planetary subsurface explorer robot that uses peristaltic crawling'. in *Aerospace Conference, 2013 IEEE*; 2013. pp. 1–9.

[13] Bekker M. *Theory of land locomotion*. University of Michigan Press; 1956.

[14] Bekker M. *Introduction to terrain-vehicle systems. Part I: The terrain. Part II: The vehicle*. University of Michigan Press; 1969.

[15] Bekker M.G. *Mechanics of locomotion and lunar surface vehicle concepts*. Sae Transactions; 1964. pp. 549–69.

[16] Meirion-Griffith G. *Advances in vehicle-terrain interaction modeling for small, rigid-wheeled vehicles operating on deformable terrain*. Illinois Institute of Technology; 2012.

[17] Wong J.Y. *Theory of ground vehicles*. John Wiley & Sons; 2008.

[18] Bekker M.G. *Off-the-road locomotion: research and development in terrame-chanics*. University of Michigan Press; 1960.

[19] Wong J.Y. 'Predicting the performances of rigid rover wheels on extra-terrestrial surfaces based on test results obtained on earth'. *Journal of Terramechanics*. 2012;**49**(1):49–61.

[20] Aguilar J., Zhang T., Qian F., *et al.* 'A review on locomotion robophysics: the study of movement at the intersection of robotics, soft matter and dynamical systems'. *Reports on Progress in Physics*. 2016;**79**(11):110001.

[21] Maladen R.D., Ding Y., Li C., Goldman D.I. 'Undulatory swim-ming in sand: subsurface locomotion of the sandfish lizard'. *Science*. 2009;**325**(5938):314–18.

[22] Maladen R.D., Ding Y., Umbanhowar P.B., Kamor A., Goldman D.I. 'Mechanical models of sandfish locomotion reveal principles of high perfor-mance subsurface sand-swimming'. *Journal of The Royal Society Interface*. 2011;**8**(62):1332–45.

[23] Ding Y., Gravish N., Goldman D.I. 'Drag induced lift in granular media'. *Physical Review Letters*. 2011;**106**(2):028001.

[24] Ding Y., Sharpe S.S., Masse A., Goldman D.I. 'Mechanics of undula-tory swimming in a frictional fluid'. *PLoS Computational Biology*. 2012;**8**(12):e1002810.

[25] Li C., Zhang T., Goldman D.I. 'A terradynamics of legged locomotion on granular media'. *Science*. 2013;**339**(6126):1408–12.

[26] Hatton R.L., Ding Y., Choset H., Goldman D.I. 'Geometric visualiza-tion of self-propulsion in a complex medium'. *Physical Review Letters*. 2013;**110**(7):078101.

[27] Zhang T., Goldman D.I. 'The effectiveness of resistive force theory in granu-lar locomotion'. *Physics of Fluids*. 2014;**26**(10):101308.

[28] Aydin Y., Chong B., Gong C. 'Geometric mechanics applied to tetrapod lo-comotion on granular media'. *Conference on Biomimetic and Biohybrid Systems*. 2017:595–603.

[29] Dunatunga S., Kamrin K. 'Continuum modelling and simulation of gran-ular flows through their many phases'. *Journal of Fluid Mechanics*. 2015;**779**:483–513.

[30] Dunatunga S., Kamrin K. 'Continuum modeling of projectile impact and penetration in dry granular media'. *Journal of the Mechanics and Physics of Solids*. 2017;**100**(3):45–60.

[31] Askari H., Kamrin K. 'Intrusion rheology in grains and other flowable materi-als'. *Nature Materials*. 2016;**15**(12):1274–9.

[32] Slonaker J., Motley D.C., Zhang Q., *et al.* 'General scaling relations for loco-motion in granular media'. *Physical Review E*. 2017;**95**(5):052901.

[33] Team R. 'Characterization of the Martian surface deposits by the Mars Pathfinder rover, Sojourner. Rover team'. *Science*. 1997;**278**(5344):1765–8.

[34] Yoshida K., Britton N., Walker J. 'Development and field testing of moon-raker: a four-wheel rover in minimal design'. ICRA13 Planetary Rovers Workshop; 2013.

[35] Karras J., Fuller C., Carpenter K. 'Puffer: pop-up flat folding explorer robot'. Google Patents; US Patent App. 15/272,239; 2017.

[36] Karras J., Fuller C., Carpenter K. 'Pop-up mars rover with textile enhanced rigid-flex PCB body'. 2017 IEEE International Conference on Robotics and Automation (ICRA). IEEE; 2017. pp. 5459–66.

[37] Ibrahim A., Aoshima S., Fukuoka Y. 'Development of wheeled rover for traversing steep slope of cohesionless sand with stuck recovery using assistive grousers'. 2016 IEEE International Conference on Robotics and Biomimetics (ROBIO). IEEE; 2016. pp. 1570–5.

[38] Inotsume H., Sutoh M., Nagaoka K. 'Modeling, analysis, and control of an actively reconfigurable planetary rover for traversing slopes covered with loose soil'. *Journal of Field Robotics*. 2013;**30**(6):875–96.

[39] Heverly M., Matthews J., Lin J. 'Traverse performance characterization for the Mars science laboratory rover'. *Journal of Field Robotics*. 2013;**30**(6):835–46.

[40] Mueller R. 'Lunabotics mining competition: inspiration through accomplishment'. Earth and Space 2012: Engineering, Science, Construction, and Operations in Challenging Environments; 2012. pp. 1478–97.

[41] Mueller R., Van Susante P. 'A review of extra-terrestrial mining robot concepts'. Earth and Space 2012: Engineering, Science, Construction, and Operations in Challenging Environments; 2012. pp. 295–314.

[42] Suescun-Florez E., Roslyakov S., Iskander M., Baamer M. 'Geotechnical properties of BP-1 lunar regolith simulant'. *Journal of Aerospace Engineering*. 2015;**28**(5):04014124.

[43] Suescun-Florez E., Roslyakov S., Iskander M., Baamer M. 'Geotechnical properties of BP-1 lunar regolith simulant'. *Journal of Aerospace Engineering*. 2015;**28**(5):04014124.

[44] Townsend I., Muscatello A., Dickson D. 'Mars ISRU pathfinder regolith autonomous operations-modeling and systems integration'. AIAASPACE and Astronautics Forum and Exposition; 2017. p. 5150.

[45] Rahmatian L., Metzger P. 'Soil test apparatus for lunar surfaces'. Earth and Space 2010: Engineering, Science, Construction, and Operations in Challenging Environments; 2010. pp. 239–53.

[46] Stoeser D., Rickman D., Wilson S. *Preliminary geological findings on the BP-1 simulant*. National Aeronautics and Space Administration, Marshall Space Flight Center; 2010.

[47] Thoesen A., McBryan T., Mick D., Green M., Martia J., Marvi H. 'Comparative performance of granular scaling laws for lightweight grouser wheels in sand and lunar simulant'. *Powder Technology*. 2020;**373**(2):336–46.

[48] Thoesen A., McBryan T., Green M., Mick D., Martia J., Marvi H. 'Revisiting scaling laws for robotic mobility in granular media'. *IEEE Robotics and Automation Letters*. 2020;**5**(2):1319–25.

[49] Thoesen A. *Helically-driven dynamics in granular media*. Arizona State University; 2019.

[50] Kobayashi T., Fujiwara Y., Yamakawa J., Yasufuku N., Omine K. 'Mobility performance of a rigid wheel in low gravity environments'. *Journal of Terramechanics*. 2010;**47**(4):261–74.

[51] Johnson J.B., Duvoy P.X., Kulchitsky A.V., Creager C., Moore J. 'Analysis of Mars exploration Rover wheel mobility processes and the limitations of classical terramechanics models using discrete element method simulations'. *Journal of Terramechanics*. 2017;**73**(E00F15):61–71.

[52] Wasfy T.M., Mechergui D., Jayakumar P. 'Understanding the Effects of a Discrete Element Soil Model's Parameters on Ground Vehicle Mobility'. *Journal of Computational and Nonlinear Dynamics*. 2019;**14**(7).

[53] Recuero A., Serban R., Peterson B., Sugiyama H., Jayakumar P., Negrut D. 'A high-fidelity approach for vehicle mobility simulation: nonlinear finite element tires operating on granular material'. *Journal of Terramechanics*. 2017;**72**(5):39–54.

[54] Lommen S., Schott D., Lodewijks G. 'DEM speedup: stiffness effects on behavior of bulk material'. *Particuology*. 2014;**12**(2):107–12.

[55] Cil M.B., Alshibli K.A. '3D assessment of fracture of sand particles using discrete element method'. *Géotechnique Letters*. 2012;**2**(3):161–6.

[56] Thoesen A., McBryan T., Mick D., Green M., Martia J., Marvi H. 'Comparative performance of granular scaling laws for lightweight grouser wheels in sand and lunar simulant'. *Powder Technology*. 2020;**373**(2):336–46.

[57] Thoesen A., McBryan T., Mick D., Green M., Martia J., Marvi H. 'Granular scaling laws for helically driven dynamics'. *Physical Review E*. 2020;**102**(3):032902.

Chapter 3

Compliant pneumatic muscle structures and systems for extra-vehicular and intra-vehicular activities in space environments

Samuel Wandai Khara[1], Alaa Al-Ibadi[2], Hassanin S. H. Al-Fahaam[2], Haitham El-Hussieny[1,3], Steve Davis[1], Samia-Nefti Meziani[1], and Olivier Patrouix[4]

In space environments, astronauts have duties that need to be addressed and some of them can be overwhelming, especially considering, in most cases, the crew members in the space station are normally a few. This can be resolved by accompanying the astronauts with assistant robots to reduce the workload. Thus, deployment of soft robots, inspired by the morphological adaptation ability existing in octopus tentacles, elephant trunks, and snakes, is promising in a very long horizon for space exploration. In this chapter, we will discuss one of the soft robotic technologies we developed based on pneumatic muscle actuators' (PMAs) principle and show how they can be effective to improve the existing robotic systems that can be implemented in space environments. These developed PMAs are promising and can serve as alternatives to the traditional rigid robotic systems, by performing dexterous activities that are tedious, repetitive, and difficult to perform.

3.1   Introduction

Recently, considerable literature has grown up around the theme of Human-Robot Collaboration (HRC) that is still in continuous development and improvement. Sectors like industry/manufacturing, households, and hospitals, even in extreme environments such as nuclear and subsea sites, are benefiting from the synergy in the

[1]Autonomous Systems and Robotics Research Centre, School of Science, Engineering & Environment, The University of Salford, Salford, M54WT, United Kingdom
[2]Computer Engineering Department, Engineering College, University of Basrah, Basrah, Iraq
[3](on-leave) Electrical Engineering Department, Faculty of Engineering (Shoubra), Benha University, Cairo, Egypt
[4]ESTIA , 92 allée Théodore Monod, Technopole Izarbel, 64210 BIDART

collaboration between the human decision-making ability and the robot's accuracy [1]. Indeed, the form of such collaboration provided by robots to humans varies significantly according to the application or the mission that is required to be achieved. For instance, in harsh and unsafe environments, where it is risky for humans to have access, a teleoperated robot could be controlled remotely via a dedicated Human–Robot Interface (HRI) [2]. Moreover, robotics manipulators could also share the working environment and work closely and safely with humans in helping them achieving tedious, repetitive, and difficult tasks [3]. Rehabilitation and assistive robots are other forms of human-robot collaboration where either wearable or non-wearable robot systems aid human patients in enhancing their irregular body movements [4]. In space applications, robots such as Robonaut 2 (R2) [5] are playing a crucial role in HRC by relieving some repetitive and tiring duties from human space explorers due to their high dexterity and remote operation.

Inspired by the morphological adaptation ability existing in octopus tentacles, elephant trunks, and snakes, soft robots have proved the potential to facilitate the exploration of challenging environments [6]. Soft robotics is a novel approach in robotics that utilizes alternative flexible materials that are deformable to increase flexibility and controllability. It is a relatively new direction that has the potential to advance the robotics field as it has many benefits such as affordability due to its low cost, lightweight, safety, and morphological adaption ability. As compared to rigid robots, soft robots have curvilinear structures with continually bending backbones that not only make them highly adaptable to the surroundings [6] but also allow sharing the working environment safely with humans [7]. Promising performance could be achieved by replacing rigid robots with soft continuum robots in fulfilling the desired tasks. However, several challenges need to be addressed. For instance, stiffness modulation is one of the crucial features that soft robots are better to have to be able to modulate their stiffness according to the required task at hand.

3.2   Robotic solutions for space environments

As early as 1991, one of the conferences held by NASA discussed Automation and Robotics for space-based systems, which discussed various proposals on how robotic systems such as manipulators could impact the space environments. Over time, it was considered that telerobotics and telepresence technologies could be utilized to reduce the crew extra-vehicular activity (EVA) and intra-vehicular activity (IVA) workloads in the international space station (ISS) [8]. The development of a ground teleoperated IVA robot was proposed to relieve astronauts from tedious and routine tasks, which could also provide ground researchers with a chance to interact and participate in space experiments.

A Dexterous Manipulator Development (DeManD) is an example of an EVA robotics for space station freedom suggested by Williams [9] to carry out specific tasks like inspection, workspace setup, and repair. Also, an IVA robotics for space station freedom was illustrated to demonstrate how it could increase productivity in conducting experiments using its automation technology. Besides, a Robotic

Intravehicular Assistant (RIVA) was considered to be implemented inside a full-scale mock-up of a space station laboratory module to assist in protein crystal growth (PCG) experiments [10], demonstrating how robotic manipulators in space would be very essential in assisting astronauts even in their IVAs, where applicable.

On the other hand, humanoid robot such as R2 [11] has been developed to provide several benefits to the crew in the ISS by its advanced manipulators. It is an improvement of the Robonaut 1 (R1) developed by DARPA and NASA, which had the motivation of carrying out EVAs on behalf of astronauts [12]. The modifications on R2 were also meant to assist astronauts in maintenance tasks with minimal human supervision and R2 can reposition itself to different worksites, providing the ISS crew with more time to conduct other activities [11]. The R2 is usually mounted on a rail to facilitate repositioning and aside from the hands, the legs as well have end effectors for holding tools to perform tasks independent of human input.

Although there is diversity of the mentioned robotics involvement in space applications, several directions are still promising and will show significant performance if robots have been incorporated. For example, traditional astronauts' spacesuit gloves are mainly designed for protection in harsh space environments while allowing astronauts to perform dexterous activities for EVA and IVA. However, existing gloves are made of many layers of insulating materials and tough fabrics to handle pressurization, warmth, and protection. For instance, the extra-vehicular passive spacesuit glove has reduced hand mobility by 10 percent due to its material that is designed to withstand pressure and thermal conditions [13]. One of the solutions to mitigate this issue, particularly in the gas-pressurized spacesuits and gloves, is to utilize the concept of mechanical counter pressure (MCP). So, spacesuits and elastic gloves are having counter pressures to meet the mobility challenges [14, 15]. The MCP concept is still undergoing research where the amount of counter- pressures needs to be in harmony with the hand movement, which mainly depends on the task. As an example, cable-driven power-assisted gloves were proposed in [16] to aid in improving the functionality of the pressurized spacesuit glove.

3.2.1 Soft robotic systems as an alternative robotic solution for space environments

R2 and RIVA are rigid robotic structures that are susceptible to challenges when flexibility in confined and limited spaces and gentle interactions with humans are considered. As NASA's early implementation of one of the first IVA-type robotic manipulator systems in ISS, they highly stated that despite benefits acquired from the robot to assist in a specific task, the central area of concern is the crew's laboratory safety [17]. Soft robotic flexible structures can resolve some of these challenges since they pose less danger when interacting with humans and are relatively at a lower cost and much power efficient with a higher power to weight ratio.

Soft robotics is a relatively new field where soft materials and soft actuators are used to develop robot manipulators and grippers that are inherently safe to operate alongside people, possessing low mass and inertia, are often low cost, and have other abilities not offered by traditional rigid robots. Thus, we will explore how dexterous soft robotic systems designed using PMA technology can offer a contributing solution to some of the mentioned challenges being addressed in space environments.

Space colonies will soon get to be established, and humans might start to settle in different places in space such as the Moon and Mars. In such environments, robots can play a crucial role, especially at the early stages when establishing the colonies in space, since the labor might be limited by the available human explorers at the site. Therefore, cost becomes a critical consideration for developing and manufacturing space robots. Both the academia and industries have been researching due to the several shortcomings faced by traditional robotic systems to develop different types of grippers, end effectors and manipulators to cut down the cost of the robotic systems and include additional abilities that allow handling and manipulation of a wide range of objects and materials [18]. Choi *et al.* [19] had a similar approach where the gripper can cost as much as 20 percent of the total cost of a whole robotic system. Meanwhile, the gripper design is limited to handling a specific object or a small range of objects. It could be expensive to manufacture grippers for varieties of objects each time and consequently could result in complex control systems. Hence, there is a need to develop systems that are low cost, multipurpose, with simple control burden wherever possible, and able to be integrated easily with existing robotics hardware. The development of new robotics designs, actuations, sensing, and control techniques is thus the different subfield contributing to the above discussions. Newer ways of designed actuators and their usefulness in space environments will be explained here.

One type of soft robots' actuation techniques is the PMA [20], which over recent years is shown to be a potential alternative to traditional actuators in existing manipulators and end effectors. PMAs have significant potential benefits due to their low cost, low weight, flexibility, and inherent softness [21].

Autonomous Systems and Robotics Research Centre at the University of Salford has been developing a soft robotic system using PMA technology. Some of the robotic systems include multi-fingered grippers designed based on novel Self-Bending Contraction Actuator (SBCA) and ring-shaped circular gripper that use the newly developed Circular PMA (CPMA). These new grippers are particularly well suited to grasping unknown objects as they naturally deform to the shape of the object being handled with their intrinsic flexibility. In the context of space applications, these soft robots are important for assisting inter-vehicular activities such as grasping of space rocks that might have irregular shapes and textures and probably also are very delicate for handling. These gripper designs offer enhanced efficiency by eliminating the need for extensive grasp preplanning, having high payloads, minimal power requirements, and reduced control complexity. More of this based on their construction and experiments is explained later. Wearable soft robotics systems designed using the Extensor-Bending PMA (EBPMA) are also discussed and we will show how they can be beneficial for activities that require force augmentation to improve human performance when using the hands. A low-cost power assistive glove using the EBPMA is developed and experimented to demonstrate its ability to augment the user's strength and reduce fatigue. Such actuation technology

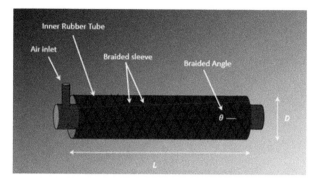

Figure 3.1   *The structure of the PMA, where L and D are the length and the diameter of the rubber muscle, respectively, at zero air pressure and θ being the braided angle, which is the angle between the vertical line and the braided strand*

for a glove can extend the abilities for a spacesuit glove worn by astronauts for either EVA or IVA.

3.3   Soft robotic systems based on PMA as an alternative to rigid robotic systems for space environments

Soft robots are commonly made from soft and deformable materials and typically are not relying on traditional actuators. Instead, soft robots use soft actuation technologies and one popular actuator is the McKibben Muscle [22] known as the PMA. This actuator is formed from lightweight, soft materials, and being pneumatic it is compliant due to the compressibility of air. Thus, this mechanical property is the reason for replacing the rigid structures used in developing space robots, for example, manipulators and end effectors used to develop R2. In addition to the reduction of weight for safety most of such robots are still mounted on rails or supporting member structures. The weight reduction also improves power efficiency and ease of mobility.

The simplicity of McKibben muscle design is the reason for its popularity among PMA basic structures [22]. This PMA type is constructed by inserting a rubber tube inside a braided sleeve with caps at each end, one end is closed and the other left open, to let pressurized air in or out, see Figure 3.1. For a contractor PMA at zero pressure, the tube and the sleeve have equal length and diameter, whereas for the extensor PMA, the sleeve is longer than the tube but of equal diameters. The braided sleeve angle governs the ability to change in length. If the braided angle is greater than 54.7 degrees (minimum energy state) the actuator will extend upon pressurization until reaching the minimum energy state. And if the braid angle is less than 54.7 degrees, any pressurization allows the contraction of the actuator.

Behavioral analysis of the two types of PMA demonstrated their unique properties. A contractor and extensor PMAs of lengths 18.4 and 16 cm, respectively, and

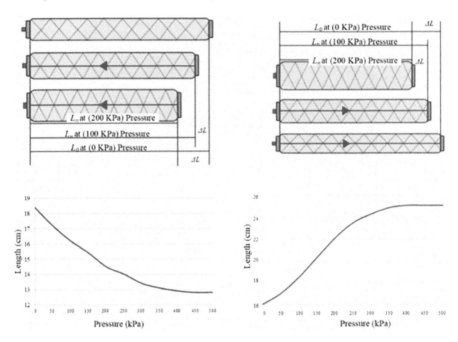

Figure 3.2   *Schematic representation of contractor and extensor PMA to the*
right and left, respectively, and their summary in length changes with
respect to pressure changes

of diameters 5 mm were used in an experiment (Figure 3.2) to analyze changes in
length upon changes in pressure between zero and 500 kPA at steps of 50 kPA (at no
load). The contractor muscle reduced in length to 12.8 cm, about 30 percent reduction, and extensor increased to 25.1 cm, a 56 percent extension.

The contractor PMAs possess similar behavior to the human muscles; they contract by thickening when pressurized. These are three distinguishable and crucial
observations for the PMAs when in use [23, 24]:

• Isotonic characteristics: Varying input pressure under constant load, Figure 3.2
  on the left.
• Isobaric characteristics: Varying load under constant pressure input, Figure 3.2
  on the right.
• Isometric performances: Constant operation ratio by changing both pressure
  and load.

An experiment to investigate the three characteristics is done by fixing the contractor PMA at one end and mounting a load M on the other end (Figure 3.2) and
applying incremental pressure from zero kPa to pressure P_1 produces a pulling force
equilibrium to the loading mass M [25]. As a result, the volume increases to V_1 and
length reduces to L_1. Additional pressure to P_2 further increases the volume to V_2

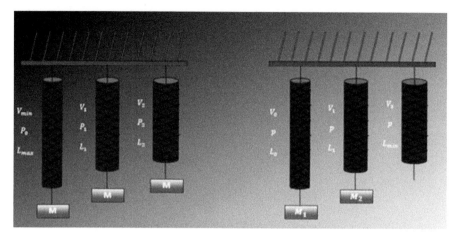

Figure 3.3   *On the left the work of the PMA using constant load test and on the (right) is the constant pressure test of the PMA.*

and a length reduction to L_2 until the maximum pressure is attained, which depends on the (sleeve and rubber) material properties of the PMA.

Figure 3.3 likewise shows the effects of variable load under constant pressure. Reduction of mass M_1 to M_2 decreases the length from L_1 to L_2.

3.3.1   Modeling of a pneumatic muscle actuator

Most engineering systems must have quite accurate models that are useful in improving their behavior when implemented as part of a system [26]. Mathematical modeling of the PMAs is an ongoing process. One of the models relates the air pressure and the length of the PMA in generating a contractile force for the case of the contracting PMA. The model is affected by several factors, such as the properties of the materials used in the construction of the PMA, its length, its diameter, the braided angle, the air pressure, and the load force. Having a better understanding of their relationship enables the achievement of accurate models especially in developing a control model for the muscle [27].

One of the derived model [28] of a contraction PMA assumes that the PMA is cylindrical in shape, a contact point between the braided sleeve and the surface of the inner tube is always present, there is negligible friction between the rubber and the braided sleeve and no considerations of the elastic forces in the inner rubber tube.

Using the parameters shown in Figure 3.4 the work input W_{in}, for a McKibben's muscle under air pressure supply is given by (3.1):

$$dW_{in} = \int_{s_i} \left(P - P_0\right) dl_i.ds_i = \left(P - P_0\right) \int_{s_i} dl_i.ds_i = P_g dV \tag{3.1}$$

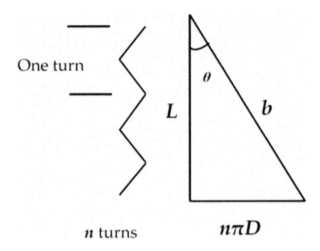

One turn

n turns

$n\pi D$

Figure 3.4   The parameters of the PMA

where P is the absolute pressure, P_0 is the environment pressure ($P_0 = 1.0336$ bar), P_g is the gauge pressure (the relative pressure), s_i is the total inner surface, dl_i is the inner surface displacement, and dV is the volume change.

Equation 3.2 shows the relation of how the change in work input W_{in} is proportional to the contractile force F and change in axial length L of the actuator.

$$dW_{in} = -FdL \qquad (3.2)$$

Assuming the actuator has no storage of energy, then the input work will be equal to work output (3.3). The work output W_{out} is a result of the shortening of the actuator's length while dissipating a contractile force due to the change in volume and pressure input (3.4):

$$dW_{out} = dW_{in} \qquad (3.3)$$
$$-FdL = P_g dV \qquad (3.4)$$
$$F = -P_g \frac{dV}{dL} \qquad (3.5)$$

$\frac{dV}{dL}$ assumes that the lengths of the strands of the braided sleeve are constant during the pressurization of the actuator. Thus, volume V of the muscle actuator is dependent on the length L and diameter D, which are given by (3.6) and (3.7).

$$L = b\cos\theta, \qquad D = \frac{b\sin\theta}{n\pi}, \qquad V = \frac{1}{4}\pi D^2 L \qquad (3.6)$$

where b and n are constant, equation 3.6 assumes a volume of the actuator is in the form a cylinder:

$$V = \frac{b^3}{4\pi n^2}\sin^2\theta\cos\theta \qquad (3.7)$$

Hence the contractile force can be expressed as follows:

$$F = -P_g \frac{dV}{dL} = -P_g \frac{\frac{dV}{d\theta}}{\frac{dL}{d\theta}} = \frac{P_g b^2 \left(3\cos^2\theta - 1\right)}{4\pi n^2} \tag{3.8}$$

The mathematical model given by (3.8) illustrates that at maximum contraction when $F = 0$ the strands of the braided sleeve will be at $\theta = 54.7$ degrees. This is one of the fundamental analyses demonstrating the basic behavior of this type of PMA. Further analysis is still on research by the robotics community to consider how friction, material deformation, elastic energy, and other factors affect energy losses in the system.

3.3.2   Characterization of contractor PMA

The characterization is a crucial step in developing an effective system. Therefore, using the force formulas for extensile and contractile muscles derived by [28, 29] gives useful insight into the performance of an actuator. However, it has been seen that despite the many models produced for pneumatic muscles they all contain errors that lead to inaccuracies. It is therefore prudent to experimentally investigate the behavior of the actuators before use. We did a study that involved a designed contractor and extensor PMA using different pressures and loading. The length of the contractor PMA was taken to be a function of air pressure and initial length. And further led to the modification of the contractile force formula [23, 29] to reduce the error between the experimental and model data. Additional study on the lengths and forces of both serial and parallel configurations of contractor PMAs was also done to observe the effects on different configurations and designs. Comparison analysis for a single extensor PMA and extensor PMA with multiple extensor PMAs connected in series was performed to determine their respective lengths and forces relationship on performance.

Figure 3.5 shows a contractor PMA with application of air pressure within the range of 0–500 kPa manually via a valve with increments of 50 kPa. The reverse

Figure 3.5   *Prototype of a contractor PMA used in the characterization experiments*

Table 3.1   Specifications of a contractor PMAs under experiment

Nominal length L_0 (cm)	Diameter D (cm)	Braided angle (θ^o)
20	1.767	31.35
30	1.752	30.02
40	1.764	30.28

process has also been done to reduce the pressure to zero and observe its hysteresis property. Table 3.1 is a summary of the different PMAs that were used in the experiment with varying lengths.

Figure 3.6 (a) shows a length change experiment for the contractor PMA's changes in length to changes in pressure. The recorded changes in length at a certain pressure level were determined by averaging the respective changes in length during the increase and decrease of air pressure. This was done to remove the hysteresis effect.

$$\dot{L} = \frac{L_{n_{dec}} + L_{n_{inc}}}{2} \tag{3.9}$$

where $0 \leq n \leq P_{max}$, $L_{n_{dec}}$, and $L_{n_{inc}}$ are the contraction and elongation lengths, respectively. The $\dot{L}$ is then used as the actuator length at respective points, Figure 3.6 (b).

3.3.3   Analysis of contractor PMA

The force modeling of contractor PMA conducted by Deimel *et al.* [29] served as the foundation of some underlying assumptions used for the modification of the model. The procedure involved experimental loading with varying weights: 0–10 kg in steps of 0.5 kg at the free end of PMAs that were of different lengths. The air

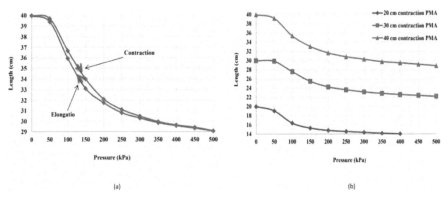

(a) (b)

Figure 3.6   (a) Changes in length to changes in air pressure and (b) Experimental data for three PMA prototypes of different initial lengths of 20, 30, and 40 cm

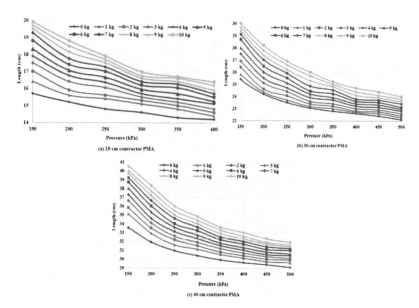

Figure 3.7   Changes in length of (a) 20 cm, (b) 30 cm, and (c) 40 cm PMA with respect to changes in pressure at fixed loadings

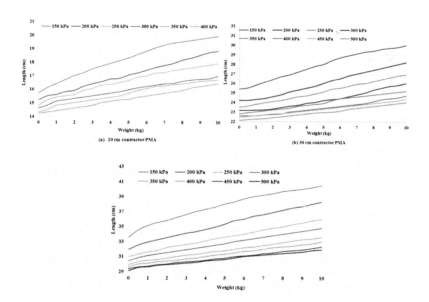

Figure 3.8   Changes in length of (a) 20 cm, (b) 30 cm, (c) 40 cm PMA with respect to changes on loadings at fixed pressures

pressure also was varied within a range of 0 kPa–P_{max}, as summarized in Figures 3.7 and 3.8.

In Figure 3.7, the contractor PMAs have similar behavior of change in length when loaded and pressurized at varying air pressure. They also showed a lower contraction ratio at higher loadings because of more downward force by the weight of the loads. The experiment recorded in Figure 3.8 showed an increase in length that is proportional to the loadings at fixed air pressure. At higher pressures, the increase in length is lower due to the generation of higher forces.

The model assumes the PMA is cylindrical in shape and there is constant contact between the inner rubber tube with the outer braided sleeve [22]. However, in practice, this is not the case and the factors vary with actuator design. A correction factor ε is therefore introduced but highly dependent on the PMA structure. In a test actuator, full contact of the rubber and sleeve occurs only above certain pressure P; in the experiment, it was found to be 45 kPa for the PMA prototype. The resulting modified force f is seen in (3.10):

$$f = \pi r^2_{0centers} \, (P - 45) \left[\alpha \left(1 - q\varepsilon\right)^2 - \beta\right], \qquad P \geq 45 \, kPa \tag{3.10}$$

where $\alpha = \frac{3}{tan^2\theta}$, $\varepsilon = \frac{L}{L_0}$, $\beta = \frac{1}{sin^2\theta}$, $q = 1 + 1e^{-0.5P}$ while θ is the braided angle, and $r_{0centers}$ is the radius of the outer braided sleeve.

Figure 3.9 (a) and (b) are plots of force against pressure for practical experiment and theoretical data using (3.11) and (3.10), respectively. The difference with the two line graphs is because of the underlying assumptions of cylindrical in shape and continuous contact of the sleeve and the rubber tube. Additionally, Figure 3.9 (b) shows how the modified formula did manage to map the theoretical plot to closely resemble the experimental plot.

$$f = \pi r^2_0 \, (p) \left[\alpha \left(1 - q\varepsilon\right)^2 - \beta\right] \tag{3.11}$$

A comparison performance experiment for multiple actuators arranged in series versus a single PMA was conducted. This involved prototypes of two 30 cm contractor PMAs in series and a single contractor PMA of 60 cm in length. The plot in Figure 3.10 shows that the force generated by the two actuators in series is like that of a single actuator of the same length. However, the serial arrangement provides a degree of redundancy, should one of the actuators faults the other will continue to operate.

Multiple actuators could also be arranged in parallel to each other. A continuum arm was constructed by arranging several PMAs in parallel. Figure 3.11 depicts a 30 cm contractor PMA arm with four actuators; one at the center and the other three are positioned 30 mm away from the arm's center to their axial centers and are separated 120° apart from the neighboring actuator.

A comparison experiment of a single contractor PMA and the continuum PMA arm with four parallel actuators was performed as recorded in Figure 3.12 (a). Both actuator systems did contract by the same amount when pressurized since each actuator was of similar in design. The slight variations between the results are caused

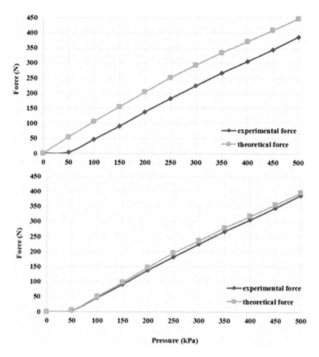

Figure 3.9   *Experimental and theoretical force analysis for a 30 cm PMA using (3.10) and (3.11)*

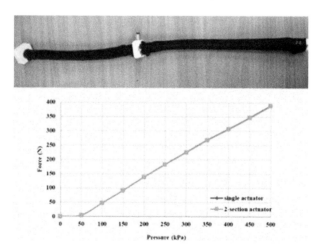

Figure 3.10   *Two 30 cm PMAs connected in series and its comparison force results to a single 60 cm PMA, showing their similar output results by having their line graph overlapping each other*

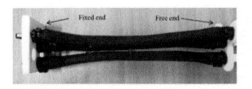

Figure 3.11   Prototype and mechanical design of the fixed and free end of the four parallel actuators for the continuum arm [30]

by friction between the four actuators. When analyzing the force output, the force generated by the parallel actuators is approximately four times that of a single PMA when each of the four actuators is pressurized at equal pressures, Figure 3.12 (b). The relation of force F produced by multiple actuators can be generalized by (3.12).

$$F = Kf \tag{3.12}$$

where K is the number of parallel muscles and f is the force of a single muscle.

The above experiments were based on pressurizing the four actuators with the same value and this results in a linear contraction of the continuum arm. However, if each of the muscles was supplied with different pressures, each would contract by differing amounts causing the actuator to bend and thus behaving like a continuum arm. This is demonstrated by supplying a fixed pressure to two of the outer actuators, while the remaining actuator is supplied with increasing pressure from 0 to 500 kPa. The bending angle δ of the four actuated arms is shown in Figure 3.13 (a) and it can be seen that the bend angle is dependent on the amount of pressure. The maximum angle δ_{max} is affected by the amount of loading as observed from the line curves in Figure 3.13 (b), with the heavy loadings resulting in a smaller maximum angle.

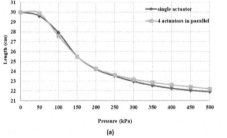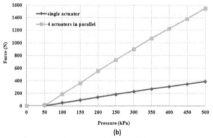

Figure 3.12   Comparison experiment for a single contractor PMA and a continuum PMA arm with four parallel actuators, (a) length-pressure characteristic and (b) force-pressure characteristic

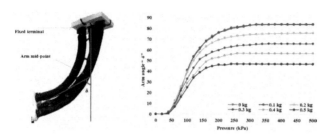

Figure 3.13　On the left, a contractor PMA continuum-arm with bending angle, and on the right a bending angles relative to loading values

3.3.4　Modeling of extensor PMA

Experiments for an extensor PMA are similar to the contractor PMA. Figure 3.14 shows changes in length, which increases with an increase in air pressure to a maximum length determined by a fixed length of the braided sleeve.

The modified force formula for an extensor PMA also includes the initial 45 kPa pressure as its zero-force value (3.13), which shows the change in direction of force with a minus sign. Figure 3.15 gives the relationship of the force generated and the pressure input for experimental and theoretical values, the variation is a result of the contact between the rubber tube and the braided sleeve.

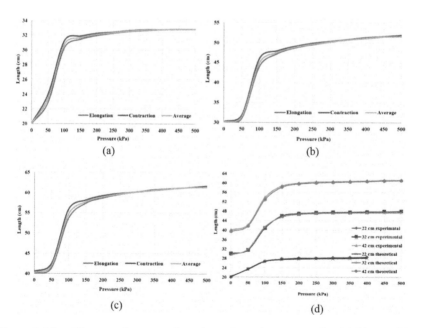

Figure 3.14　Change of actuator length against air pressure for (a) 20 cm, (b) 30 cm, (c) 49 cm PMA, and (d) practical and theoretical lengths against pressure changes

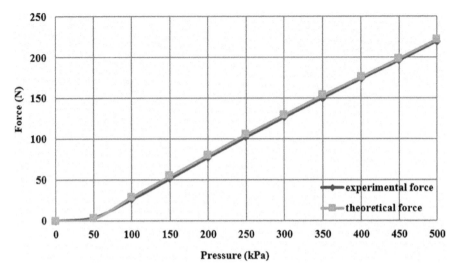

Figure 3.15   Practical and the presented theoretical force for a 30 cm extensor PMA

$$f = -\pi r_0^2(p - 45)[\alpha(1 - q\varepsilon)^2 - \beta], \; p \geq 45kPa \qquad (3.13)$$

Experimental analysis recorded in Figure 3.16 shows changes in length due to pressure changes for the extensor PMA under fixed loadings while Figure 3.17 has the pressure values fixed with varied loadings. The observations show that the elongation due to loadings is much less at higher pressures.

An extensor PMA continuum arm with four actuators that can bend in all directions is designed as shown in Figure 3.18 (b) (whose fundamental working principle is discussed in the Section 3.4). Each of the four actuators is of length 30 cm, Figure 3.18 (a), one at the center and the other three are 3 cm away from the center of the arm and are 120° apart from the neighboring actuator.

When the air pressure is equally distributed to the four PMAs, the continuum arm extends but bends towards the direction of PMA with the highest distribution. The bending angle is dependent on the amount of weight attached at the free end. Figure 3.19 (a) shows the most acute angle is attained at lower loadings.

3.4   PMA designs that can be adapted as robotic manipulators for space

The work in this section involves design concepts using the techniques discussed in Section 3.4. These unique designs have promising results of how they can be an alternative to the rigid manipulators, since they are extremely flexible and possess infinite degrees of freedom, making them ideal to do tasks that are challenging to achieve with rigid manipulators. Their simplicity in construction and lesser weight

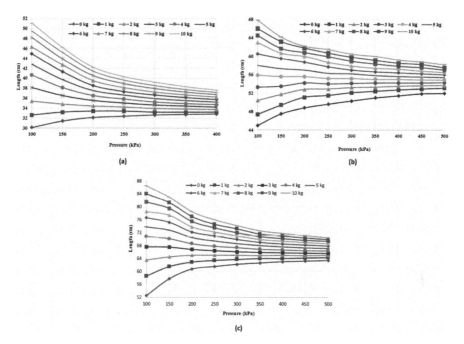

Figure 3.16   Changes in length of (a) 20 cm, (b) 30 cm, and (c) 40 cm PMAs at varying loads with fixed pressures

are good achievements when accounting for transporting robotic system to space and when mounting them on rails in the space station. These systems have not tested yet in low-gravity settings to emulate space environments and are currently under investigation as part of the FAIR-SPACE project.

Both the contractor and extensor PMAs discussed in Section 3.3 generate linear motion or force that acts in one direction. However, some modification produces novel actuator designs that can produce bending motions, some that can contract or extend in length, a ring design for gripping action. This section gives the detailed explanations of self-bending PMAs, double-bending PMA, extending and contract-ing PMA, and CPMA.

3.4.1   Self-bending contraction actuator and extensor-bending pneumatic artificial muscles

Bending motion of the PMAs can be achieved by preventing one side of the actua-tor from changing length when the opposite side extends or contracts. Two types of bending muscles include an EBPMA and the SBCA.

The EBPMA is created by longitudinal fixing along one side of an extensor PMA as depicted in Figure 3.20. This fixing takes the form of an inextensible thread that prevents one side of the muscle from extending when pressurized. This is the same principle used in the construction of the extensor PMA continuum-arm with

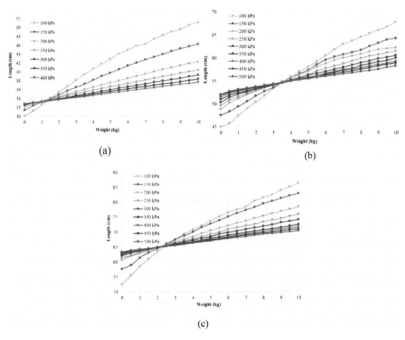

Figure 3.17   Changes in length of (a) 20 cm, (b) 30 cm, and (c) 40 cm PMAs at varying loads with fixed pressures

four parallel PMAs discussed earlier. If the actuator is pressurized, the free side of the muscle extends and because the thread on the opposite side limits changes in length, a bending motion occurs as depicted in Figure 3.20.

The SBCA is a biologically inspired design that imitates how bones support surrounding muscles. The bending actuator prototype is formed from a contraction PMA that has a thin flexible rod of 2 mm width inserted in between the rubber tube and braided sleeve, as seen in Figure 3.21 (a). The rod prevents one side of the

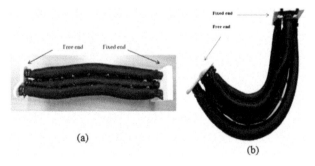

Figure 3.18   (a) A four extensor PMA continuum-arm (b) at a certain bending angle when pressurized [30]

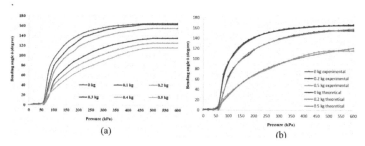

(a)   (b)

Figure 3.19   *Result analysis for the four extensor PMA continuum-arm. (a) The bending angle against the pressure at different load conditions and (b) comparison of experimental and theoretical bending angles at three different load conditions.*

actuator from contracting when pressurized while the other side contracts freely, Figure 3.21 (b). Just like the EBPMA, this mismatch between displacements causes a bending motion. Comparing the payload capacity made by these two bending designs is an ongoing experiment at the time of writing this chapter.

3.4.2   Double-bending pneumatic muscle actuator

The bending muscle presented above are only capable of bending in a single direction; however, by reinforcing different sections of the muscle it is possible to create more complex bending motions, for example, the prototype shown here can bend in two directions. It is based on the SBCA by placing two flexible rods on opposite

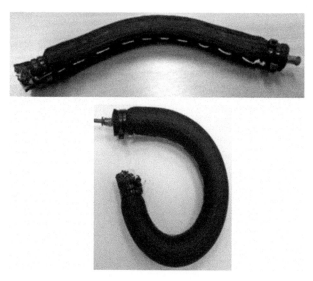

Figure 3.20   An EBPMA at bending when pressurized at 300 kPa [30]

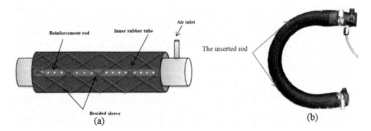

Figure 3.21   *(a) SBCA with a flexible rod inserted between the braided sleeve and rubber tube and (b) 30 cm SBCA pressurized at 300 kPa [31]*

sides of the PMA, a distance apart on either half of the length, resulting in a bending behavior shown in Figure 3.22 (c), which forms an S shape when pressurized.

Identical reinforcement rods of similar material and size result in a symmetrical shape for the upper and lower half of the actuator, Figure 3.22 (b). Despite the simplicity of the design and minimal usage of resources, the actuator can move in both the horizontal and vertical directions using one actuator. The calculation of the vertical or horizontal length, L_v and L_h, of the actuator is given by (3.14).

$$L_v = \left(\lambda_1^2 + L_{h1}^2 - 2\lambda_1 L_{h1}\cos\alpha_1\right)^{\frac{1}{2}} + \left(\lambda_2^2 + L_{h2}^2 - 2\lambda_2 L_{h2}\cos\alpha_2\right)^{\frac{1}{2}}$$

$$L_h = \left(\lambda_1^2 + L_{v1}^2 - 2\lambda_1 L_{v1}\cos\gamma_1\right)^{\frac{1}{2}} + \left(\lambda_2^2 + L_{v2}^2 - 2\lambda_2 L_{v2}\cos\gamma_2\right)^{\frac{1}{2}} \qquad (3.14)$$

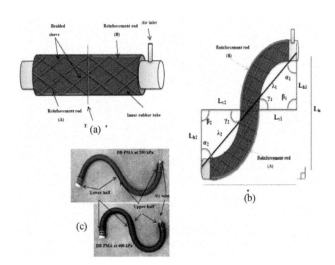

Figure 3.22   *(a) The structure of the double-bend pneumatic muscle actuator (DB-PMA), (b) the bending behavior and the geometrical analysis of the DB-PMA, and (c) DB-PMA at different pressure values of 200 and 400 kPa*

where λ_1 and λ_2 are the two rods diagonals while α_1, α_2, γ_1, and γ_2 are the angles defined in Figure 3.22 (b).

This design means that by appropriately positioning the reinforcement rods, a complex bending motion can be achieved. As such it could be used to develop soft fingers capable of forming a bending shape desirable to certain grasping actions, for example, two metal rods, each one is at the formed trough.

3.4.3 Extensor-contraction pneumatic muscle actuator

The two types of PMA discussed earlier can independently provide either pulling or pushing forces. However, each type is only capable of producing either of the forces but not both. This means that they must be used in pairs or to work together with another mechanism such as a spring to provide reversed motion. Unfortunately, in applications where space is limited, it might be desirable to combine the two PMAs into one single device that can provide the two forces in a more compact and convenient package. Hence, an extensor-contractor PMA (ECPMA) has been developed by inserting a contractor muscle inside an extensor muscle. All pneumatic muscles contain a "dead space" inside them since the force produced by the muscle is not based on the inside volume of the actuator but the surface area of the inner tube. This means this inner volume can be exploited by inserting another PMA without affecting the actual forces the outer PMA can provide.

The ECPMA has modified end caps to host the contractor element inside the extensor muscle: two inlets to supply air in each of the respective muscle types as shown in Figure 3.23 (a). From the ECPMA's resting length, if the extensor PMA is pressurized the actuator will extend to its maximum length limited by the outer sleeve. When the contraction PMA is pressurized, it contracts past the extensor PMA resting length until reaching maximum reduction in length. Therefore, depending upon the relative pressures within the extensor and contraction PMA, the actuator can lengthen or shorten and produce tensile or compressive forces.

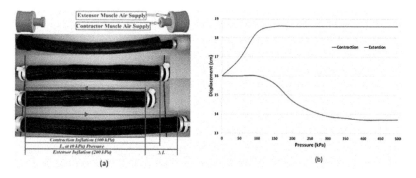

Figure 3.23   *(a) The modified caps with two inlets for air supply and three state of the ECPMA depending on pressure input (initial length, contraction, extension). (b) Change in length of ECPMA when pressurizing the two muscles independently*

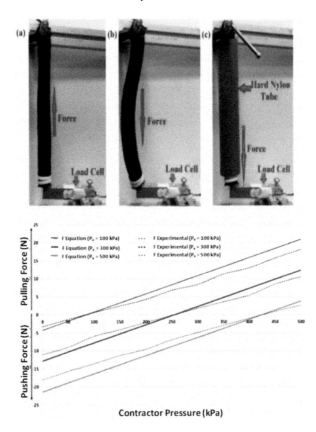

Figure 3.24   Dimensions of the CPMA on the left and on the right, the actual experiments for its validation

Figure 3.23 (b) shows the experimental results obtained for the new actuator. It has a resting length of 16 cm and when the extensor PMA is pressurized, it extends to 18.6 cm. When the contraction PMA is pressurized, the actuator shortens to approximately 13.7 cm. Therefore, unlike a conventional PMA, this variation can both extend and contract depending on relative pressure in the two PMAs.

Figure 3.24 shows how the force measurements in the ECPMA was experimentally conducted. The actuator was mounted between a fixed surface and load cell and it was pressurized to predefined testing pressures, and the respective forces on the load cell were measured and recorded. The pressure in the contraction PMA was then increased incrementally, which reduced the actuator length, and the pulling force was measured by the load cell when the pulling force created by the contraction PMA exceeded the pushing force by the extensor. The load cell also measured the generated pushing force, which increased in proportion to the increase in pressure input. Thus, the actuator can produce both a pushing force and a pulling force

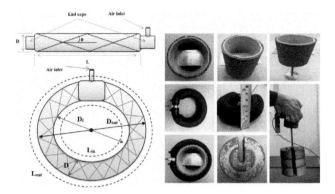

Figure 3.25   Dimensions of CPMA and actual experiments for its validation [32]

on the load cell. The transition point between these two is dependent on the relative pressures in the types of PMA.

3.4.4   Circular pneumatic muscle actuator

Another interesting PMA design concept is the CPMA, which is inspired by facial muscles around eyes and mouth. It is designed to shrink its inner circumference and increase the diameter of the actuator. The is achieved by joining the ends of a contraction PMA to form a ring as seen in Figure 3.25. Increasing the air pressure reduces the circumference of the actuator, which in turn reduces the inner and outer diameter of the CPMA.

The reduction of the inner diameter generates a radial force towards the center of the ring, which in turn is applied to any object placed within it. In Figure 3.25 the CPMA is being used to grasp a circular object. When pressurized, the actuator tightens around the object and the resultant force in conjunction with friction between the actuator and object helps for a secure grasp.

3.5   PMA applications in developing novel grippers, manipulators, and power assistive glove for space environments

In this section, some design concepts of soft robotic grippers and manipulators are explored using the above pneumatic muscle actuation technologies. These grippers have been designed with improvements in mind of the limitations offered by some of the already existing technologies in providing similar services. As pointed out earlier, weights of components making up a robotic system can become a challenge by itself, especially if efficiency in power consumption or motion control is to be considered. Irregular objects are also a challenge when being handled by the end effectors of robots. And in some cases, the rigid end effectors are not suitable to handle delicate objects, especially the handling of space rocks during experiments is critical. Thus, the following alternative soft grippers and manipulators are designed to be compliant and light in weight, of which they can be adapted for improving the existing robotics systems in IVAs in the space station or future space colonies. The power assistive glove discussed will be demonstrating how the technology can be useful for augmenting force to human fingers, to aid in better grasping and handling of objects. This particularly aims at the spacesuit glove used by astronauts since the pressurization and stuffing of many layers affects the mobility of the fingers. Therefore, in addition to the limitation by design, the astronaut's hands also get the natural muscle fatigue especially amplified by countering the spring back effects of the inner pressure contained by the glove, which in turn causes a reduction in performance. In the future, the power assistive space suit gloves could then be able to assist astronauts while performing their IVA and EVAs.

The Section 3.4 has presented a range of designs derived from the conventional operation of PMAs, and they have been used to develop a three- and six-fingered gripper, extension-circular gripper, and varied manipulators.

3.5.1   Three fingers gripper base on SBCA

The SBCA described in Section 3.3.1 can be used to construct a multi-finger gripper to aid in compliant grasping. A prototype soft gripper is constructed using three identical 14 cm long SBCAs that act as the fingers. The actuator only flexes when pressurized. Thus, an elastomeric material is placed on the rear of each finger to spread out the actuators when deflated. The fingers can spread 20 cm apart and touch each other on closure when pressurized, which makes them suitable to grasp a wide range of object sizes.

A rigid and noncompliant gripper requires some elements of grasp planning and precision control for each finger when handling certain objects. However, this design relieves such requirements as its ability to automatically bend around and deform to suit the shape of the objects irrespective of the complexity in shape. This ability means there is no need for grasp preplanning or sophisticated control; the gripper can operate well even with open-loop control. When activated, the finger bends and makes contact with the target object and continues to wrap itself around it. Each finger in the prototype has a maximum bending angle of 72 degrees, which allows them to touch each other at the center of the gripper. This property allows the soft gripper to handle small-sized objects of less than 10 mm (e.g. a pen) or much larger complex-shaped parts.

A further modification to the design included the addition of three shorter fingers as seen in Figure 3.26. This has the benefit of being able to grasp objects more securely and increases the grasping force as well. Experimental results that used a prototype soft gripper weighing 0.34 kg could achieve a maximum payload of 3.6 kg.

The potential applications of this gripper in space environments can be very useful in the handling of unknown and undefined objects like space rocks, which may vary in physical properties: weight, shape, and fragility. This is something that traditional grippers struggle with: demanding extensive preplanning, which is difficult to achieve without a geometric model of the object to be grasped or extensive sensor,

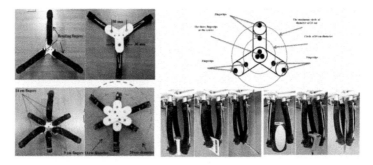

Figure 3.26   A three and six-fingers gripper based on SBCA handling different objects and a diagram representation of the three-finger design's fingertip at different positions within its workspace

especially visual data. Thus, the grasping of objects can be slow and inefficient. Therefore, the potential applications of this gripper design can be for extra-vehicular to capture space debris. By its very nature, this debris is often of unknown shape and size and this gripper would be able to grasp the debris with relative ease as the fingers deform and conform to the object being grasped automatically and rapidly without the need for extra operations or extensive planning. Similarly, the gripper could also be suited to the collection of samples on other terrestrial bodies, which again will likely have unknown geometry, and especially when teleoperation can be limited due to communication latencies.

3.5.2   Extension-circular gripper

A very different soft gripper has been created using the CPMA highlighted in Figure 3.27. Unlike many common soft grippers, this design does not have flexible

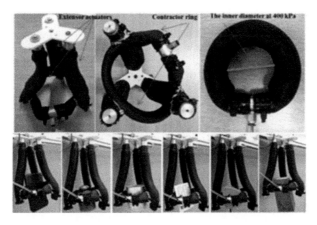

Figure 3.27   The structure of the extension-circular gripper also showing its ability to grasp objects of different shapes and center axis [32]

fingers, instead it operates by deforming the shape of the gripper around the object to be grasped. The prototype soft gripper is constructed from three 18 cm extensor PMA and one 30 cm long CPMA with a maximum diameter of 7.8 cm. The extensor PMA provides the ability to direct and extend the gripper to the target object to be grasped while the CPMA does the actual grasping action by contracting around the target object and applying a compressive force to its entire periphery.

The extensor muscles allow the gripper to extend from 18 cm in length to 24 cm. Therefore, to grasp an object the gripper is placed above the target and then the extension muscles are activated to extend the gripper. This pushes the circular gripping element towards the target object. The benefit of this design is that the soft nature of the gripper means that should the target object has an odd pose, the natural compliance of the extensor muscles will cause the gripper to flex to automatically align the circular element with the target object appropriately. This is usually a challenging task for rigid grippers since they lack bending ability. Once located as desired, the CPMA is pressurized and reduces to a minimum diameter of 4.45 cm, which is a 43 percent reduction. This applies compressing force to the outer edges of the target object leading to a firm grasp.

This extension-circular gripper has an advantage over a multi-finger in that it has infinite contact points with the target object while a gripper with fingers is limited to the number of available fingers being used to grasp. The size of the object that the gripper can handle is determined by the maximum and minimum diameter of the CPMA, which for the case of our prototype means target object must be between 4.45 and 7.8 cm across. However, the design is fully scalable, and larger and smaller designs can be created to match a range of target objects.

3.5.3   Three CPMAs gripper

The previous gripper design consists of a single CPMA that limits the grasp force of the gripper, especially if the object to be grasped is long and large, meaning that the grasp is only applied over a relatively small part of the target object's length. The extension-circular gripper has been modified by increasing the number of CPMAs to three, raising the grasping payload and increasing the gripping surface area. In this prototype Figure 3.28, the extensor muscle can vary from 27 to 38.1 cm, which

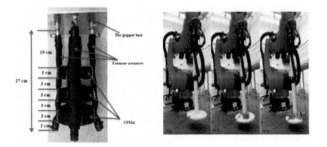

Figure 3.28   *The dimensions and the structure of the three CPMAs gripper and a grasp lifting experiment at different loadings [30]*

is a 41 percent extension, whereas the CPMA grasping diameter ranges from 8 to 4.3 cm at 0 to 400 kPa, respectively, which is a 46 percent reduction in diameter. The weight of the gripper is 0.8 kg and it could lift a 6 cm wide cylinder of 40 kg as demonstrated in Figure 3.28.

Just like the soft gripper consisting of fingers, the main application for this type of gripper in space environments would be best suited where the geometry of the target objects is unknown. These could be samples of space rocks collected or parts of debris needing retrieval. The main advantage of the gripper is the fact that grasp force is spread over a large area instead of specific localized points and this makes it particularly well suited to handling delicate and easily damaged samples.

3.5.4 Soft robot manipulators

In addition to producing the soft grippers using PMA designs, the concept can be extended to create soft robotics manipulators. In Figure 3.29 we see contraction and extensor PMA are combined to form two continuum manipulators. As was described in Section 3.2, a continuum manipulator can be formed by arranging multiple muscles in parallel. Depending upon the relative pressures in each of the muscles, the manipulator will flex and bend. The manipulators do not have individual discrete joints but instead, the entire body of the system flexes. Figure 3.29 also shows a

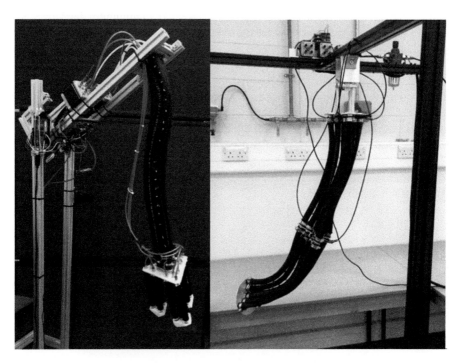

Figure 3.29   *Two manipulator designs that can flex in all direction of a workspace to handle objects using their attached grippers at the free end [31]*

single-section and a two-section manipulator. The multi-section version has a greater workspace as each section moves independently of the other. Both systems have a high power-to-weight ratio, each weighing less than 1 kg but can carry payloads of up to 7 kg.

An alternative design of continuum manipulator can be created using the SBCA described previously, see Figure 3.30. Here the direction of bending is not a function of the differential pressure in the muscles but is determined by the construction of the self-bending actuators. This means the system has a more limited workspace, but it is mechanically simpler, has lower cost and is much less complex for a control model. This design is well suited where a specific motion must be of high repeatability, for example, a pick and place type motion. Though with appropriate design modifications for the system, complex behaviors can still be achieved, as seen in Figure 3.30, where two manipulators are working together to relocate a sample cylindrical object.

Soft manipulators have several potential applications in space environments. Their lightweight makes them less costly to get to orbit as payload weight has a significant effect on launch cost. The systems are also potentially at a lower cost to manufacture than the traditional robots due to the materials used. However, such systems are still a long way from being produced commercially. In the future, this could reduce the cost of building and deploying remote robotic systems.

Soft arms also have the potential to offer safer alternatives to conventional robot arm in terms of both the safety of astronauts and other mission hardware. The soft arms are lightweight, have low inertia, and are compliant, which mean should they collide with an astronaut there is fewer chances of an injury. Also, because they deform on contact with objects the contact forces become spread over a greater area, meaning less likelihood of causing damages to other objects when a collision occurs.

Soft robot manipulator could work safely alongside astronauts to aid in completing tasks through collaborative working. Soft manipulators could even be mounted to an astronaut's suit to operate as an additional limb and enhance their productivity during EVAs. One of the other benefits is that as the soft continuum arms do not have discrete joints and can deform around obstacles, they can access locations where traditional manipulators struggle, like a constrained path to an object. This enhances the range of tasks that robots could be used to perform in an uncertain space environment.

Figure 3.30   A single bending continuum arm and the three fingers gripper with two of them performing a collaborative task

3.5.5 Power assistive soft glove

The current pressurized spacesuit design astronauts wear to perform EVAs pose challenges in ease of completing certain tasks. This is particularly the case for the hands where multiple protective layers and internal air pressure in the spacesuit limit the mobility of the hand tiring, hence much of the dexterity that the human hand possessed is lost. Soft robotics technology has the potential to assist in this area as actuators can be used to provide forces to the user's finger to augment the power to the hand muscles especially the fingers and reduce fatigue thereof.

Figure 3.31 shows a prototype force augmentation glove that uses the EBPMA. The EBPMA is used to form a hand exoskeleton to assist the bending of finger joints to perform multiple gripping and pinching movements. When pressurized, the pneumatic muscles bend and in turn apply a force to the user's fingers to amplify the power of the hand. According to [33] males and females have an average pinching force of 43 and 38 N, respectively, and experiment performed using the soft glove showed it could boost the pinching forces of a user by 40–45 percent.

To demonstrate that the force augmentation glove can reduce user fatigue, an experiment was performed using EMG (electromyography) signals taken from the user's muscles, Figure 3.32. In the experiment, the test subjects were required to grasp a range of objects both with and without the glove and their muscle activity was measured using the EMG sensors. It was observed that there was a significant impact when the glove aided in grasping, and the electrical activity in the test subject's muscles reduced. This experiment did indicate the huge impact the power assistive glove allows a user to perform a task using less of their muscle force than when not using the device. Therefore, with such a glove a reduction in fatigue over time can be experienced.

If this soft force augmentation technology were introduced into astronaut's spacesuits, it could have the potential to reduce fatigue during EVAs as the actuators would help to overcome the resistance to motion caused by the pressurized glove. The technology also has the potential to augment the astronaut's strength allowing them to perform tasks they otherwise may not able to with their natural strength.

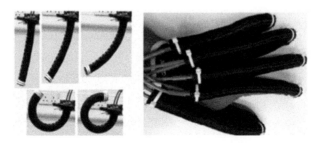

Figure 3.31   *A controllable stiffness and bendable actuator and its application on soft robotic glove [32]*

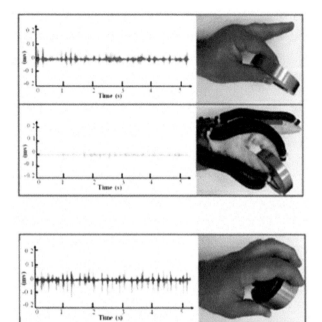

*Figure 3.32   Study of the impact on human muscles when using the power
            assistive glove to pinch or grasp an object*

3.6   Recommendations and future works

This chapter has introduced and discussed some of the soft robotics systems undergoing development and has looked at how this technology could potentially be used
in space application in the future. While to date no soft robotic system has been used
in space it is likely that this rapidly advancing field of research will see the application in space in the future. Soft robots are constructed in a very different manner and
from different materials than traditional robots giving them unique behavior and
abilities that conventional robots cannot provide. Thus, they are highly likely to be
a viable solution for space-based challenges that cannot be addressed using existing
robot technologies.

Some of the discussed advantages offered by soft robotics that make them
potentially useful in space applications include:

Low weight – Soft robots are typically formed from lightweight material that
reduces overall system mass. In space applications, system mass is a critical factor

that determines the launch and mission cost. Anything that reduces the mass of robot hardware is likely to provide financial saving. Soft robots also often have a high power-to-weight ratio meaning they can often achieve the same feats as much heavier robots do.

Low-cost materials – Although little soft robotic technology is currently commercially available, most systems being developed are low cost when compared to the already established robotics technology. This factor also has the potential to reduce mission cost.

Inherent safety – Soft robots are typically inherently safe due to the use of soft materials, their low weight and inertia, and their compliant actuation. This means soft robots could work collaboratively and close to an astronaut without presenting danger and also handle fragile and irregular objects. This is something much more difficult to achieve with conventional robots, especially in an unstructured environment.

Deformable structures – Soft robots do not have discrete joints and they are often able to flex any part of their structure. Thus they can achieve much more complex motions than a traditional serial link manipulator can. This gives soft manipulators the ability to reach into and around obstacles and reach places a traditional robot cannot and increase the range of orbit and remote manipulation tasks that can be performed. The deformable nature of soft robots also provides advantages when grasping objects. Traditional grippers often need extensive preplanning before grasping an object, but many soft grippers can be used in an open-loop manner and will naturally deform around an object and create a secure grasp.

Wearable – Soft robotics technology can easily be incorporated into lightweight, low-cost wearable devices such as exoskeleton and assistive devices. Incorporating soft actuators into an astronaut's spacesuit could enhance their abilities or reduce fatigue. Similarly, a soft body-mounted manipulator could provide an astronaut with an additional limb to help increase their productivity.

Despite the potential future advantages that soft robotic technology offers towards working and exploration in space, the technology still has several challenges that need to be overcome.

Power sources – Many of the soft robotics systems developed, and indeed those described in this chapter, are pneumatic-based systems. It is the compressibility of air that gives many soft robots their desirable behavior. In a terrestrial environment, the pneumatic system is widespread with air being compressed as and when required. This is not possible in the vacuum of space and air recycling systems or another actuator technology will need to be developed.

Sensing – As we have seen soft robots continuously flex and deform, and this means traditional sensors are inappropriate to monitor their motion and behavior. Advances are therefore needed in soft sensors.

Mechanical resilience – To date, soft robots have seen little application outside of the laboratory environment and so they have not been designed to withstand the rigors of the environment in which they operate. Space is an extreme environment and presents a further challenge. Research and development are needed to increase the resilience of soft robots and this will require advances in materials and the development of self-repairing technology.

Modeling and control – Advances are required in the modeling and control of soft robots as the already well-established techniques developed for traditional robots are not directly applicable.

Alternative soft technologies – The pneumatic technology is just an example of a soft robotic system that is still under research. However, this is not the only concept of soft systems that can be used to improve and develop futuristic robots. Other technologies that are promising include SMA (Shape Memory Alloys), FEA (Fluidic Elastomeric Actuators), SMP (Shape Morphing Polymers), DEAP (Dielectric-Electrically Actuated Polymers), and E/MA (Magnetic/Electro-Magnetic Actuators).

References

[1]   Gombolay M.C., Gutierrez R.A., Clarke S.G., Sturla G.F., Shah J.A., *et al.* 'Decision-making authority, team efficiency and human worker satisfaction in mixed human–robot teams'. *Autonomous Robots*. 2015;**39**(3):293–312.

[2]   Di Castro M., Ferre M., Masi A. 'CERNTAURO: a modular architecture for robotic inspection and telemanipulation in harsh and semi-structured environments'. *IEEE Access*. 2018;**6**:37506–22.

[3]   Zanchettin A.M., Ceriani N.M., Rocco P., Ding H., Matthias B. 'Safety in human-robot collaborative manufacturing environments: metrics and control'. *IEEE Transactions on Automation Science and Engineering*. 2016;**13**(2):882–93.

[4]   Rahman M.H., Rahman M.J., Cristobal O.L., Saad M., Kenné J.P., Archambault P.S. 'Development of a whole arm wearable robotic exoskeleton for rehabilitation and to assist upper limb movements'. *Robotica*. 2015;**33**(1):19–39.

[5]   Reynolds E. *Nasa's new humanoid robot to tackle space missions*. [online] Wired.co.uk. 2020. Available from https://www.nasa.gov/mission_pages/station/multimedia/robonaut_photos.html [Accessed 20 Feb 2020].

[6]   Calisti M., Giorelli M., Levy G., *et al.* 'An octopus-bioinspired solution to movement and manipulation for soft robots'. *Bioinspiration & Biomimetics*. 2011;**6**(3):036002.

[7]   Althoff M., Giusti A., Liu S.B., Pereira A. 'Effortless creation of safe robots from modules through self-programming and self-verification'. *Science Robotics*. 2019;**4**(31):eaaw1924.

[8]   Swaim P., Thompson C., Campbell P. 'The Charlotte (TM) intra-vehicular robot'. *Artificial Intelligence, Robotics, and Automation for Space Symposium*. 1994:157–62.

[9]   Williams R. *Automation and robotics for space-based systems-1991*. Hampton, VA: National Aeronautics and Space Administration, Langley Research Center; 1992.

[10] Morris A., Barker L. *Technology Demonstration of Space intravehicular Automation and Robotics*; 1994.

[11] robonaut.jsc.nasa.gov. Robonaut. [online]. n.d. Available from https://robonaut.jsc.nasa.gov/R1/index.asp [Accessed 9 Aug 2020].

[12] Ahlstrom T., Curtis A., Diftler M., *et al*. Robonaut 2 on the International Space Station: Status update and preparations for IVA mobility. In *Aiaa Space 2013 Conference and Exposition*; 2013. p. 5340.

[13] General Disclaimer. [online]. n.d. Available from https://ntrs.nasa.gov/archive/nasa/casi.ntrs.nasa.gov/19760003644.pdf [Accessed 3 Aug 2020].

[14] Waldie J.M.A.T. 'Compression under a mechanical counter pressure space suit glove'. *Journal of Gravitational Physiology: A Journal of the International Society for Gravitational Physiology [online]*. 2002:93–7.

[15] Tanaka K., Danaher P., Webb P., Hargens A.R. 'Mobility of the elastic counterpressure space suit glove'. *Aviation, Space, and Environmental Medicine*. 2009;**80**(10):890–3.

[16] Sanner R., Sorenson B., Fogleman M., Hq N., Kosmo M., Jsc N. NASA research announcement phase h final report for the development of a power assisted space suit glove [online]. 1997. Available from https://ntrs.nasa.gov/archive/nasa/casi.ntrs.nasa.gov/19980007975.pdf [Accessed 3 Aug 2020].

[17] Konkel C.R., Powers A.K., Dewitt J.R. IVA the robot: Design guidelines and lessons learned from the first space station laboratory manipulation system. In *Fourth Annual Workshop on Space Operations Applications and Research (SOAR'90): Proceedings of a Workshop Sponsored by the National Aeronautics and Space Administration*, Vol. 1; Washington, DC, the US Air Force, Washington, DC, and Cosponsored by the University of New Mexico, Albuquerque, New Mexico, and Held in Albuquerque, New Mexico, June 26-28, 1990, National Aeronautics and Space Administration; 1991, January. p. 9.

[18] Fantoni G., Santochi M., Dini G., *et al*. 'Grasping devices and methods in automated production processes'. *CIRP Annals*. 2014;**63**(2):679–701.

[19] Choi H., Koç M. 'Design and feasibility tests of a flexible gripper based on inflatable rubber pockets'. *International Journal of Machine Tools and Manufacture*. 2006;**46**(12):1350–61.

[20] Caldwell D.G., Tsagarakis N., Medrano-Cerda G.A., Schofield J., Brown S. 'A pneumatic muscle actuator driven manipulator for nuclear waste retrieval'. *Control Engineering Practice*. 2001;**9**(1):23–36.

[21] Martell J., Gini G. 'Robotic hands: design review and proposal of new design process'. *Image*. 2007;**180**:9270.

[22] Jamwal P.K., Xie S.Q. 'Artificial neural network based dynamic modelling of indigenous pneumatic muscle actuators'. Paper presented at the 2012 IEEE/ASME International Conference on Mechatronics and Embedded Systems and Applications (MESA); 2012.

[23] Andrikopoulos G., Nikolakopoulos G., Manesis S. 'Advanced nonlinear PID-based antagonistic control for pneumatic muscle actuators'. *IEEE Transactions on Industrial Electronics*. 2014;**61**(12):6926–37.

[24]   Tondu B. 'Modelling of the McKibben artificial muscle: a review'. *Journal of Intelligent Material Systems and Structures.* 2012;**23**(3):225–53.

[25]   Trivedi D., Rahn C.D., Kier W.M., Walker I.D. 'Soft robotics: biological inspiration, state of the art, and future research'. *Applied Bionics and Biomechanics.* 2008;**5**(3):99–117.

[26]   Tatlicioglu E., Walker I.D., Dawson D.M. 'Dynamic modelling for planar extensible continuum robot manipulators'. Paper presented at the 2007 IEEE International Conference on Robotics and Automation; 2007.

[27]   Kelasidi E., Andrikopoulos G., Nikolakopoulos G., Manesis S. 'A survey on pneumatic muscle actuators modeling'. Paper presented at the 2011 IEEE International Symposium on Industrial Electronics (ISIE); 2011.

[28]   Ching-PingChou., Hannaford B. 'Measurement and modeling of McKibben pneumatic artificial muscles'. *IEEE Transactions on Robotics and Automation.* 1996;**12**(1):90–102.

[29]   Tondu B., Lopez P. 'Modeling and control of McKibben artificial muscle robot actuators'. *Control Systems, IEEE.* 2000;**20**(2):15–38.

[30]   Al-Ibadi A., Nefti-Meziani S., Davis S. 'Controlling of pneumatic muscle actuator systems by parallel structure of neural network and proportional controllers (PNNP)'. *Frontiers in Robotics and AI, section Soft Robotics.* 2020;**7**.

[31]   Al-Ibadi A., Nefti-Meziani S., Davis S. 'The design, kinematics and torque analysis of the self-bending soft contraction actuator'. *Actuators.* 2020;**9**(2):33.

[32]   Al-Ibadi A., Nefti-Meziani S., Davis S., Theodoridis T. 'Novel design and position control strategy of a soft robot arm'. *Robotics.* 2018;**7**(4):72.

[33]   Kadowaki Y., Noritsugu T., Takaiwa M., Sasaki D., Kato M 'Development of soft power-assist glove and control based on human intent'. *Journal of Robotics and Mechatronics.* 2011;**23**(2):281–91.

Chapter 4

Biologically-inspired mechanisms for space applications

Craig Pitcher[1], Mohamed Alkalla[1], Xavier Pang[1], and Yang Gao[1]

Natural organisms are constantly having to adapt in order to overcome the challenges posed by their environment, with the most beneficial traits being continuously improved and refined over millions of years of evolution [1]. This long refinement process takes place at all scale levels, from the nano to the macroscopic, resulting in materials and processes far superior to human solutions to similar problems. As such, studying these natural techniques and adapting them for man-made applications can lead to innovations and new improvements in current and future technologies.

Biomimetics is defined as the study and understanding of the mechanisms and processes found in the natural world as a model for solving or providing new approaches to engineering problems. While the term biomimetics is relatively new, humans have taken inspiration from nature throughout history from developing methods for improving early domesticated life to artificial silk made by the Chinese over 3 000 years ago and Leonardo da Vinci's bird-inspired flying machines [2]. The techniques evolved by organisms to overcome the challenges posed by nature are typically not fully compatible or optimised for a particular engineering problem. Biomimetics therefore involves taking inspiration from natural solutions, as opposed to directly copying them, and a number of different methodologies and tools for developing biologically-inspired concepts have been established [3, 4].

Today, biomimetics is used in a wide variety of fields, resulting in countless innovations, materials and concepts, such as superhydrophobic surfaces, Velcro, aircraft design and artificial intelligence [1, 5, 6]. Biomimetics has also been applied to space exploration, with the characteristics of several natural systems having properties that could be ideal for implementation into a range of space applications. Numerous mechanisms for use in space have been developed, some by the European Space Agency (ESA)'s Advanced Concepts Team [7] and others through independent research. While summaries of some biologically-inspired space mechanisms have been made in previous discussions of biomimetics [8, 9], this chapter aims to

[1]STAR LAB, Surrey Space Centre, University of Surrey, Guildford, GU2 7XH, United Kingdom

present an up-to-date and comprehensive discussion of the many designs and concepts considered for use in various fields related to space exploration.

There are two criteria that must be met for concepts to be included in this study. First, the source of biological inspiration will need to have been either examined in depth in the concept design process, or relevant and detailed studies of the specific biological system must have been referenced. This is used to eliminate concepts, of which several were found, that took inspiration at a surface level only. Second, as well as having a traceable link to a biological inspiration, the mechanisms will also need to have been intended for use as a space application at some point during the concept's evolution, whether it be only during initial or later iterations, or throughout the entire design. This is intended to discount concepts that used similar versions of the mechanisms discussed to achieve the same goals, but do not have either a biological origin or an intended use in space.

4.1   Subsurface exploration

Accessing the subsurface of an extraterrestrial body using a drilling system has been one of the goals of many planetary exploration missions, from Apollo 15's manually-operated Lunar Surface Drill [10] to Rosetta's Sample, Drill and Distribution System [11] and Curiosity's Powder Acquisition Drill System [12]. These missions have typically used the rotary or rotary-percussive drilling techniques, which are commonly used in terrestrial applications. However, the rotary technique requires a large overhead force to push the drill into the substrate [13] and the rotary-percussive system is heavy and complex. Other missions have utilised the percussive drilling technique [14] in the design of self-propelling moles, such as Beagle 2's PLUTO [15] and InSight's HP³ [16].

As minimising the total mass is one of the most critical aspects of mission design, biologically-inspired solutions have been explored with the aim of developing small, low-mass and energy-efficient alternatives to conventional drilling techniques. Two biological mechanisms that have been explored for use in subsurface exploration systems are insect ovipositors and the peristaltic motion of earthworms.

4.1.1   Ovipositor drilling

The mechanisms used by the ovipositors of the females of locust species and the wood wasp to drill into sand and wood, respectively, in order to lay their eggs are markedly different [9]. The locust's ovipositor contains a pair of valves hinged upon their apodemes. These contain intrinsic muscles to the valves, allowing them to open and close without needing muscles within the abdomen wall. When inserted into the substrate, the valves are opened, as seen in frames 5–9 of Figure 4.1. The valves dig into and grip the surrounding substrate, pulling the abdomen down, while another pair of valves higher up also open in order to clear away the debris [18].

A numerical simulation modelled the ovipositor's digging mechanism and performance, and calculated the resistive force experienced in various granular and cohesive soils, as well as the penetration of the ovipositor per cycle of opening

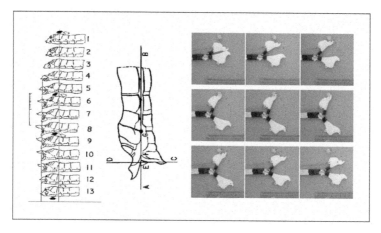

*Figure 4.1   Diagram of the locust ovipositor mechanics (left) and pictures
showing the macro-scale physical model (right) [17]*

and closing the valves. A simplified plastic physical prototype was concurrently
developed, also shown in Figure 4.1, which used wires to simulate the action of the
muscles and tendons to open the valves [17].

The ovipositor of the wood wasp is split into two halves that are reciprocated in
opposing directions by the abdomen muscles. These halves are lined with backwards-
facing teeth, as shown in Figure 4.2, which are designed to have little resistance
when penetrating into the wood, but engage it as they are pulled upwards, creating a
tensile force. This is then added to the initial force provided by the wasp's abdomen,
assisting the ovipositor's penetration [20]. The ability to self-generate penetration
forces would lower the masses needed to produce the overhead force required for
drilling, making it a promising technique for planetary subsurface exploration.

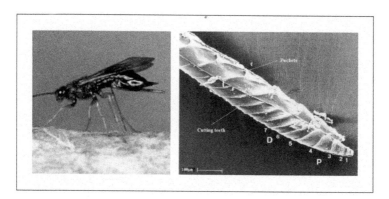

*Figure 4.2   The wood wasp (left) [19], reproduced by permission of Yang Gao,
and a cross-section of its ovipositor (right) [20], reproduced by
permission of Julian Vincent*

First envisioned as part of a micro-penetrator and sampling system, the cutting forces and speeds of several plastic designs lined with backwards-facing teeth were examined, with the two halves reciprocated by a pin-crank mechanism [21]. A feasibility study of a steel prototype demonstrated its ability to drill into soft rocks, with an energy efficiency comparable to other conventional drilling techniques [19].

4.1.1.1   Dual-reciprocating drill

The results of these studies inspired the development of the Dual-Reciprocating Drill (DRD). The first design, shown in Figure 4.3, used an external test rig to convert the rotary motion of a motor into reciprocation of two halves. Several test campaigns examined the effects of various design and operational parameters on the DRD's performance in regolith [22–24]. The system later evolved into a compact design [25], also shown in Figure 4.3, in which the actuation mechanism was integrated within the drill heads. Lateral movements of the drill heads were also implemented, creating complex burrowing motions that were shown to improve drilling performance [26], confirming the importance of lateral forces seen in previous tests [27].

Two new system prototypes are currently in development, each designed to be integrated onto a rover and tested in field trials. These designs will also incorporate novel sampling and actuation mechanisms, one of which has been inspired by a fish's caudal fin. This propelling element, used by the fish as it retracts and relaxes its body muscles to create forward momentum [28], led to the design of an oscillation mechanism for creating lateral movements of the drill heads [29].

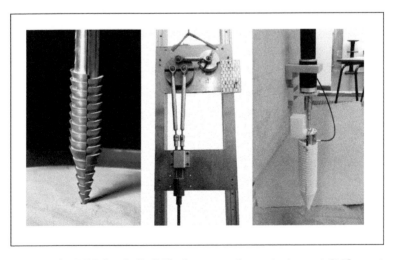

Figure 4.3   *The DRD bit (left) [22], the external test rig (centre) [23], reprinted with permission from Elsevier, and the integrated actuation mechanism design (right)*

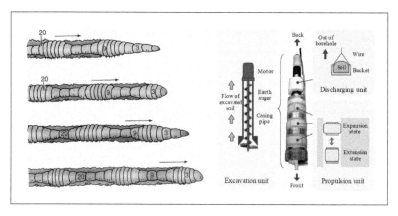

Figure 4.4   Diagram of the contraction and relaxation of earthworm muscles (left), reprinted from [31] with permission from Elsevier, and a schematic of the excavation and propulsion units in the peristaltic mole concept LEAVO (right) ©2018 IEEE. Reprinted, with permission, from [32]

4.1.2   Peristaltic motion

The sequential contraction and relaxation of rings of muscle to create a wave of locomotion along a body is known as peristaltic motion. This can be seen in the intestine, where food is transported towards the colon [30], and is used by soft-bodied animals such as worms as a method of burrowing or locomotion, as shown in Figure 4.4 [31]. The use of peristalsis for subsurface exploration has been proposed for a mole concept. This design, also seen in Figure 4.4, has a body consisting of several contraction-extension units. As each segment contracts, they subsequently expand outwards into the surrounding regolith, with the increased friction holding them in place, which allows the thinner segments to move. This contraction propagates from front to back, pulling the rear segments and pushing the front segments forwards [33]. The mole's mechanisms, including an internal discharging system for excavating the surrounding regolith, were tested [34], and the design evolved into the LEAVO concept, which explored the benefits of a flexible excavation unit [32].

4.2   Surface mobility inspired by animals

In the design of a rover's surface locomotion system, traversal of both rough terrain, with rocks and crevasses of various sizes, and low-cohesion soil, in which traction is required to overcome resistances to motion such as sinkage, is taken into account [35]. Despite challenges such as poor performance across rough terrain, every mobile rover to date has used wheeled systems for locomotion, as tracks can be power-hungry and legged systems are difficult to control. Attempts to maximise rover mobility have resulted in innovations in suspension system and wheel designs. These include ExoMars' wheel-walking mechanism, in which the rover produces different walking gaits by rotating each wheel about its axis [36], and the rocker-bogie suspension system used by the Spirit and Opportunity rovers [37].

Several animal mobility mechanisms have been used as the inspiration for novel robotic concepts that aim to have a greater manoeuvrability and range than current systems in both spacecraft and extraterrestrial environments. One such concept that was briefly explored is a hybrid wheeled design for a surface exploration robot. This uses the undulatory movements of sandfish lizards [38] to keep a constant level of slip as the robot traverses granular environments [39]. Other animal mobility concepts discussed in this section are based upon gecko and spider foot adhesion, various walking gaits and hopping animals.

4.2.1   Gecko and spider adhesion

Climbing robots are often used in terrestrial applications when direct human access is either very difficult or dangerous [40]. These have the potential to traverse space environments well beyond the capabilities of typical mobile rovers, such as on the surfaces of spacecraft and over steep planetary features. However, terrestrial adhesion techniques are unsuitable for use in space, as suction only works on smooth, clean surfaces, magnetic attraction requires a ferromagnetic surface and grasping robots are unable to climb smooth surfaces [41]. As a result, several space robotic concepts have explored dry adhesion techniques used by geckos and spiders.

The gecko's ability to stick to and climb vertical, smooth surfaces has long been the focus of scientific study. Their feet contain around 500 000 keratinous hairs, or setae, each one-tenth the diameter of a human hair and containing hundreds of projections ending in spatula-shaped protrusions, as shown in Figure 4.5. Coupled with its toe uncurling and peeling behaviours, the gecko uses a preload force normal to the surface and orientates the hairs to create extremely close contact on even molecularly smooth surfaces [42]. The setae surface geometry produces an adhesion

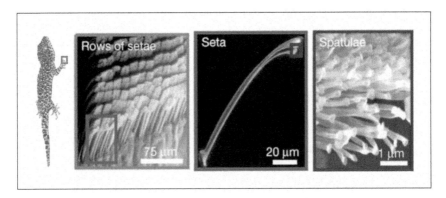

Figure 4.5   Microscopic images of the setae in the gecko's feet and their spatula-shaped endings. Reprinted by permission from Springer Nature Customer Service Centre GmbH: [42] ©2000

Figure 4.6   *Microscopic images showing the setae hairs in the scopula of the jumping spider foot (left) and the broad endings of the setules covering the setae (centre) [44] ©IOP Publishing. Reproduced with permission. All rights reserved. Picture of the Abigaille-III robot (right), reprinted from [45] with permission from Elsevier*

force created by van der Waals dispersion forces of 10 N/cm^2, allowing the gecko to be equally capable of dry adhesion and sticking to highly hydrophobic surfaces [43].

Similarities have been noted between the attachment systems of the gecko and some species of spider. The tip of the jumping spider's foot contains a cluster of setae known as a scopula. These are covered with around 78 000 setules that broaden at the tips, as can be seen in Figure 4.6. Relying on van der Waals forces, the setules in a single foot can produce an adhesive force of up to 3.2×10^{-3}N [44].

4.2.1.1   Waalbot

Waalbot, named after the van der Waals forces, was devised with a potential application of spacecraft hull inspection and repair. Several prototypes were initially developed, each using a synthetic adhesive suitable for use on climbing robots consisting of polymer micro- and nano-fibres [46]. Two of these concepts, the Rigid Gecko Robot and Compliant Gecko Robot, were designed to mimic the gait of the gecko and could climb acrylic surfaces with a slope angle of up to 65° [47].

Another Waalbot concept uses a specialised wheel-leg pair design, as shown in Figure 4.7. Each wheel has three feet with polymer adhesive pads connected by a passive revolute joint, enabling either one or two feet to be in contact with the surface, depending on the point of rotation. When two feet are attached, a motor torque presses on the front foot while pulling away the rear. The prototype was capable of climbing in any direction on a smooth, acrylic surface up to a maximum angle of 110°, and was able to steer left and right with a small turning circle [41]. Waalbot II further improved upon the design by using polyurethane fibrillar foot-pads, with a fibre stem diameter of 57 μm and height of 113 μm, and a passive peeling mechanism was implemented to prevent immobilisation [48].

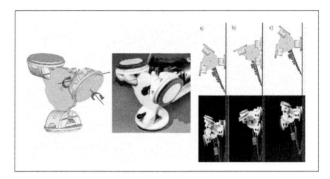

Figure 4.7   CAD models and pictures of the Waalbot tri-foot design with adhesive pads (left), showing how the feet stick to walls as it climbs a vertical surface (right) ©2006 IEEE. Reprinted, with permission, from [41]

4.2.1.2   Abigaille

Abigaille-I is a six-legged robot developed with the aim of being able to traverse any type of terrain or material and to eventually be capable of operating in a space environment. Each leg has a polydimethylsiloxane (PDMS) synthetic dry adhesive inspired by the gecko and spider setae bonded to it. This consists of arrays of microscale posts 10 μm wide and 60 μm tall, which are then further treated with a process known as micromasking to create nanoscale-compliant structures [49]. The Abigaille-II and Abigaille-III evolutions of the robot, the latter of which is also shown in Figure 4.6., use a hierarchical PDMS adhesive structure, wherein three layers consisting of the plastic foot and PDMS macro- and micro-posts are bonded together with silicone [45].

4.2.1.3   Legged excursion mechanical utility rover

The Legged Excursion Mechanical Utility Rover (LEMUR) was developed with the aim of performing assembly, inspection and maintenance of small-scale objects in space, using six limbs equipped with three-digit grippers. Initially intended to have an axi-symmetric design akin to octopuses and starfish, limitations caused by testing the first prototype in Earth gravity resulted in a design requiring an alternating tripod gait similar to walking insects [50]. While LEMUR IIa was a natural evolution of the concept, LEMUR IIb instead focused on planetary surface exploration. This used a four-limbed design with the aim of creating a system capable of free-climbing rough terrain and slopes up to and including vertical rock faces and overhangs [51].

LEMUR 3, shown in Figure 4.8, was designed both for crawling along the exterior of the International Space Station (ISS) with synthetic adhesive grippers and for climbing vertical cliffs and cave ceilings on the Moon and Mars using microspine grippers. Each adhesive gripper uses eight 10 cm² tiles made of silicone hairs 20 μm thick and 80 μm tall, allowing it to support a normal weight of up to 150 N. Using

Figure 4.8   Pictures of the gecko adhesive in its attached and detached states, a gripper attached to a solar panel (left) and LEMUR 3 climbing a cliff wall (right) ©2017 IEEE. Reprinted, with permission, from [52]

van der Waals forces to stick to the surface, the gripper detaches itself by shearing in a direction that causes the hairs to peel up. Mobility is achieved by releasing, moving, lowering, preloading and engaging each limb one at a time. Once all four limbs have moved, the end effectors shift the robot's body forward [52].

4.2.1.4   Additional concepts
Other gecko-based concepts include a crawling robot adapted for the assembly of orbital truss structures, which uses a multijointed, symmetrical leg configuration with an adhesive microstructure on the inside of each leg. This enables the robot to alter its shape for different trusses, using a combination of mechanical clamping and adhesion forces to grip the structures [53]. Another is the IBSS_Gecko_6, a four-legged climbing robot that uses a gecko-inspired multilayered footpad with multiple toes, which was able to climb vertical surfaces under conditions mimicking zero gravity [54]. Lastly, the ACROBOT concept was designed to be able to fit within a 1.5 in. gap behind equipment racks on the ISS. Consisting of a linked front and rear module, each with two-layered gecko pads, a shear load is applied to bend the layers to create adhesion. The robot moves in a motion similar to the inchworm, with the front module turning off its adhesion and extending forwards before adhering again to the surface, after which the process is repeated for the rear module [55].

4.2.2   Legged locomotion
Although the inherent complications associated with legged rovers mean that no such designs have yet been used in a space environment, there are several robotic concepts that have been investigated. These use legged locomotion and actuation mechanisms inspired by the gait and physiology of spiders and scorpions.

4.2.2.1   Abigaille
The mechanical design of the Abigaille series of robots discussed in Section 4.2.1.2 was also inspired by spiders. Each foot has three compliant and passive degrees of freedom (DOFs), allowing them to maximise contact with surfaces at different

angles. The feet can then be detached by moving the legs outwards in a manner similar to that seen in spiders [49]. The legs also have six DOFs and are symmetrically positioned around the prosoma, enabling the robot to transition from one surface angle to another, which was demonstrated with Abigaille-II [56]. Abigaille-III was optimised for vertical climbing, with an improved dexterity demonstrated in its ability to attach its feet to surfaces at different heights from its body, allowing it to traverse vertically over uneven surfaces. To detach the feet, a motor rotates a cam into and out of contact with the surface, disengaging the adhesive [45].

4.2.2.2   SCORPION

The SCORPION series of concepts focused on the development of an eight-legged robot capable of moving in difficult terrains, such as the slopes of craters found on the Moon and Mars. The gait pattern used is a simplified approximation of that seen in scorpions and many arthropods [57]. The four legs on the left-hand and right-hand sides of the body from front to back are labelled as L1–L4 and R1–R4, respectively. Two groups of legs step in a phased alternating tetrapod motion, i.e. group R4–L3–R2–L1 step together, and after a phase shift group L4–R3–L2–R1 step together. The legs in each group are also individually phased, with the movement of L3 happening slightly after the movement of R4, and so on [58]. The ambulation control was also achieved by combining two biologically-inspired principles: the Central Pattern Generator (CPG), in which the rhythmic behaviour of animals is generated by the central nervous system [59], and the Reflex. The CPGs control the legs under normal conditions without the need for feedback, while the Reflex is only activated when required in order to overcome obstacles [60].

The development of this concept culminated in field tests of the SCORPION IV [61], which also implemented an evolved control system known as the Posture-CPG-Reflex. This approach was also used in ARAMIES, a four-legged rover [62] whose ground contact, stumbling correction and step control processes were inspired by an analysis of the locomotion of cats [63]. The same biologically-inspired control approach was also used in the SpaceClimber robot. This was designed to traverse obstacles up to 40 cm tall, loose regolith and slopes with an incline of up to 40° [64], though a series of experiments performed in a simulated lunar environment noted that adjustments to the design were needed to enable it to avoid slippage [65].

4.2.2.3   Additional concepts

Another study proposed an eight-legged design based upon spiders for traversing outside of spacecraft, using both the same phased alternating tetrapod motion discussed in Section 4.2.2.2 and spider-inspired adhesion. Each leg also has femur, tibia and metatarsus segments, with the size ratios based upon those observed in several spider species [66]. The kinematics required to achieve the desired gait and angle of contact between the metatarsus tips for reversible adhesion was simulated, before being successfully demonstrated in a microgravity environment [67].

An active articulation mechanism was also developed from the hinge joints of spider legs, which are themselves actively extended by a hydraulic mechanism.

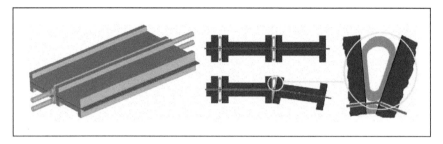

Figure 4.9   Diagrams of the Smart Stick held within an elastic joint (left) and the joint bending caused by the pressurised elliptical section expanding (right) [69]. Reproduced by permission of AISB

Although the joints have no extensor muscles, a torque is produced by a high internal pressure, which is caused by the empty space between the muscles and skeleton being filled with pressurised haemolymph [68]. A hydraulic mechanism for moving joints in a lightweight structure known as Smart Stick was developed from this system. Here, fluid flows through a miniaturised tube that has had sections plastically deformed into an elliptical shape. Acting upon hydraulic principles, the increased pressure in the elliptical section causes it to expand. When placed within the joint of a structure, the expansion of the Smart Stick will cause the system to bend without any external loads being applied, as shown in Figure 4.9 [70].

4.2.3   Hopping locomotion

Hopping mechanism concepts for use in planetary surface exploration scenarios have been inspired by both desert locusts and kangaroos. One such design aimed to bridge the gap between mechanisms mimicking the gait of the locust and planetary hopper concepts. As the locust prepares to jump, its hind femora bend until they are approximately horizontal. Energy is stored in elastic tissues known as the semilunar processes and released as the locust jumps [71]. The system in question used preloaded springs in a small, simplified single-leg mechanism to mimic the locust's muscles and semilunar processes. The dynamics of the pre-launch, launch and aerial stages were each modelled with the design's calculated performance found to be comparable to that of locust data. The study also explored the use of thrusters during the aerial phase and the feasibility of scaling the design up to the size of typical hopping rovers [72].

Another hopping concept examined the legs of a kangaroo, which consist of the foot, shank and thigh [73]. Whereas kangaroo-inspired robots have typically used open-chain linkages and several actuators, a minimally actuated, intermittent hopping robot with a compact design was proposed, which mimics the kangaroo by using a bipedal, synchronous hopping motion. This has an asymmetrical geared six-bar linkage system with a single DOF, and uses springs to store and release the energy within the robot, as shown in Figure 4.10. The kinematics and dynamics of the mechanism were analysed, and a small prototype model was built to confirm its

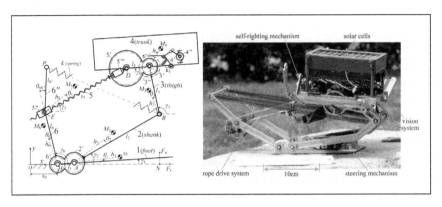

*Figure 4.10   Schematic of the kangaroo-inspired hopping robot (left) and the
second experimental prototype (right) [74]*

hopping capabilities [75]. A larger prototype was later constructed, which included
a self-righting mechanism and a rough sole to reduce slippage. This design demonstrated an ability to independently jump and upright itself after landing, and it was
able to jump a distance of 1 m horizontally and 0.3 m vertically [74].

4.3   Object capture

Capturing devices in space is typically required for on-orbit servicing, debris
removal and the retrieval of tumbling or uncontrolled satellites. There are numerous factors that have to be taken into account when devising a capture mechanism,
including manipulator design, approach trajectories and payload impact [76]. The
gecko adhesion discussed in Section 4.2.1 and kangaroo legs are two biological
sources that have been used to provide novel solutions to different aspects of capture
mechanism design.

4.3.1   Adhesive grippers

Several concepts for grasping objects have used gecko-inspired adhesion. One such
design for gripping smooth surfaces has two tiles, each with a PDMS microwedge
dry adhesive oriented in opposition to the other. The tiles are connected by tendons,
to which a load tendon is also attached. This is pulled when the grippers make contact with a surface, as shown in Figure 4.11, which produces a shear load in the tile
tendons, engaging the adhesives. The grippers are released by removing the tension in the load tendon, which detaches the adhesives [77]. A similar method was
applied to another concept for grasping curved spinning objects, in which the tiles
are attached to a hinged wrist mechanism, allowing the gripper to passively align
with the target surface [79].

This was further expanded upon in the design of an integrated gripper instrument, also shown in Figure 4.11, which uses a nonlinear passive wrist to produce

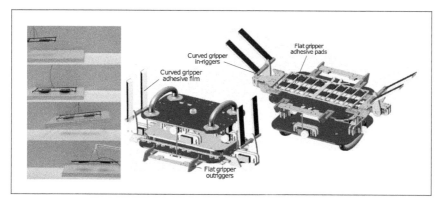

Curved gripper
in-riggers

Curved gripper
adhesive film

Flat gripper
adhesive pads

Flat gripper
outriggers

Figure 4.11   *Picture of the pulled load tendon attaching adhesive pads to*
a surface (left) [77] ©2016, reprinted by permission of SAGE
Publications, Inc. CAD models of the integrated gripper
instrument's adhesive pads (right), from [78]. Reprinted with
permission from AAAS

a tension force for engaging and detaching both flat and curved grippers. The full
design was able to grasp and manipulate a number of objects in microgravity, while
handheld flat grippers were also tested on the ISS [78]. Other studies also include the
development of a docking mechanism using four adhesive units attached to a load-
sharing component mounted on a robotic arm [80], as well as an exploration of the
impact processes between an adhesive robot and a target flat surface [81].

4.3.2   Kangaroo vibration suppression

In on-orbit servicing operations, post-capture vibrations tend to propagate from
moving payloads onboard the target, such as reaction wheels and driving mechanisms, or from the impact of capturing the object itself. The latter is particularly
significant when dealing with non-cooperative targets such as space debris [82].
To mitigate post-capture vibrations, one study took inspiration from the limb structures of birds. An asymmetric X-shaped structure was built, with varying rod lengths
and horizontal and vertical inner springs representing the bird's limbs and muscles,
respectively. The Bio-Inspired Quadrilateral Shape (BIQS) design was proposed for
use in on-orbit operations. This uses a symmetrical structure with only a horizontal
spring, considered similar to that of a kangaroo [83], as shown in Figure 4.12. The
system demonstrated a vibration isolation performance better than typical spring-
mass-damper systems [84], and experiments with an optimised design produced a
damping performance consistent with numerical models [86]. To achieve the multi-
direction vibration isolation required for on-orbit capturing, the structure was applied
to the legs of the Stewart platform mechanism, resulting in the X-shape Structure-
based Stewart Isolation Platform (XSSIP), also shown in Figure 4.12. When inserted
between a spacecraft's robotic arm and capture mechanism, the system showed

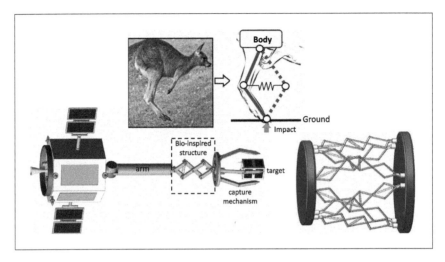

Figure 4.12   Models of the on-orbit capturing mechanism using the kangaroo-based (top) BIQS (left) [84] and the XSSIP (right) vibration suppression systems [85], reprinted with permission from Elsevier

favourable vibration isolation performance compared to the conventional Stewart platform [85].

4.4   Mobility inspired by plants

The mechanisms that have taken inspiration from animals for increasing mobility have often led to concepts that tend to have a number of similarities. As plants have very varied and unorthodox methods of locomotion, with areas of interest including seed dispersal, tendril movement and root growth, the different concepts inspired by plants have notably unique features.

4.4.1   Seed dispersal

The potential for implementing a number of seed dispersal methods in space applications has been explored [87]. The *Erodium cicutarium L.* plants have a self-burial mechanism, in which the seed's awn changes shape based on the humidity, resulting in a coiling and uncoiling motor action [88]. The seeds' ability to bury themselves in various planetary regoliths was examined as part of a study into the design of a self-burying probe concept known as SeeDriller [89]. A parachute design based upon the primary and secondary branching fibres that make up the structure of the parachute seed of the *Tragopogon dubius* plant [90] was also briefly considered.

4.4.1.1   Mars Tumbleweed

A seed dispersal method explored in more detail was the Mars Tumbleweed concept. Observations of the Mars Pathfinder mission noted that the airbags cushioning

Figure 4.13   *Pictures of the tumbleweed (left) and Dandelion design (centre) used in aerodynamics testing [93] and the Dandelion with whisk-shaped legs concept (right) [94]. Source: NASA*

its landing allowed it to travel a much farther distance than the accompanying Sojourner rover was able to achieve during its lifetime. From this, the possibility of a Tumbleweed rover using surface winds to propel itself along the Martian surface was considered. The feasibility of a number of concepts, and the characteristics of the Russian thistle tumbleweed, were subsequently discussed [91]. Simulations examining the interactions between spherical tumbleweeds, with radii of 3 and 5 m, and the Martian surface have shown that it is indeed possible for such a design to travel long distances over rock fields [92].

Although the thistle's non-spherical shape and offset centre of mass were not used in any of the concepts, the Dandelion design, shown in Figure 4.13, took inspiration from the thistle's ability to harness the wind with its lightweight branch structure. Closely resembling a dandelion, the concept used a symmetrical array of legs protruding from a central source, with padded ends to prevent sinkage [91]. The aerodynamic properties of several designs were tested alongside the tumbleweed, and the Dandelion with inverted pads was considered promising [93]. An evolution of the concept was later created, in which the legs were replaced with a curved shape resembling that of a whisk, as seen in Figure 4.13 [94].

4.4.2   Vine and tendril climbing

Continuum robots have largely been inspired to some extent by natural invertebrate structures such as elephant trunks, octopus tentacles and tongues [95]. These consist of a compliant, continuous backbone structure that enables them to bend, and often extend, at any point along their length. This allows them to navigate and reach places that traditional robots with rigid joints are unable to. Numerous prototypes and commercial designs using various continuum architectures have been developed [96], and the kinematics and dynamics of continuum robot modelling have been the subject of considerable study [97].

Another form of continuum robot has been inspired by the long, thin structures found in plant vines and anteater tongues [97]. Vines are able to climb objects either by using growths protruding from the main stem, known as tendrils, to wrap around nearby support structures, or by coiling around a body and establishing discrete points of contact that anchor it to the object [98]. This anchoring would allow a vine-inspired design to extend its body length beyond that of typical continuum robots, without needing a corresponding increase in the internal structural support at its base to accommodate the added length [99]. A robot with a small diameter-to-length ratio and controllable bending has potential for use in space applications, such as reaching inaccessible areas aboard spacecraft.

4.4.2.1   Tendril

The Tendril is a long, thin continuum robot concept designed for spacecraft inspection applications. It has a diameter-to-length ratio of 1:100, which is much smaller than typical continuum designs. The majority of the body consists of extension springs that provide passive bending, while compression springs produce active bending by actuating sets of tendons that run through the structure [100]. Though the coupling between the springs proved to be a source of significant error, resulting in the robot being very hard to control [97], later iterations of the Tendril improved the performance by using a hybrid concentric tube design [96, 101].

Another vine-inspired continuum prototype was developed, which also used the concentric tube backbone design of the Tendril. This concept included small fixed hooks attached throughout the body, mimicking the prickles or roots seen on some vines, which were used to aid contact as the robot wrapped around and braced against the target structure. These added spines enabled the robot to increase its reach, stability and accuracy compared to a smooth design [102]. A later concept, shown in Figure 4.14, utilised circumnutation, where the tip of the vine moves in an elliptical pattern as it searches for support structures [104]. The robot's ability

Figure 4.14   Picture of a vine coiling around a tree (left) ©2015 IEEE. Reprinted, with permission, from [99]. The vine tendril concept (centre) and a close-up of the spines attached to the robot body (right) ©2018 IEEE. Reprinted, with permission, from [103]

to explore unknown environments was then demonstrated on a mock-up of the ISS [103].

4.4.3 Plant root growth

Plant roots grow and dig into the soil through the processes of cell division and expansion just behind the root apex, which is caused by a force produced by a turgor pressure as water flows into the cell [105]. Information is also sent to each apex to direct the cell elongation towards mineral-rich soil. These root growth processes were implemented on a deployable probe known as SeedBot, which can anchor spacecraft on low-gravity bodies. This uses a modular design similar in shape to penetrators, with each module able to be extended by an osmotic axial piston. Between each segment is a steering stage, allowing the probe to bend in the direction chosen by the apex module [106, 107].

The collective behaviour of the spread of roots as they extend towards a goal is also governed by rules of repulsion, alignment and attraction. This was considered for use in swarm behaviours [108], which are discussed in Section 4.8.1, as well as in guidance systems for a fleet of probes descending from a satellite towards a planetary body and for penetrators exploring the subsurface [109].

4.5 Artificial muscle actuators

Although robotic actuators such as motors and gearboxes can be likened to the muscles of biological systems, they are considerably heavier and less efficient. The complex microscopic ionic mechanism that drives muscles is incredibly difficult to mimic [110]; however, it is possible to create muscle-like movement through the use of specific polymers. These are low density, mechanically flexible materials that change shape or size when stimulated by a range of sources.

The development of electrically-stimulated electroactive polymers (EAPs) allowed muscle characteristics to be transferred to robotic actuators. These have the added benefits of large actuation displacement, high strain capability and inherent vibration damping [111], with several candidate EAP materials exhibiting muscle-like properties and forces [112]. For future areas of development, both the significantly higher energy density of muscles and their ability to use different composite materials, visco-elasticity and variations at the macro and micro level to perform different tasks were highlighted [113]. The ionic and dielectric classes of EAP have both been highlighted as having potential applications for space [114].

4.5.1 Ionic polymer-metal composites

Ionic EAPs are able to produce large displacements when stimulated with low voltages, although disadvantages include slow response times and often requiring a wet environment to operate [115]. Several ionic EAP materials have demonstrated tolerances to hazardous space environmental factors including freezing, vacuum and radiation [114]. One of these materials, the ionic polymer-metal composite (IPMC),

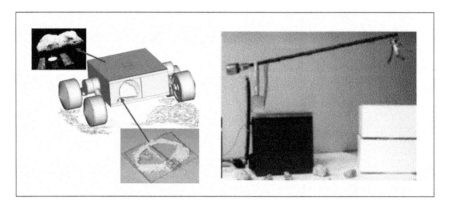

*Figure 4.15   The IPMC-actuated dust wiper concept for the MUSES-CN
           Nanorover, and a picture of the robotic arm and four-finger IPMC
           gripper [111]. Reproduced with permission from ASCE*

was used in the design of an actuator for a wiper that would remove dust from the infrared camera of the MUSES-CN Nanorover, shown in Figure 4.15. It was demonstrated that the IPMC could operate in temperatures as low as −140°C and in vacuum, though issues with permanent deformation and coating permeability were noted [116]. IPMCs have also been considered for use in satellite pointing systems [117] and as a four-finger gripper, as seen in Figure 4.15. This would be mounted aboard a miniature robotic arm, itself actuated by an electro-statically stricted polymer, and would operate like a human hand to grab objects [111].

4.5.2   Dielectric elastomers

The high-voltage requirements of dielectric elastomers (DEs), a subset of electric EAPs, allow for the development of actuators capable of precise displacements when low voltages are applied [118]. This was considered for use in the positioning of rigid telescope mirror segments, with the DE actuator able to produce a subwavelength accuracy [119]. Other proposed space mechanisms using DEs include shape control of flexible mirrors for astronomy and remote sensing [119], an actuation system for a robotic arm [120] and a jumping mechanism for a spherical Mars rover [121].

Another study also explored the use of DEs in two distributed actuation mechanism concepts for transporting lightweight objects that take inspiration from the peristaltic motion discussed in Section 4.1.2. The first uses a series of buckling actuators, each consisting of an elastomeric membrane that is displaced upwards when an electric field is applied. These are lined up along a path and sequentially activated, creating a travelling wave. The second mimics the peristaltic movement of the intestine, with a sequentially activated segmented tubular structure similar to that used in the LEAVO design, to transport objects internally [122].

4.6 Aerial mobility

Mars' low atmospheric density makes flight significantly more difficult than on Earth. Flying vehicles on Mars would need to have either a high Mach number or a very low Reynolds number in the region of 50 000 for wings and 15 000 for propellers. Such a small Reynolds number presents difficulties in transitioning from a laminar to a turbulent flow, which will cause laminar separation of the boundary layer if not achieved, resulting in a loss of lift [123]. Studies of insects have shown that they have a flight mechanism that enables them to exploit low Reynolds number environments. As their wings flap, they generate a horseshoe-shaped vortex that wraps around the wings in the early down and upstroke, which consists of several vortices including a leading-edge vortex (LEV). While the LEV behaves differently depending on the insect, as seen in Figure 4.16, it is believed to be a universal mechanism for producing lift for all insects at any Reynolds number up to 10^5, beyond which it becomes less effective [124]. Lift generation is also increased by the wings' flexible structures, with the maximum force achievable believed to be at flapping speeds near the wings' structural resonance. The generated force can also be increased or reduced through a rapid pitching rotation when performed before or after the end of the wing stroke, respectively, in a manner similar to the Magnus effect [126].

4.6.1 Wing-flapping mechanisms

The use of insect wing-flapping mechanisms for Mars flight has been explored in ornithopter designs [127], with two notable concepts being NASA's Entomopter and Marsbee, which are shown in Figure 4.17. The Entomopter uses the hawk moth as a baseline model for the wing aerodynamics. It also takes inspiration from another technique used by insects, where energy is temporarily stored in muscles or resilin,

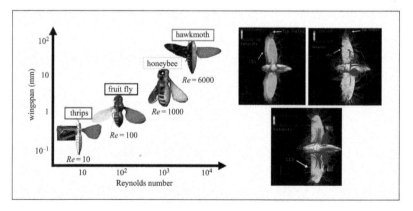

Figure 4.16   Graph of the Reynolds numbers of various insects (left) [124] and numerical results of the leading-edge vortices on (clockwise from top left) the thrip, fruit fly and hawk moth (right) [125], reproduced by permission of Wei Shyy and AIAA Journal

Figure 4.17   Concept model of the Entomopter (left) [123] and a picture of the Marsbee concept (right) [128]. Source: NASA

to recover flapping energy from torsional resonance in the fuselage [123]. The flapping wings are powered by a Reciprocating Chemical Muscle, which produces a high energy density without the need for atmospheric oxygen [129]. The Entomopter was envisioned as a short-range reconnaissance vehicle for collecting data and samples, which could then return to a refuelling rover in order to perform several flights. The use of multiple Entomopters employed in a swarm was also considered [130], and this mission structure was later implemented into the Marsbee concept, in which a cluster of vehicles, each weighing around 10 g, is deployed from a mobile base. A prototype robot was developed and tested in proof of concept experiments [128], with the design driven by the bumblebees' minute wing-to-body mass ratio of 0.52 per cent. An aerodynamic analysis suggested that, were the wing scaled up to 2–4.5 times its normal size, it would be able to hover in the Mars environment with a penalty mass increase of only 8–22 per cent [131].

4.7   Navigation systems for mobility

Given the significant telecommunications delay in all but Earth orbit and lunar missions, robotic systems must be able to autonomously make decisions that will allow them to safely navigate their surroundings. The processes required to do this are similar to the behaviours of insects that, despite the small size of their brains, have a remarkable ability to solve complex navigation problems, be it flying through the air or traversing along the ground. As a result, a number of methods have been proposed for implementing insect navigation techniques into various systems with the aim of increasing mobility for both surface and aerial traversal of planetary environments.

4.7.1   Natural and invasive interfacing

The processes required for planetary rovers to autonomously make decisions that will allow them to safely navigate unknown, rough terrain and overcome or avoid obstacles can be compared to the techniques used by desert ants and honeybees. These both use a constantly running path integration system as their primary means

Figure 4.18   *Pictures of the naturally-interfaced silkworm-controlled robot (left) [135] and an invasive multi-electrode array inserted into the coxa of a cockroach (right) [136], reprinted with permission from Elsevier*

of navigation, taking cues from a skylight compass and neural odometer, and learn a number of routes through landmark-based navigation. The information from both systems is then used by ants in a procedural and context-dependant way [132], while honeybees combine the information into a hierarchical navigation system [133]. However, replicating and implementing these biological processes into a robotic system is very challenging, and so an alternative hybrid mechanism has been proposed. Rather than taking inspiration from animal navigation, the rover's movement will be taken directly from a biological source embedded within the control loop of the robot. This is achievable via two methods, which can be used separately or together [134].

Natural interfacing involves giving stimuli to an animal that is kept in a cockpit-like area, with its movements then transmitted by sensors to a robotic system. An example of this is the silkworm-controlled robot shown in Figure 4.18 [135]. The invasive technique, on the other hand, involves direct contact between the living tissue and controller. This can be achieved using non-neural approaches, such as through muscle stimulation via multi-electrode arrays, as can be seen in Figure 4.18 [136], or by taking information from neuron interfaces. Here, electrodes stimulate and monitor cultured neurons or brain tissue either still connected to, or pre-grown and removed from, the body. Though the behaviour of dissociated neuronal cultures is still imperfect, the performance of such systems is considered very promising [137, 138]. A non-invasive application of this technique is the brain–machine interface, which monitors and translates neural activity into commands for external devices and has been envisaged for use in support activities for astronauts [139, 140].

4.7.1.1   Insect/machine hybrid controller

Integrating insect intelligence mechanisms into the control system of a rover will allow it to produce complex responses and decision-making potentially well beyond the capabilities of current systems, thus resulting in a greater level of mobility [141]. The Insect/Machine Hybrid Controller (IMHC) is a concept that combines natural and neural interfacing techniques for use in an autonomous planetary explorer. Low-level behaviours such as obstacle avoidance are carried out solely by the robotic system, while high-level behaviours, including path planning and decision-making, will require input from an insect tethered inside a cockpit. Inputs from the rover of the extraterrestrial environment are transmitted via both natural and neural interfaces and translated into nature-analogue signals for the insect to interpret [134]. From this, the insect is able to create an internal map of the environment, and the motor responses it produces are detected by a variety of sensors, which are then processed by the controller into movement of the system [142].

4.7.2   Honeybee optics

Insect eyes are close together and immobile, with fixed focus optics and low spatial resolution, meaning that they cannot gauge distances to objects through depth perception as vertebrates do. To determine how insects are able to navigate, honeybees were trained to fly through tunnels with moveable patterned walls. When the walls were given a constant motion, the position of the bee's flight would change, as shown in Figure 4.19 [144]. This is the result of optic flow, a process of image motion in which distance is measured by integrating the movement of images of the environment across the retina of the eye [143]. By maintaining a constant image motion velocity, honeybees use optic flow to stabilise their flight, control their speed and height, estimate object distances and negotiate narrow gaps [145]. A number of studies have attempted to increase the mobility of flying robotic concepts by

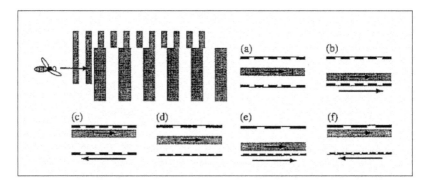

Figure 4.19   *Diagram showing how the position of the paths travelled by a bee through a tunnel are dependent on the motion of the patterned walls [143]. Reproduced with permission of the Licensor through PLSclear*

developing guidance systems that implement optic flow mechanisms, and the uses in both manned and autonomous spacecraft have been suggested [146].

4.7.2.1 Bio-inspired engineering of exploration systems

One of the earliest studies to implement insect navigation into a space application is the Bio-inspired Engineering of Exploration Systems (BEES) concept. Using multiple biological sources to develop a suite of autonomous flying navigation systems [147], BEES was envisioned as a multidisciplinary series of small biomorphic explorers, with subclasses including flight systems and planetary surface and sub-surface explorers [148]. One such system was a lightweight video camera onboard a flying vehicle that can process optic flow even in Mars twilight conditions, allowing for terrain navigation and hazard avoidance at very low altitudes [149].

Another BEES design took inspiration from the median ocellus of dragonflies for a flight stabilisation and attitude referencing system, which was successfully implemented and demonstrated on a flying platform concept [150]. Ocelli are small eyes on the dorsal and forward regions of the heads of a number of insects. While these typically function as a single optical horizon sensor that detects changes in light intensity, informing the insect of movements of its head away from a level position, the dragonfly ocellus is able to produce a focused image on the retina. Its elliptical shape enables it to both resolve details in the vertical direction and detect horizontally-orientated structures, allowing the ocellus to find the absolute position of the horizon [151].

A concurrent BEES study also envisaged the merging of insect-based technologies with an imaging system inspired by mammals [149]. The inner plexiform layer in the retinas of mammals has a vertical stack of neural strata, wherein parallel representations of the visual world are organised. A study of rabbit retinas showed that a neuronal interaction between strata results in the formation of at least ten unique outputs, from which visual data can be extracted and sent to higher visual centres [152]. A multilayer Cellular Neural/non-linear Network (CNN) capable of modelling the processes of the retina strata was developed [153]. This was then implemented in a small package capable of performing a wide range of image-processing operations, and the feasibility of using the CNN for recognising features in a natural scene was examined [149].

4.7.3 Optic flow landing

Optic flow sensing has also been considered for use in autonomous systems designed to improve the mobility of a landing spacecraft. One of the most hazardous phases of a planetary exploration mission is the landing. The spacecraft not only needs to reduce its speed but must also have selected a landing site that has both a minimal hazard density and a high scientific interest. Given the challenges present in developing a processing system that can extract the relevant data and use it to send appropriate steering commands, almost every unmanned spacecraft that has landed on an extraterrestrial body has relied on remote sensing data obtained by orbiters. The exception is the Chang'e-3 spacecraft, which successfully used an autonomous hazard detection and avoidance technique when approaching the lunar surface [154].

4.7.3.1   Elementary motion detectors

The motion of an insect is sensed by Elementary Motion Detectors (EMDs), from which a visual processing centre known as the lobula plate analyses the optic flow and transmits the resulting instructions to the neurons driving the insect's muscles [146]. The use of artificial EMDs in a simple, lightweight system was explored for a spacecraft landing in a simulated visual lunar environment. The outputs of an optic flow sensor pointed towards the surface were compared by the EMD to the surface images provided by the simulation. This allowed the spacecraft descent to be robustly regulated with acceptable approach speeds as it reached the low gate point, the distance at which the landing site is obscured by dust clouds [155]. Optic flow landing techniques using EMDs were further explored with an analysis of optimal descents with regard to mass consumption [156] and a study of fuel-optimised trajectories created using suboptimal descents. These could be achieved with either two optic-flow sensors pointing at 90° and 135° to the local horizontal [157] or by using a non-gimbaled sensor configuration [158].

4.7.3.2   Additional concepts

Other techniques for aiding spacecraft landing using optic flow include a visual motion sensor based on the retina lenslets of the fly. This was developed and tested onboard an unmanned helicopter, where it demonstrated an ability to measure optic flow in complex environments as may be seen by a planetary lander [159]. Another concept used a seven-lens compound optical sensor inspired by the compound eye of the common housefly. This is composed of many light-sensitive cells with overlapping fields of view that each produces a quasi-Gaussian response to visual stimuli, allowing the eye to detect very small movements. The sensor can be used to detect small displacements in land masses and is envisioned for use in hazard analysis of fault lines when considering sites for landing spacecraft and building structures [160]. Wide-field-integration of optic flow, in which a large range of optic flow is used for motion estimation, was also considered as a back-up sensor for spacecraft performing near-ground manoeuvres around asteroids [161].

Another navigation technique is Time-To-Contact (TTC), which is used by pigeons to regulate their landing approach speed [162]. TTC is calculated by dividing the pigeon's height by its vertical velocity, which is measured by the increasing expansion of observed ground features. This has been combined with optic flow measurements to create successful landing strategies when the information from images is delayed or noisy [163]. Finally, birds of prey use optic flow created by the motion of their target to produce high-resolution tracking and to stabilise their bodies while moving. This was implemented in an autonomous flyer concept that orbits a descending payload, taking images and storing the data for future use [164].

4.8 Multi-agent spacecraft system architectures

The majority of the biomimetic mechanisms discussed so far have been considered for use on a single instrument or system. This section explores the design of the spacecraft as a whole and details how taking inspiration from the behaviour of large groups of animals and various cellular processes has resulted in unique system architectures that can improve the mobility and reliability of missions involving small spacecraft.

4.8.1 Swarm intelligence

Swarm intelligence in nature is defined as large groups of animals that appear to work as a single fluid entity, while each individual organism has a simple specified role. This arises from the self-organised interactions with nearby individuals and the local environment, which when combined together enables the swarm to produce complex behaviours as it acts towards a goal, seemingly without supervision [165]. Swarms from several animal groups have been studied extensively, including foraging ants [165], flocks of birds [166], such as that seen in Figure 4.20, termite nest-building [168] and coordinated locust swarms [169]. These have provided the basis for intelligent swarm technology, in which individual units within a group have independent intelligence. Applications include agent swarms, which are used in computer modelling and the study of complex systems, and swarm robotics [170].

Swarms are an attractive concept for a wide range of small satellite mission scenarios. By utilising a large number of identical, simple units built using commercial off-the-shelf (COTS) technology, this would provide a great degree of robustness without requiring extreme redundancy techniques. Swarms also have a much greater degree of manoeuvrability, as they can provide a large area coverage, are able to explore areas that are difficult for traditional spacecraft to reach, and sub-sections can break away to investigate objects of interest before rejoining the main group [170]. Additional benefits include a lower total mass than a single large spacecraft, potentially reducing the mission cost and a low processing power per unit [171].

Figure 4.20   A starling murmuration as an example of swarm intelligence [167]

The major disadvantage of satellite swarm design is the challenging prospect of establishing collective control, which has been the subject of a number of studies [172–174].

4.8.1.1   Autonomous Nano Technology Swarm

The Autonomous Nano Technology Swarm (ANTS) project sought to utilise swarm technologies inspired by the task specialisation and sociality in insect colonies for spacecraft and planetary rovers [175]. The ANTS mission concepts include the Saturn Autonomous Ring Array (SARA) [176], a 1 000-strong swarm with specialised sub-swarms for performing in situ exploration of Saturn's rings, the Super Miniaturized Addressable Reconfigurable Technology (SMART) [170], a swarm of interconnectable tetrahedron robots that can create complex robotic structures in order to adapt to different environmental challenges, and the Prospecting Asteroid Mission (PAM).

ANTS/PAM envisions launching a swarm of 1 000 pico-satellites, each carrying a specialised instrument, into the asteroid belt. Teams of three spacecraft classes are sent to different asteroids, as shown in Figure 4.21. Here, the data-gathering workers are coordinated by rulers, while messengers communicate to the other classes and Earth. Each team requires total autonomy to, with artificial intelligence software that can reconfigure their operations at each asteroid, optimise efficiency, plan orbital changes to avoid high-risk collisions and recover from damage [177].

4.8.1.2   Additional concepts

Several other concepts have also explored the use of swarm intelligence. One study examined the possibility of using swarms to traverse the surface of Mars and search for caves, using a method inspired by the strategy employed by honeybees as they look for a new hive [178]. Other designs have utilised swarm intelligence in robotic

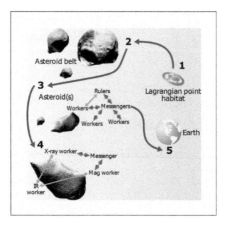

Figure 4.21   Overview of the ANTS/PAM mission concept ©2004 IEEE. Reprinted, with permission, from [177]

system applications, such as the manoeuvring and self-assembly of multiple units to create a single physical structure in the YETE2/SIW concept [179], and the deployment of a debris capture mechanism, which uses a net that is held and controlled by multiple autonomous units [180]. Other studies have also included the development of swarm navigation and exploration strategies [181], as well as the use of human operators for overseeing the swarm communication and strategy in a planetary exploration scenario [182].

4.8.2   Cellular spacecraft architecture

The living cells that make up a multicellular organism have a huge variety of functions, compositions and processes, such as replication, information transfer and DNA repair. Artificial cells have been developed and used as models for the study and application of different aspects of cell biology [183, 184], and two of these processes have also been considered for implementation into spacecraft architectures.

4.8.2.1   Cell apoptosis

Programmed cell death, also known as apoptosis, is a key factor in physiological homeostasis, in which some cells are genetically encoded to die. This occurs when biochemical signals provided by various sources, such as hormones and growth factors, are reduced or stopped, as shown in Figure 4.22. By maintaining an equilibrium between cell growth and death, apoptosis regulates the size of cell populations and tissues, preventing cancers, viral infections and immune deficiencies [187].

Autonomic computing uses the principles of apoptosis to create a self-managing computer system. By being aware of the internal and external conditions, and having the ability to detect and adapt to changing circumstances, an autonomic system will be able to optimise its performance, protect itself and recover from faults [185]. Apoptotic computing has been considered for use in agent-based swarm systems,

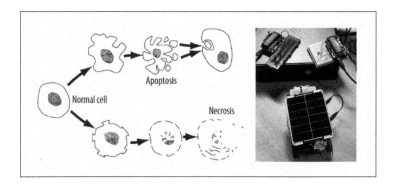

Figure 4.22   Diagram of cell apoptosis and necrosis (left) ©2011 IEEE. Reprinted, with permission, from [185]. An apoptotic mechanism in a proof of concept (right) ©2019 IEEE. Reprinted, with permission, from [186]

such as the ANTS concept discussed in Section 4.8.1.1. If a worker becomes damaged or begins causing emergent behaviour detrimental to the mission, its ruler can remove the stay alive signal, allowing the swarm to reconfigure and adapt [185, 186]. A proof of concept hardware was developed to demonstrate the use of autonomic computing in disposing of satellites within a swarm if they are deemed to be endangering the mission, and showed the successful deployment of apoptosis when different failure conditions were met, as seen in Figure 4.22 [186].

4.8.2.2 Satellite stem cell

Small, low-budget satellite missions such as CubeSats often have rapid development times. This tends to come at the cost of system-level testing, which has been highlighted as a key factor in the significantly low reliability rate of such missions [188, 189]. Typical techniques for improving reliability, such as component screening and functional redundancies, are often too costly for small satellites, resulting in single-string architectures with many potential single points of failure [190].

Simple multicellular organisms use decentralised task management to coordinate body functions, by using cellular peer-to-peer messaging as opposed to a central nervous system. Some vertebrates, including the zebrafish and newt, are able to redifferentiate mature cells to repair damage [191]. This is similar to the k-out-of-n satellite architecture, in which multifunctional subsystems are placed into a global pool. Though this method has superior reliability compared to other architectures, the implementation of these subsystems is very difficult [192].

The Satellite Stem Cell (SSC) concept shown in Figure 4.23 is a simplified model of the differentiation process in a biological cell. A proposed satellite architecture consists of multiple SSCs connected together, with the system tasks distributed autonomously across the cells. Discrete programmable elements, acting as proteins, will actively carry out a task or remain in either a dormant or a redundant state. If an element fails, its task is then redistributed to a free protein, if one is available. The system was simulated on a microsatellite set-up and it demonstrated an increased reliability compared to the COTS architecture [193].

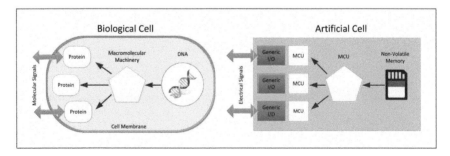

*Figure 4.23   Simplified differentiation in a biological cell and a proposed
            artificial implementation. Reprinted from [193] with permission
            from Elsevier*

4.9   Hibernation for human spaceflight

One of the major challenges for human spaceflight missions to Mars and beyond is maintaining the health of the astronauts. Physiological issues caused by the microgravity environment include the redistribution of body fluids, a decrease in strength and atrophy of muscles of up to 50 and 30 per cent, respectively, and a bone density decrease of 1–2 per cent per month, with some astronauts being unable to fully recover pre-flight bone density levels [194]. The astronauts will also be exposed to a greater risk of cellular damage to DNA, cancer and fatal doses of radiation from coronal mass ejection events [195]. The psychological issues present in long-term missions have also been highlighted in a simulated 520-day manned mission to Mars, with the participants experiencing notable drops in benevolence and stimulation, as well as an increase in tension, fatigue and physical withdrawal from social contact [196].

Life support systems in long-term missions, which include food, water, entertainment and atmosphere provisions, could take up to 40 per cent of the total wet mass. One proposed method for reducing this mass is to minimise crew activity by inducing a state similar to hibernation [197]. This is a type of natural hypometabolism used by mammals to survive extreme environments, in which the animal drastically reduces its body temperature, metabolic activity, heart rate and energy consumption, and can last from a few hours to several weeks [198]. Attempts have been made to induce hypometabolism in non-hibernating mammals using various techniques [199].

Hibernation would effectively eliminate the psychological challenges of long-term missions, and radiation shielding would only need to be focused on the area where the crew are located. It may also be possible to use hibernating black bears as a model for preventing bone atrophy through disuse [200]. Strategies for artificially inducing a mild hypometabolic state and integrating hibernacula into spacecraft life support systems have been considered for long-term missions [197], and the benefits of inducing synthetic torpor through controlled metabolic depression have also been explored [201]. During hibernation, nutrition and hydration can be provided by a liquid solution fed via an intra venous line into the body [202, 203]. One mission strategy that has been examined in detail uses a combination of Therapeutic Hypothermia (TH) to lower body temperature and Total Parenteral Nutrition (TPN) to provide the liquid nutrition solution [202].

4.10   Summary and future

This chapter has presented a comprehensive review of the biologically-inspired mechanisms that have been developed and suggested for use in space applications. A chronological summary of the concepts discussed is given in Table 4.1.

Immediately noticeable is the wide variety of both biological inspirations and intended applications for the mechanisms discussed. While there are several concepts that have quite a logical origin, such as the designs based upon spider legs

Table 4.1   Summary of the mechanisms discussed

Biological inspiration	Mechanism	Intended space application	Stage of development
Locust ovipositor	Drill	PSub* exploration	Prototype design
Wood wasp ovipositor	DRD	PSub exploration	3rd gen. D & T†
Peristaltic motion	LEAVO	PSub exploration	Prototype D & T
Sandfish lizard	Hybrid wheels	PSurf‡ traversal	M & S§
Gecko foot hairs (GFH)	Waalbot	SC¶ traversal	2nd gen. D & T
GFH, spider legs	Abigaille	All surface traversal	3rd gen. D & T
GFH	LEMUR	SC, PSurf traversal	3rd gen. D & T
GFH	Crawler	SC truss traversal	Prototype D & T
GFH	IBSS_Gecko_6	Vertical surface climbing	Prototype D & T
GFH, inchworm gait	ACROBOT	SC small gap traversal	Prototype D & T
Scorpion gait control	SCORPION	Crater traversal	4th gen. D & T
Cat gait control	ARAMIES	Crater traversal	Prototype D & T
Gait control	SpaceClimber	Crater traversal	Prototype D & T
GFH, spider legs	Legged robot	SC traversal	M & S
Spider leg joints	Smart Stick	Structure hydraulics	Prototype D & T
Desert locust jumping	Hopper	PSurf traversal	M & S
Kangaroo hopping	Hopper	PSurf traversal	2nd gen. D & T
GHF	Gripper	Debris capture	Prototype D & T
GHF	Gripper	Object capture, docking	M & S
GHF	Integrated gripper	Object capture	Testing on ISS
GHF	Docking	Capture, servicing	M & S
Kangaroo legs	BIQS	Vibration isolation	M & S
Kangaroo legs	XSSIP	Vibration isolation	M & S
Self-burying seed	SeeDriller	PSub exploration	Concept D & T
Parachute seed	Parachute	Sensor parachute	Concept proposal
Thistle tumbleweed	Mars Tumbleweed	PSurf traversal	Concept D & T
Tentacles, tongues	Tendril	SC small gap traversal	2nd gen. D & T
Vines	Continuum robot	SC traversal	2nd gen. D & T
Plant roots	SeedBot	PSub exploration	Concept design
Plant roots, swarms	Probes	PSub, landing guidance	Concept study
Muscles	IPMC actuator	Camera dust removal	Concept D & T
Muscles	IPMC pointing	Satellite attitude control	M & S
Muscles	IPMC gripper	Object grasping	Concept D & T
Muscles	DE actuator	Mirror position, shape	Concept D & T
Muscles	DE actuator	SC servicing robotic arm	Concept D & T
Muscles	DE actuator	Mars surface hopper	M & S
Peristaltic motion	DE actuators	Object transportation	Concept proposal
Flapping wings	Ornithopter	Mars flight exploration	Prototype testing
Hawk moth wings	Entomopter	Mars flight exploration	Mission scenario
Bumblebee wings	Marsbee	Mars flight exploration	Mission scenario
Natural interfacing	Mimetic robot	PSurf traversal	Concept D & T

(Continues)

Table 4.1   Continued

Biological inspiration	Mechanism	Intended space application	Stage of development
Neural interfacing	Brain-machine	Remote control	Concept proposal
Hybrid interfacing	IMHC	PSurf traversal	Concept proposal
Insect optic flow (OF)	BEES	PTerr** navigation	Concept D & T
Dragonfly ocellus	BEES	PTerr navigation	Concept D & T
Mammal retinas	CNN	PTerr navigation	Concept D & T
OF	EMD	Landing guidance	M & S
Insect OF	Lander	Landing guidance	Concept test
Housefly eye	Imager	Landing site analysis	Concept D & T
OF	Imager	Asteroid manoeuvres	Concept D & T
Pigeon TTC	Lander	Landing guidance	M & S
Birds of prey OF	Imager	Landing recorder	Concept test
Swarms	ANTS/SARA	Saturn exploration	Concept proposal
Swarms	ANTS/SMART	Connectable robots	Prototype design
Swarms	ANTS/PAM	Asteroid exploration	Concept proposal
Honeybee searching	Swarm	Mars caves search	M & S
Swarms	YETE2/SIW	Component self-assembly	Concept design
Swarms	Aggregator	Debris removal	Concept proposal
Swarms	Swarm	Navigation strategies	M & S
Swarms	Swarm	PTerr exploration	Concept D & T
Cell apoptosis	Satellite death	Satellite removal	Proof of concept
Multicellular life	SSC	Architecture reliability	Concept test
Hibernation	Hypometabolism	Human stasis	Concept proposal
Hibernation	Synthetic torpor	Human stasis	Concept proposal
Hibernation	TH, TPN	Human stasis	Mission design

*PSub: Planetary Subsurface.
†D & T: Design and Testing.
‡PSurf: Planetary Surface.
§M & S: Modelling and Simulation.
¶SC: Spacecraft.
**PTerr: Planetary Terrain.

and swarm intelligence, for small satellite missions, there are a number of concepts taken from unlikely sources, such as peristaltic wave motion, tumbleweeds and cell apoptosis. It can also be seen that, while some concepts have gone through several iterations, with testing of second-and even third-generation designs, the majority have yet to progress beyond either initial prototype demonstrations or modelling and simulation studies. This is to be expected, given both the relative infancy of biomimetics and the natural discontinuation that occurs in many novel technology studies.

Though a number of biomimetic concepts have been used in terrestrial applications, such as the kingfisher-inspired shape of the Shinkansen bullet train nose [204], architecture for cooling buildings based on termite nests [205] and wind turbine blade designs inspired by humpback whales [206], so far none have been implemented in current space systems. One of the greatest challenges for new concepts, be they biologically inspired or not, is to achieve a technology readiness level that will allow them to be considered as viable alternatives to current systems. The cost and relative infrequency of launches, especially for more complex missions such as planetary exploration, naturally result in only a very select few designs being used. The cost of failure, and the consequent importance of reliability, will often mean that established technologies that have seen extensive use will be favoured over novel, untested concepts. This is exacerbated in fields such as robotics, which will often require the use of complex systems whose reliability will be much less certain [9].

A technology demonstration of adhesive grippers aboard the ISS [78] was the only example found of a biologically-inspired system that has been tested in space. A significant step forward for biomimetics will be the first mission to feature a technology demonstration of a fully constructed mechanism. This could be achieved through further tests performed aboard the ISS, which can be expanded to include other lightweight mechanisms such as the legged climbers, continuum robots and artificial muscle actuators, or through the verification of software such as the multicellular architecture and swarm intelligence systems in small satellite missions. However, this can only be accomplished if the development of these novel ideas continues towards a point where system prototypes can be built and thoroughly tested.

The field of biomimetics for space applications can also expand simply by continuing to explore new sources of inspiration. The mechanisms that animals and plants use to survive and thrive in their environments are nearly limitless, and as such there is a vast well of untapped potential. Even if concepts initially intended for use in space never reach the stage of being considered for a mission, it will often be possible to adapt them for use in terrestrial applications, presenting solutions and innovations that would not have been considered or thought possible using traditional techniques.

References

[1] Bar-Cohen Y. 'Introduction: nature as a source of inspiring innovation' in Bar-Cohen Y. (ed.). *Biomimetics: Nature-Based Innovation*. CRC Press; 2011. pp. 1–34.

[2] Vincent J.F.V., Bogatyreva O.A., Bogatyrev N.R., Bowyer A., Pahl A.-K. 'Biomimetics: its practice and theory'. *Journal of the Royal Society Interface*. 2006;**3**(9):471–82.

[3] Wanieck K., Fayemi P.-E., Maranzana N., Zollfrank C., Jacobs S. 'Biomimetics and its tools'. *Bioinspired, Biomimetic and Nanobiomaterials*. 2017;**6**(2):53–66.

[4] Menon C., Lan N., Sameoto D. 'Towards a methodical approach to implement biomimetic paradigms in the design of robotic systems for space applications'. *Applied Bionics and Biomechanics*. 2009;**6**(1):87–99.

[5] Bhushan B. 'Biomimetics: lessons from nature – an overview'. *Philosophical Transactions of the Royal Society of London. A*. 1893;**2009**(367):1445–86.

[6] Hwang J., Jeong Y., Park J., Lee K., Hong J., Choi J. 'Biomimetics: Forecasting the future of science, engineering, and medicine'. *International Journal of Nanomedicine*. 2015;**10**(1):5701–13.

[7] Summerer L. 'Thinking tomorrows' space – research trends of the ESA advanced concepts team 2002–2012'. *Acta Astronautica*. 2014;**95**(December (2)):242–59.

[8] Ayre M. 'Biomimetics applied to space exploration' in Collins M.W., Brebbia C.A. (eds.). *Design and Nature II*. WIT Press; 2004. pp. 603–12.

[9] Menon C., Broschart M., Lan N. 'Biomimetics and robotics for space applications: challenges and emerging technologies'. *IEEE Int Conf Robot Autom - Workshop on Biomimetic Robotics*. Roma, Italy; 2007. pp. 1–8.

[10] Zacny K. 'Lunar drilling, excavation and mining in support of science, exploration, construction, and in situ resource utilization (ISRU)' in Badescu V. (ed.). *Moon. Prospective Energy and Material Resources*. Berlin Heidelberg: Springer-Verlag; 2012. pp. 235–65.

[11] Di Lizia P., Bernelli-Zazzera F., Ercoli-Finzi A., *et al*. 'Planning and implementation of the on-comet operations of the instrument SD2 onboard the Lander Philae of Rosetta mission'. *Acta Astronautica*. 2016;**125**:183–95.

[12] Abbey W., Anderson R., Beegle L., *et al*. 'A look back: The drilling campaign of the curiosity rover during the mars science laboratory's prime mission'. *Icarus*. 2019;**319**(6153):1–13.

[13] Bar-Cohen Y., Zacny K. *Drilling in extreme environments*. Wiley VCH; 2009.

[14] Badescu M., Kassab S., Sherrit S., Aldrich J., Bao X., Bar-Cohen Y. 'Ultrasonic/sonic driller/corer as a hammer-rotary drill'. *Sensors and Smart Structures Technologies for Civil, Mechanical and Aerospace Systems*. San Diego, USA; 2007. p. 65290S–1.

[15] Richter L., Coste P., Gromov V.V., *et al*. 'Development and testing of subsurface sampling devices for the beagle 2 Lander'. *Planetary and Space Science*. 2002;**50**(9):903–13.

[16] Marshall J.P., Hudson T.L., Andrade J.E. 'Experimental investigation of insight HP³ mole interaction with martian regolith simulant'. *Space Science Reviews*. 2017;**211**(1-4):239–58.

[17] Menon C., Vincent J., Lan N., Bilhaut L., Ellery A., Gao Y. 'Bio-inspired micro-drills for future planetary exploration'. *CANEUS*. Toulouse, France: ASME; 2006. pp. 117–28.

[18] Vincent J. 'How does the female locust dig her oviposition hole?' *Physiological Entomology*. 1975;**50**(3):175–81.

[19] Gao Y., Ellery A., Sweeting M.N., Vincent J. 'Bioinspired drill for planetary sampling: literature survey, conceptual design, and feasibility study'. *Journal of Spacecraft and Rockets*. 2007;**44**(3):703–9.

[20] Vincent J., King M. 'The mechanism of drilling by wood wasp ovipositors'. *Biomimetics*. 1995;**3**(4):187–201.

[21] Gao Y., Ellery A., Jaddou M., Vincent J., Eckersley S. 'A novel penetration system for in situ astrobiological studies'. *International Journal of Advanced Robotic Systems*. 2005;**2**(4):29–86.

[22] Gouache T.P., Gao Y., Coste P., Gourinat Y. 'First experimental investigation of dual-reciprocating drilling in planetary regoliths: proposition of penetration mechanics'. *Planetary and Space Science*. 2011;**59**(13):1529–41.

[23] Pitcher C., Gao Y. 'Analysis of drill head designs for dual-reciprocating drilling technique in planetary regoliths'. *Advances in Space Research*. 2015;**56**(8):1765–76.

[24] Alkalla M., Gao Y., Bouton A. 'Customizable and optimized drill bits bio-inspired from wood-wasp ovipositor morphology for extraterrestrial surfaces'. *IEEE/ASME Int Conf Adv Intell Mechatron*. Hong Kong, China; 2019. pp. 430–5.

[25] Gao Y., Frame T.E.D., Pitcher C. 'Piercing the extraterrestrial surface: integrated robotic drill for planetary exploration'. *IEEE Robotics & Automation Magazine*. 2015;**22**(1):45–53.

[26] Pitcher C., Gao Y. 'First implementation of burrowing motions in dual-reciprocating drilling using an integrated actuation mechanism'. *Advances in Space Research*. 2017;**59**(5):1368–80.

[27] Gouache T., Gao Y., Frame T., Coste P., Gourinat Y. 'Identification of the forces between regolith and a reciprocating drill-head: perspectives for the exploration of martian regolith'. *62nd International Astronautical Congress*. Cape Town, South Africa; 2011. pp. 1280–9.

[28] Lauder G., Tangorra J. 'Fish locomotion: biology and robotics of body and fin-based movements'. *Robot Fish*. Berlin Heidelberg: Springer; 2015. pp. 25–49.

[29] Pitcher C., Alkalla M., Pang X., Gao Y. 'Development of the third generation of the Dual-Reciprocating drill'. *Biomimetics*. 2020;**5**(3):38.

[30] Smith M., Morton D. 'The small intestine'. *The Digestive System*. Elsevier Ltd; 2010. pp. 107–27.

[31] Yekutieli Y., Flash T., Hochner B. 'Biomechanics: hydroskeletal'. *Encyclopedia of Neuroscience*. Elsevier Ltd; 2009. pp. 189–200.

[32] Fujiwara A., Nakatake T., Tadami N., Isaka K., Yamada Y., Sawada H. 'Development of both-ends supported flexible auger for lunar earthworm-type excavation robot leavo'. *IEEE/ASME Int Conf Adv Intell Mechatron*. Auckland, New Zealand; 2018. pp. 924–9.

[33] Kubota T., Nagaoka K., Tanaka S., Nakamura T. 'Earth-worm typed drilling robot for subsurface planetary exploration'. *IEEE ROBIO*. Sanya, China; 2007. pp. 1394–9.

[34] Omori H., Murakami T., Nagai H., Nakamura T., Kubota T. 'Development of a novel bio-inspired planetary subsurface explorer: initial experimental study by prototype Excavator with propulsion and Excavation units'. *IEEE/ASME Transactions on Mechatronics*. 2013;**18**(2):459–70.

[35] Ellery A., Patel N., Richter L., Bertrand R., Dalcomo J. 'ExoMars rover chassis analysis & design'. *The 8th Int Symp on Artificial Intelligence, Robotics and Automation in Space*. Munich, Germany; 2005. pp. 1–8.

[36] Patel N., Slade R., Clemmet J. 'The ExoMars rover locomotion subsystem'. *Journal of Terramechanics*. 2010;**47**(4):227–42.

[37] Reina G., Foglia M. 'On the mobility of all-terrain rovers'. *Industrial Robot: An International Journal*. 2013;**40**(2):121–31.

[38] Maladen R.D., Ding Y., Li C., Goldman D.I. 'Undulatory swimming in sand: subsurface locomotion of the sandfish lizard'. *Science*. 2009;**325**(5938):314–18.

[39] Lopez-Arreguin A.J.R., Montenegro S. 'Towards bio-inspired robots for underground and surface exploration in planetary environments: an overview and novel developments inspired in sand-swimmers'. *Heliyon*. 2020;**6**(6):e04148.

[40] Silva M., Machado J., Tar J. 'A Survery of technologies for climbing robots adhesion to surfaces'. *6th Int Conf on Computational Cybernetics*. Stará Lesná, Slovakia; 2008. pp. 127–32.

[41] Murphy M., Tso W., Tanzini M., Sitti M. 'Waalbot: an agile small-scale wall climbing robot utilizing pressure sensitive adhesives'. *Int Conf on Intelligent Robotics and Systems*. Beijing, China; 2006. pp. 3411–16.

[42] Autumn K., Liang Y.A., Hsieh S.T., *et al.* 'Adhesive force of a single gecko foot-hair'. *Nature*. 2000;**405**(6787):681–5.

[43] Autumn K., Sitti M., Liang Y.A., *et al.* 'Evidence for van der Waals adhesion in gecko setae'. *Proceedings of the National Academy of Sciences*. 2002;**99**(19):12252–6.

[44] Kesel A.B., Martin A., Seidl T. 'Getting a grip on spider attachment: an AFM approach to microstructure adhesion in arthropods'. *Smart Materials and Structures*. 2004;**13**(3):512–18.

[45] Henrey M., Ahmed A., Boscariol P., Shannon L., Menon C. 'Abigaille-III: a versatile, bioinspired hexapod for scaling smooth vertical surfaces'. *Journal of Bionic Engineering*. 2014;**11**(1):1–17.

[46] Menon C., Murphy M., Angrilli F., Sitti M. 'Waalbots for space applications'. *55th International Astronautical Congress*. Vancouver, Canada; 2004. pp. 1972–82.

[47] Menon C., Sitti M. 'A biomimetic climbing robot based on the gecko'. *Journal of Bionic Engineering*. 2006;**3**(3):115–25.

[48] Murphy M.P., Kute C., Mengüç Y., Sitti M. 'Waalbot II: adhesion recovery and improved performance of a climbing robot using fibrillar adhesives'. *The International Journal of Robotics Research*. 2011;**30**(1):118–33.

[49] Menon C., Li Y., Sameoto D., Martens C. 'Abigaille-I: towards the development of a spider-inspired climbing robot for space use'. *2nd IEEE/RAS-EMBS Int Conf Biomed Robot Biomechatron*. Scottsdale, USA; 2008. pp. 384–9.

[50] Kennedy B., Agazarian H., Cheng Y., *et al.* 'Lemur: legged excursion mechanical utility rover'. *Autonomous Robots*. 2001;**11**(3):201–5.

[51]   Kennedy B., Okon A., Aghazarian H., *et al.* 'Lemur IIb: a robotic system for steep terrain access'. *Industrial Robot: An International Journal.* 2006;**33**(4):265–9.

[52]   Parness A., Abcouwer N., Fuller C., Wiltsie N., Nash J., Kennedy B. 'Lemur 3: a limbed climbing robot for extreme terrain mobility in space'. *IEEE Int Conf robot Autom.* Singapore; 2017. pp. 5467–73.

[53]   Tang T., Hou X., Xiao Y., Su Y., Shi Y., Rao X. 'Research on motion characteristics of space truss-crawling robot'. *International Journal of Advanced Robotic Systems.* 2019;**16**(1):172988141882157–17.

[54]   Wang Z., Wang Z., Dai Z., Gorb S. 'Bio-inspired adhesive footpad for legged robot climbing under reduced gravity: multiple toes facilitate stable attachment'. *Applied Sciences.* 2018;**8**(1):114.

[55]   Kalouche S., Wiltsie N., Su H., Parness A. 'Inchworm style gecko adhesive climbing robot'. *IEEE/RSJ Int Conf on Intelligent Robots and Systems.* Chicago, USA; 2014. pp. 2319–24.

[56]   Li Y., Ahmed A., Sameoto D., Menon C. 'Abigaille II: toward the development of a spider-inspired climbing robot'. *Robotica.* 2012;**30**(1):79–89.

[57]   Bowerman R.F. 'The control of walking in the scorpion'. *Journal of Comparative Physiology ? A.* 1975;**100**(3):183–96.

[58]   Wilson D. 'Stepping patterns in tarantula spiders'. *The Journal of Experimental Biology.* 1967;**47**(1):133–51.

[59]   Delcomyn F. 'Neural basis of rhythmic behavior in animals'. *Science.* 1980;**210**(4469):492–8.

[60]   Klaassen B., Linnemann R., Spenneberg D., Kirchner F. 'Biomimetic walking robot scorpion: control and modeling'. *Robotics and Autonomous Systems.* 2002;**41**(2-3):69–76.

[61]   Spenneberg D., Frank K. 'The Bio-inspired scorpion robot: design, control & lessons learned'. *Climbing and Walking Robots towards New Applications.* InTech; 2007. pp. 197–218.

[62]   Spenneberg D., Bosse S., Hilljegerdes J., Kirchner F., Strack A., Zschenker H. 'Control of a bio-inspired four-legged robot for exploration of uneven terrain'. *9th Workshop on Advanced Space Technologies for Robotics and Automation.* Noordwijk, The Netherlands; 2006. p. 2.3.1.2.

[63]   Forssberg H. 'Stumbling corrective reaction: a phase-dependent compensatory reaction during locomotion'. *Journal of Neurophysiology.* 1979;**42**(4):936–53.

[64]   Bartsch S., Birnschein T., Cordes F., Kühn D., Kampmann P., Hilljegerdes J. 'SpaceClimber: Development of a Six-Legged Climbing Robot for Space Exploration'. *41st Int Symp on Robotics.* Munich, Germany; 2010. pp. 1265–72.

[65]   Bartsch S., Birnschein T., Römmermann M., Hilljegerdes J., Kühn D. 'Development of the six-legged walking and climbing robot SpaceClimber'. *Journal of Field Robotics.* 2012;**29**(3):506–32.

[66] Yoshida H., Tso I., Severinghaus L. 'The spider family Theridiidae (Arachnida: Araneae) from orchid Island, Taiwan: descriptions of six new and one newly recorded species'. *Zoological studies*. 2000;**39**(2):123–32.

[67] Gasparetto A., Vidoni R., Seidl T. 'Passive control of attachment in legged space robots'. *Applied Bionics and Biomechanics*. 2010;**7**(1):69–81.

[68] Parry D., Brown R. 'The hydraulic mechanism of the spider leg'. *The Journal of experimental biology*. 1959;**36**(2):423–33.

[69] Menon C., Lira C. 'Spider-inspired embedded actuator for space applications'. *AISB '06: Adaptation in Artificial Systems*. vol. 2. Bristol, UK; 2006. pp. 85–92.

[70] Menon C., Lira C. 'Active articulation for future space applications inspired by the hydraulic system of spiders'. *Bioinspiration & Biomimetics*. 2006;**1**(2):52–61.

[71] Bennet-Clark H.C. 'The energetics of the jump of the locust Schistocerca gregaria'. *The Journal of experimental biology*. 1975;**63**(1):53–83.

[72] Punzo G., McGookin E.W. 'Engineering the locusts: hind leg modelling towards the design of a bio-inspired space hopper'. *Proceedings of the Institution of Mechanical Engineers, Part K: Journal of Multi-body Dynamics*. 2016;**230**(4):455–68.

[73] Alexander R.M., Vernon A. 'The mechanics of hopping by kangaroos (Macropodidae'. *Journal of Zoology*. 1975;**177**(2):265–303.

[74] Bai L., Ge W., Chen X., Chen R. 'Design and dynamics analysis of a bio-inspired intermittent hopping robot for planetary subsurface exploration'. *International journal of advanced robotic systems*. 2012;**4**(1):1–11.

[75] Bai L., Ge W., Chen X., Meng X. 'Hopping Capabilities of a Bio-inspired and Minimally Actuated Hopping Robot'. *Int Conf on Electronics, Communications and Control*. Ningbo, China; 2011. pp. 1485–9.

[76] Yoshida K., Dimitrov D., Nakanishi H. 'On the Capture of Tumbling Satellite by a Space Robot'. *IEEE/RSJ Int Conf on Intelligent Robots and Systems*. Beijing, China; 2006. pp. 4127–32.

[77] Hawkes E.W., Jiang H., Cutkosky M.R. 'Three-Dimensional dynamic surface grasping with dry adhesion'. *The International Journal of Robotics Research*. 2016;**35**(8):943–58.

[78] Jiang H., Hawkes E.W., Fuller C., *et al.* 'A robotic device using gecko-inspired adhesives can GRASP and manipulate large objects in microgravity'. *Science Robotics*. 2017;**2**(7):eaan4545–11.

[79] Estrada M., Hockman B., Bylard A., Hawkes E., Cutkosky M., Pavone M. 'Free-flyer acquisition of spinning objects with gecko-inspired adhesives'. *IEEE Int Conf on Robotics and Automation*. Stockholm, Sweden; 2016. pp. 4907–13.

[80] Trentlage C., Stoll E. 'A Biomimetic docking mechanism for controlling uncooperative satellites on the ELISSA free-floating laboratory'. *3rd Int Conf on Advanced Robotics and Mechatronics*. Singapore; 2018. pp. 77–82.

[81]   Yu Z., Shi Y., Luo A., Tao J. 'Research on landing experiment of gecko ro-
       bot with bio-inspired dry adhesive foot'. *IEEE Int Conf on Information and
       Automation*. Wuyi Mountain, China; 2018. pp. 336–40.

[82]   Cyril X., Misra A.K., Ingham M., Jaar G.J. 'Postcapture dynamics of a
       spacecraft-manipulator-payload system'. *Journal of Guidance, Control, and
       Dynamics*. 2000;**23**(1):95–100.

[83]   Wu Z., Jing X., Bian J., Li F., Allen R. 'Vibration isolation by explor-
       ing bio-inspired structural nonlinearity'. *Bioinspiration & Biomimetics*.
       2015;**10**(5):056015.

[84]   Dai H., Jing X., Wang Y., Yue X., Yuan J. 'Post-capture vibration suppres-
       sion of spacecraft via a bio-inspired isolation system'. *Mechanical Systems
       and Signal Processing*. 2018;**105**(2):214–40.

[85]   Wang X., Yue X., Dai H., Yuan J. 'Vibration suppression for post-capture
       spacecraft via a novel bio-inspired Stewart isolation system'. *Acta
       Astronautica*. 2020;**168**(1–2):1–22.

[86]   Sun C., Hou X. 'Passive impact/vibration control and isolation performance
       optimization for space noncooperative target capture'. *International Journal
       of Advanced Robotic Systems*. 2020;**17**(1):172988141989538.

[87]   Pandolfi C., Izzo D. 'Biomimetics on seed dispersal: survey and insights for
       space exploration'. *Bioinspiration & Biomimetics*. 2012;**8**(2):025003.

[88]   Pandolfi C., Comparini D., Mancuso S. 'Self-burial mechanism of ero-
       dium cicutarium and Its potential application for subsurface exploration'.
       Biomimetics and Biohybrid Systems. Berlin Heidelberg: Springer; 2012. pp.
       384–5.

[89]   Mancuso S., Mazzolai B., Comparini D., Popova L., Azzarello E., Masi
       E. 'Subsurface investigation and interaction by self-burying bio-inspired
       probes'. *ESA Advanced Concepts Team*. 2014:12–6401.

[90]   Pandolfi C., Casseau V., Fu T., Jacques L., Izzo D. 'Tragopon dubius, con-
       siderations on a possible biomimetic transfer'. *Biomimetic and Biohybrid
       Systems*. Berlin Heidelberg: Springer; 2012. pp. 386–7.

[91]   Antol J., Calhoun P., Flick J., Hajos G., Kolacinski R., Minton D. 'Low cost
       Mars surface exploration: the Mars Tumbleweed. NASA Langley research
       center. TM-2003-212411'. 2003.

[92]   Hartl A.E., Mazzoleni A.P. 'Terrain modeling and simulation of a
       Tumbleweed Rover traversing Martian rock fields'. *Journal of Spacecraft
       and Rockets*. 2012;**49**(2):401–12.

[93]   Antol J., Calhoun P., Flick J., Hajos G., Keyes J., Stillwagen F. 'Mars
       Tumbleweed: FY2003 conceptual design assessment. NASA Langley re-
       search center. TM-2005-213527'. 2005.

[94]   Calhoun P., Harris S., Raiszadeh B., Zaleski K. 'Conceptual design and
       dynamics testing and modeling of a mars tumbleweed rover'. *43rd AIAA
       Aerospace Sciences Meeting and Exhibit*. Reno, USA; 2005. p. 247.

[95]   Kier W.M., Smith K.K. 'Tongues, tentacles and trunks: the biomechanics
       of movement in muscular-hydrostats'. *Zoological Journal of the Linnean
       Society*. 1985;**83**(4):307–24.

[96] Tonapi M., Godage I., Walker I. 'Next Generation Rope-like Robot for In-Space Inspection'. *IEEE Aerospace Conf.* Big Sky, USA; 2014. p. 2037.

[97] Walker I. 'Robot strings: long, thin continuum robots'. *IEEE Aerospace Conf.* Big Sky, USA; 2013. p. 2182.

[98] Goriely A., Neukirch S. 'Mechanics of climbing and attachment in twining plants'. *Physical Review Letters.* 2006;**97**(18):184302.

[99] Walker I. 'Biologically inspired vine-like and tendril robots'. *Science and Information Conference.* London, UK; 2015. pp. 714–20.

[100] Mehling J., Diftler M., Chu M., Valvo M. 'A minimally invasive tendril robot for in-space inspection'. *1st IEEE/RAS-EMBS Int Conf Biomed Robot Biomechatron.* Pisa, Italy; 2013. pp. 690–5.

[101] Tonapi M.M., Godage I.S., Vijaykumar A.M., Walker I.D. 'A novel continuum robotic cable aimed at applications in space'. *Advanced Robotics.* 2015;**29**(13):861–75.

[102] Wooten M., Walker I. 'A novel vine-like robot for in-orbit inspection'. *45th international conference on environmental systems.* Bellevue, USA; 2015. p. 21.

[103] Wooten M., Frazelle C., Walker I., Kapadia A., Lee J. 'Exploration and inspection with vine-inspired continuum robots'. *IEEE Int conf robot autom.* Brisbane, Australia; 2018. pp. 5526–33.

[104] Wooten M., Walker I. 'Vine-inspired continuum tendril robots and circumnutations'. *Robotics.* 2018;**7**(3):58.

[105] Clark L.J., Whalley W.R., Barraclough P.B. 'How do roots penetrate strong soil?' *Plant and Soil.* 2003;**255**(1):93–104.

[106] Seidl T., Mugnai S., Corradi P., Mondini A., Mattoli V., Azzarello E. 'Biomimetic Transfer of Plant Roots for Planetary Anchoring'. *59th International Astronautical Congress.* Glasgow, Scotland; 2008. pp. 7930–41.

[107] Dario P., Laschi C., Mazzolai B., Corradi P., Mattoli V., Mondini A. 'Bioinspiration from Plants' Roots. ESA Advanced Concepts Team. 06/6301'. 2008.

[108] Simões L., Cruz C., Ribeiro R., Correia L., Seidl T., Ampatzis C. 'Path Planning Strategies Inspired by Swarm Behaviour of Plant Root Apexes. ESA Advanced Concepts Team. 09-6401'. 2011.

[109] Sabatini M., Palmerini G., Chiwiacowsky L. 'Root's like natural behaviors applied to guidance algorithms for space exploration missions'. *61st International Astronautical Congress.* Prague, Czech Republic; 2010. pp. 811–9.

[110] Bar-Cohen Y., Anderson I.A. 'Electroactive polymer (EAP) actuators—background review'. *Mechanics of Soft Materials.* 2019;**1**:5.

[111] Bar-Cohen Y. 'Electroactive polymers as artificial muscles - capabilities, potentials and challenges'. *4th Int Conf and Expo on Robotics for Challenging Situations and Environments.* Albuquerque, USA: ASCE; 2000. pp. 188–96.

[112] Meijer K., Rosenthal M., Full R. 'Muscle-like actuators? A comparison between three electroactive polymers'. *8th Annual Int Symp on Smart Structures and Materials*. Newport Beach, USA; 2003. pp. 7–15.

[113] Meijer K., Bar-Cohen Y., Full R. 'Biological Inspiration for musclelike actuators of robots'. *Biologically Inspired Intelligent Robots*. SPIE; 2003. pp. 25–45.

[114] Punning A., Kim K.J., Palmre V., *et al.* 'Ionic electroactive polymer artificial muscles in space applications'. *Scientific Reports*. 2015;4(1):6913.

[115] Kruusmaa M., Fiorini P. 'Electroactive polymers in space: design considerations and possible applications'. *9th ESA Workshop on Advanced Space Technologies for Robotics and Automation*. Noordwijk, The Netherlands; 2006. p. 1.3.2.3.

[116] Bar-Cohen Y., Leary S., Yavrouian A., Oguro K., Tadokoro S., Harrison J. 'Challenges to the application of IPMC as actuators of planetary mechanisms'. *7th annual Int Symp on smart structures and materials*. Newport Beach, USA; 2000. pp. 140–6.

[117] Menon C., Izzo D. 'Satellite pointing system based on EAP actuators'. *Towards autonomous robotic systems*. London, UK; 2005. pp. 173–80.

[118] Branz F., Sansone F., Francesconi A. 'Design of an innovative Dielectric Elastomer actuator for space applications'. *SPIE Smart Structures and Materials + Nondestructive Evaluation and Health Monitoring*. San Diego, USA; 2014. p. 90560Z.

[119] Kornbluh R., Flamm D., Prahlad H., Nashold K., Chhokar S., Pelrine R. 'Shape control of large lightweight mirrors with dielectric elastomer actuation'. *SPIE smart structures and materials*. San Diego, USA; 2003. pp. 143–58.

[120] Branz F., Francesconi A. 'Experimental evaluation of a dielectric elastomer robotic arm for space applications'. *Acta Astronautica*. 2017;133(4–10):324–33.

[121] Carpi F., Tralli A., Rossi D., Gaudenzi P. 'Martian jumping rover equipped with electroactive polymer actuators: a preliminary study'. *IEEE Transactions on Aerospace and Electronic Systems*. 2007;43(1):79–92.

[122] Menon C., Carpi F., De Rossi D. 'Concept design of novel bio-inspired distributed actuators for space applications'. *Acta Astronautica*. 2009;65(5-6):825–33.

[123] Colozza A., Englar R., Michelson R., Naqvi M., Beringer L., Deacey T. *Planetary Exploration Using Biomimetics - An Entomopter for Flight on Mars. NASA Institute for Advanced Concepts. NAS5-98051*; 2002.

[124] Liu H., Ravi S., Kolomenskiy D., Tanaka H. 'Biomechanics and biomimetics in insect-inspired flight systems'. *Philosophical Transactions of the Royal Society B: Biological Sciences*. 2016;371(1704):20150390.

[125] Shyy W., Liu H., Wings F. 'Flapping wings and aerodynamic lift: the role of leading-edge vortices'. *AIAA Journal*. 2007;45(12):2817–19.

[126] Shyy W., Kang C.-kwon., Chirarattananon P., Ravi S., Liu H. 'Aerodynamics, sensing and control of insect-scale flapping-wing flight'. *Proceedings of*

the Royal Society A: Mathematical, Physical and Engineering Sciences. 2016;**472**(2186):20150712.

[127] Liu H., Aono H., Tanaka H. 'Bioinspired air vehicles for Mars exploration'. *Acta Futura.* 2013;**6**:81–95.

[128] Kang C., Fahimi F., Griffin R., Landrum D., Mesmer B., Zhang G. *Marsbee – Swarm of Flapping Wing Flyers for Enhanced Mars Exploration: NASA Innovative Advanced Concepts*; 2019.

[129] Michelson R., Naqvi M. 'Extraterrestrial flight (entomopter-based Mars surveyor)'. *RTO/AVT von Karmen Institute for fluid dynamics.* Brussels, Belgium; 2003. pp. 1–17.

[130] Bar-Cohen Y., Colozza A., Badescu M., Sherrit S., Bao X. 'Biomimetic flying swarm of entomopters for Mars extreme terrain science investigations'. *concepts and approaches for Mars exploration.* Houston, USA; 2012. p. 4075.

[131] Bluman J., Kang C., Landrum B., Fahimi F., Mesmer B. 'Marsbee - can a bee fly on mars?' *55th AIAA Aerospace Sciences Meeting.* Grapevine, USA; 2017. pp. 7202–14.

[132] Wehner R. 'The desert ant's navigational toolkit: procedural rather than positional knowledge'. *Navigation.* 2008;**55**(2):101–14.

[133] Menzel R., Giurfa M. 'Cognitive architecture of a mini-brain: the honeybee'. *Trends in Cognitive Sciences.* 2001;**5**(2):62–71.

[134] Di Pino G., Seidl T., Benvenuto A., Sergi F., Campolo D., Accoto D. 'Interfacing insect brain for space applications'. *Int Rev Neurobiol.* **86**. Elsevier; 2009. pp. 39–47.

[135] Ando N., Kanzaki R. 'Using insects to drive mobile robots—hybrid robots bridge the gap between biological and artificial systems'. *Arthropod Structure & Development.* 2017;**46**(5):723–35.

[136] Spence A.J., Neeves K.B., Murphy D., *et al.* 'Flexible multielectrodes can resolve multiple muscles in an insect appendage'. *Journal of Neuroscience Methods.* 2007;**159**(1):116–24.

[137] Reger B.D., Fleming K.M., Sanguineti V., Alford S., Mussa-Ivaldi F.A. 'Connecting brains to robots: an artificial body for studying the computational properties of neural tissues'. *Artificial Life.* 2000;**6**(4):307–24.

[138] Warwick K., Xydas D., Nasuto S., *et al.* 'Controlling a mobile robot with a biological brain'. *Defence Science Journal.* 2010;**60**(1):5–14.

[139] Rossini L., Izzo D., Summerer L. 'Brain-Machine interfaces for space applications'. *31st Annual Int Conf of the IEEE-EMBS.* Minneapolis, USA; 2009. pp. 520–3.

[140] de Negueruela C., Broschart M., Menon C., del R. Millán J., Millán J. 'Brain–computer interfaces for space applications'. *Personal and Ubiquitous Computing.* 2011;**15**(5):527–37.

[141] Benvenuto A., Sergi F., Di Pino G., Campolo D., Accoto D., Guglielmelli E. 'Conceptualization of an insect/machine hybrid controller for space applications'. *2nd IEEE/RAS-EMBS Int Conf Biomed Robot Biomechatron.* Scottsdale, USA; 2008. pp. 306–10.

[142]   Benvenuto A., Sergi F., Di Pino G., *et al.* 'Beyond biomimetics: towards insect/machine hybrid controllers for space applications'. *Advanced Robotics*. 2009;**23**(7-8):939–53.

[143]   Srinivasan M., Chahl J., Nagle M., Zhang S. 'Embodying natural vision into machines' in Srinivasan M., Ventakesh S. (eds.). *From Living Eyes to Seeing Machines*. Oxford University Press; 1997. pp. 249–65.

[144]   Srinivasan M.V., Lehrer M., Kirchner W.H., Zhang S.W. 'Range perception through apparent image speed in freely flying honeybees'. *Visual Neuroscience*. 1991;**6**(5):519–35.

[145]   Baird E., Srinivasan M., Zhang S., Lamont R., Cowling A. 'Visual control of flight speed and height in the honeybee'. *9th Int Conf on simulation of adaptive behavior*. Rome, Italy; 2006. pp. 40–51.

[146]   Franceschini N. 'Towards automatic visual guidance of aerospace vehicles: from insects to robots'. *Acta Futura*. 2009;**3**:15–34.

[147]   Thakoor S., Moro J.M., Chahl J., Hine B., Zornetzer S. 'Bees: exploring Mars with bioinspired technologies'. *Computer*. 2004;**37**(9):38–47.

[148]   Thakoor S. 'Bio-inspired engineering of exploration systems'. *J Space Mission Archit*. 2000;**2**(1):49–79.

[149]   Thakoor S., Chahl J., Srinivasan M.V., *et al.* 'Bioinspired engineering of exploration systems for NASA and DOD'. *Artificial Life*. 2002;**8**(4):357–69.

[150]   Chahl J., Thakoor S., Le Bouffant N., *et al.* 'Bioinspired engineering of exploration systems: a horizon sensor/attitude reference system based on the dragonfly ocelli for Mars exploration applications'. *Journal of Robotic Systems*. 2003;**20**(1):35–42.

[151]   Stange G., Stowe S., Chahl J., Massaro A. 'Anisotropic imaging in the dragonfly median ocellus: a matched filter for horizon detection'. *Journal of comparative physiology*. 2002;**188**(6):455–67.

[152]   Roska B., Werblin F. 'Vertical interactions across ten parallel, stacked representations in the mammalian retina'. *Nature*. 2001;**410**(6828):583–7.

[153]   Bálya D., Roska B., Roska T., Werblin F.S. 'A CNN framework for modeling parallel processing in a mammalian retina'. *International Journal of Circuit Theory and Applications*. 2002;**30**(2-3):363–93.

[154]   Jiang X., Li S., Tao T. 'Innovative hazard detection and avoidance strategy for autonomous safe planetary landing'. *Acta Astronautica*. 2016;**126**(1):66–76.

[155]   Valette F., Ruffier F., Viollet S., Seidl T. 'Biomimetic optic flow sensing applied to a lunar Lander scenario'. *IEEE Int Conf on robotics and automation*. Anchorage, USA; 2010. pp. 2253–60.

[156]   Izzo D., Weiss N., Seidl T. 'Constant-optic-flow Lunar landing: optimality and guidance'. *Journal of Guidance, Control, and Dynamics*. 2011;**34**(5):1383–95.

[157]   Sabiron G., Burlion L., Rahaijaona T., Ruffier T. 'Optic flow-based nonlinear control and sub-optimal guidance for Lunar landing'. *IEEE ROBIO*. Bali, Indonesia; 2014. pp. 1241–7.

[158]   Sabiron G., Raharijaona T., Burlion L., Kervendal E., Bornschlegl E., Ruffier F. 'Suboptimal lunar landing GNC using nongimbaled

optic-flow sensors'. *IEEE Transactions on Aerospace and Electronic Systems*. 2015;**51**(4):2525–45.

[159] Raharijaona T., Sabiron G., Viollet S., Franceschini F., Ruffier F. 'Bio-Inspired landing approaches and their potential use on extraterrestrial bodies' in Badescu V. (ed.). *Asteroids*. Berlin Heidelberg: Springer; 2013. pp. 221–46.

[160] Frost S.A., Yates L.A., Kumagai H.S. 'Bioinspired optical sensor for remote measurement of small displacements at a distance'. *Biomimetics*. 2018;**3**(4):E3430 Oct 2018.

[161] Kobayashi N., Oishi M., Kinjo Y., Hokamoto S. 'Experimental verification of wide-field-integration of optic flow for state estimation'. *Transactions of the Japan Society for Aeronautical and Space Sciences, Aerospace Technology Japan*. 2016;**14**(ists30):Pd_63–Pd_68.

[162] Lee D., Davies M., Green P., van der Weel F. 'Visual control of velocity of approach by pigeons when landing'. *The Journal of Experimental Biology*. 1993;**180**:85–104.

[163] De Croon G., Alazard D., Izzo D. 'Controlling spacecraft landings with constantly and exponentially decreasing time-to-contact'. *IEEE Transactions on Aerospace and Electronic Systems*. 2015;**51**(2):1241–52.

[164] Beard R., Lee D.J., Quigley M., Thakoor S., Zornetzer S. 'A new approach to future Mars missions using bioinspired technology innovations'. *Journal of Aerospace Computing, Information, and Communication*. 2005;**2**(1):65–91.

[165] Bonabeau E., Théraulaz G. 'Swarm smarts'. *Scientific American*. 2000;**282**(2):72–9.

[166] Reynolds C.W. 'Flocks, herds and schools: a distributed behavioral model'. *ACM SIGGRAPH Computer Graphics*. 1987;**21**(4):25–34.

[167] Humphreys O. 'Starlings in the Scottish borders'. *PA Images*. 2013.

[168] Hildmann H., Almeida M., Kovacs E., Saffre F. 'Termite algorithms to control collaborative swarms of satellites'. *The 14th Int Symp on Artificial Intelligence, Robotics and Automation in Space*. Madrid, Spain; 2018. pp. 1–8.

[169] Ariel G., Ayali A. 'Locust collective motion and its modeling'. *PLoS Computational Biology*. 2015;**11**(12):e1004522.

[170] Truszkowski W., Rouff H.C.H., Karlin J., Rash J., Hinchey N., Sterritt R. '*Swarms in space missions*'. *Autonomous and autonomic systems: with applications to NASA intelligent spacecraft operations and exploration systems*. London: Springer-Verlag; 2009. pp. 207–21.

[171] Verhoeven C.J.M., Bentum M.J., Monna G.L.E., Rotteveel J., Guo J. 'On the origin of satellite swarms'. *Acta Astronautica*. 2011;**68**(7-8):1392–5.

[172] Sabatini M., Palmerini G.B. 'Collective control of spacecraft swarms for space exploration'. *Celestial Mechanics and Dynamical Astronomy*. 2009;**105**(1-3):229–44.

[173] Peng H., Li C. 'Bound evaluation for spacecraft swarm on libration orbits with an uncertain boundary'. *Journal of Guidance, Control, and Dynamics*. 2017;**40**(10):2690–8.

[174] Yun X., Zhaokui W., Yulin Z. 'Bounded flight and collision avoidance control for satellite clusters using intersatellite flight bounds'. *Aerospace Science and Technology*. 2019;**94**:105425.

[175] Curtis S., Truszkowski W., Rilee M., Clark P. 'ANTS for the human exploration and development of space'. *IEEE Aerospace Conf Proc*. Montana, USA; 2003. p. 2.0309.

[176] Clark P., Rilee M., Curtis S., Cheung C. 'In situ surverying of saturn's rings'. *Lunar Planet Sci XXXV*. League City, USA; 2004. p. 1100.

[177] Rouff C. 'NASA's swarm missions: The challenge of building autonomous software'. *IT Professional*. 2004;**6**(5):47–52.

[178] Kisdi Áron., Tatnall A.R.L. 'Future robotic exploration using honeybee search strategy: example search for caves on Mars'. *Acta Astronautica*. 2011;**68**(11-12):1790–9.

[179] Redah A., Mikschl T., Montenegro S. 'Physically distributed control and swarm intelligence for space applications'. *The 14th Int Symp on Artificial Intelligence, Robotics and Automation in Space*. Madrid, Spain; 2018. pp. 1–6.

[180] Ivanov S., Konstantinov B., Tzokov S., Ivanov T., Kanev T., Zlateva V. 'Space debris identification, classification and aggregation with optimized satellite swarms'. *Innovative ideas on Micro/Nano-Satellite missions and systems*. **4**. International Academy of Astronautics; 2017. pp. 137–51.

[181] Staudinger E., Shutin D., Manß C., Ruiz A., Zhang S. 'Swarm technologies for future space exploration missions'. *The 14th Int Symp on Artificial Intelligence, Robotics and Automation in Space*. Madrid, Spain; 2018. pp. 1–14.

[182] St-Onge D., Kaufmann M., Panerati J., Ramtoula B., Cao Y., Coffey E. 'Planetary exploration with robot teams: implementing higher autonomy with swarm intelligence'. *IEEE Robotics & Automation Magazine*. 2019.

[183] Xu C., Hu S., Chen X. 'Artificial cells: from basic science to applications'. *Materials Today*. 2016;**19**(9):516–32.

[184] Salehi-Reyhani A., Ces O., Elani Y. 'Artificial cell mimics as simplified models for the study of cell biology'. *Experimental Biology and Medicine*. 2017;**242**(13):1309–17.

[185] Sterritt R. 'Apoptotic computing: programmed death by default for computer-based systems'. *Computer*. 2011;**44**(1):59–65.

[186] Palmer R., Sterritt R. 'Autonomic & apoptotic computing prototype: providing pre-programmed death of cubesats for avoiding space junk'. *IEEE Int Conf on Space Mission Challenges for Information Technology*. Pasedena, USA; 2019. pp. 78–86.

[187] Vermes I., Haanen C., Reutelingsperger C. 'Flow cytometry of apoptotic cell death'. *Journal of Immunological Methods*. 2000;**243**(1–2):167–90.

[188] Swartwout M. 'The first one hundred CubeSats: a statistical look'. *Journal of Small Satellites*. 2013;**2**(2):213–33.

[189] Langer M., Bouwmeester J. 'Reliability of CubeSats - statistical data, developers' beliefs and the way forward'. *30th Annual AIAA/USU Conference on Small Satellites*. Logan, USA; 2016. pp. X–2.

[190] Erlank A., Bridges C. 'A multicellular architecture towards low-cost satellite reliability'. *NASA/ESA Conference on Adaptive Hardware and Systems*. Montreal, Canada; 2015. pp. 1–8.

[191] Jopling C., Boue S., Izpisua Belmonte J.C. 'Dedifferentiation, transdifferentiation and reprogramming: three routes to regeneration'. *Nature Reviews Molecular Cell Biology*. 2011;**12**(2):79–89.

[192] Erlank A., Bridges C. 'The satellite stem cell architecture'. *IEEE symposium series on computational intelligence*. Athens, Greece; 2016. pp. 1–8.

[193] Erlank A.O., Bridges C.P. 'Reliability analysis of multicellular system architectures for low-cost satellites'. *Acta Astronautica*. 2018;**147**(3):183–94.

[194] Williams D., Kuipers A., Mukai C., Thirsk R. 'Acclimation during space flight: effects on human physiology'. *Canadian Medical Association Journal*. 2009;**180**(13):1317–23.

[195] Paris A. 'Physiological and psychological aspects of sending humans to Mars: challenges and recommendations'. *Journal of the Washington Academy of Sciences*. 2014;**100**(4):3–20.

[196] Sandal G.M., Bye H.H. 'Value diversity and Crew relationships during a simulated space flight to Mars'. *Acta Astronautica*. 2015;**114**(1):164–73.

[197] Ayre M., Zancanaro C., Malatesta M. 'Morpheus - Hypometabolic Stasis in Humans for Long Term Space Flight'. *Journal of the British Interplanetary Society*. 2004;**57**:325–39.

[198] French A. 'The patterns of mammalian hibernation'. *American scientist*. 1988;**76**(6):568–75.

[199] Malatesta M., Biggiogera M., Zancanaro C. 'Hypometabolic induced state: a potential tool in biomedicine and space exploration'. *Reviews in Environmental Science and Bio/Technology*. 2007;**6**(1–3):47–60.

[200] Donahue S.W., McGee M.E., Harvey K.B., Vaughan M.R., Robbins C.T. 'Hibernating bears as a model for preventing disuse osteoporosis'. *Journal of Biomechanics*. 2006;**39**(8):1480–8.

[201] Nordeen C.A., Martin S.L. 'Engineering human stasis for long-duration spaceflight'. *Physiology*. 2019;**34**(2):101–11.

[202] Schaffer M., Bradford J., Talk D. 'A feasible, near-term approach to human-stasis for long-duration deep space science missions'. *67th International Astronautical Congress*. Guadalajara, Mexico; 2016. pp. 3731–47.

[203] Choukèr A., Bereiter-Hahn J., Singer D., Heldmaier G. 'Hibernating astronauts—science or fiction?' *Pflügers Archiv - European Journal of Physiology*. 2019;**471**(6):819–28.

[204] Kim S.-J., Lee J.-H. 'Parametric shape modification and application in a morphological biomimetic design'. *Advanced Engineering Informatics*. 2015;**29**(1):76–86.

[205]   Turner J., Soar R. 'Beyond biomimicry: what termites can tell us about realizing the living building'. *1st Int Conf on Industrialized, intelligent construction*. Leicester, UK; 2008. pp. 233–48.

[206]   Abate G., Mavris D. 'Performance analysis of different positions of leading edge tubercles on a wind turbine blade'. *Wind Energy Symp*. Kissimmee, USA; 2018. p. AIAA 2018–1494.

Part II

Sensing, perception and GNC

Chapter 5

Autonomous visual navigation for spacecraft on-orbit operations

Arunkumar Rathinam[1], Zhou Hao[1], and Yang Gao[1]

Space robotic missions with increased levels of autonomy are being pursued in wide-array of orbital applications including on-orbit servicing (OOS), on-orbit assembly (OOA), and active debris removal (ADR). In these missions, the spacecraft is expected to perform most guidance and navigation tasks such as far-, mid- and close-range rendezvous, relative navigation, and proximity operations with minimal human-in-loop. This goal brings the focus towards vision-based spacecraft navigation utilising the state-of-the-art technologies in the field of computer vision especially the deep-learning algorithms for the pose estimation. This chapter explores major deep-learning approaches suitable for spacecraft pose estimation along with the discussion on different software simulation tools that are currently used for rendering realistic images of the target in orbit to train and validate the deep-learning models, and finally the ground-based testbed is used for validating the close-proximity operations.

5.1 Introduction

Orbital robotic operations have a rich history and been around since the first deployment of the Shuttle Remote Manipulator System (SRMS) by the Space Shuttle Colombia in 1981. The SRMS is a 50 ft long manipulator arm with 6-Degrees of Freedom (DoF) mounted on the shuttle orbiter and used for deploying, manipulating, and retrieving the payloads in space [1]. In recent years, lower launch costs and minimum entry barriers have made space more accessible and this led to increased congestion in orbit. Most satellites launched into orbit will last for the entire mission life-cycle, while at the end of life, they are either moved to the graveyard orbit or left to re-enter the Earth's atmosphere. However, a few satellites may run into anomalies, malfunctions, or may run out of fuel before their full life span. These malfunctioned satellites may become non-cooperative and become a threat to existing space infrastructure. There is a need for sustainable solutions to either restore service

[1]STAR LAB, Surrey Space Centre, University of Surrey, Guildford, UK

through in-situ repair to extend the life-cycle or remove them from orbit to mitigate risks for other space infrastructures. Autonomous orbital robotic spacecraft systems with the capability to perform a multitude of functions offer the perfect solution. Research and development in robotics and space systems have significantly raised the technology readiness level of the autonomous systems over the last two decades. Autonomous on-orbit robotic operation is a key element in the development of a space frontier that will enable a sustainable space exploration in the future. The three major orbital robotic operations [2] for the next generation autonomous space systems are OOS, OOA, and ADR.

OOS is the process of inspection, maintenance, and repair of a system as an in-space operation and has been around since the first servicing of the Skylab space station in the 1970s. OOS involves tasks such as life extension and maintenance of satellites, rescue and recover satellites if any deployment failure, and assisting astronauts with extravehicular activities (EVAs). A brief timeline of the developments of systems and spacecraft aimed for robotic OOS is presented in Figure 5.1. For OOS, the autonomous capability is a critical component that enables the system to perform relative navigation, inspection, maintenance, and repair of the satellites without human-in-loop. In the past two decades, demonstration missions have been carried out to explore the reusability of space systems and develop more autonomous robotic systems. The idea of reusability was studied in the past in different contexts ranging from servicing the satellite to extend the lifetime in orbit to re-purposing the retired satellites to create new space elements. Orbital servicing helps to improve or retrofit the existing systems to the modern requirements and to deploy new capabilities in orbit thus extending the spacecraft's lifetime while bringing down the total operational costs. Recently, first commercial orbital servicing of a geostationary satellite the IntelSat-901 was carried out by SpaceLogistics (subsidiary of Northrop Grumman) using the MEV-1 (Mission Extension Vehicle) satellite platform. During mission operations, MEV-1 docked with the IntelSat satellite and re-positioned it to the designated spot. MEV-1 continues to provide in-orbit station-keeping services for the Intelsat, thus extending the satellite's lifetime for another five years until 2025 [4]. MEV-1 is designed to perform dock and undock multiple times, thus potentially enabling it to service multiple satellites, if necessary. Multiple missions are in pipeline through combinations of public and private entities that will enable a

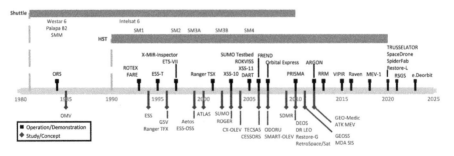

Figure 5.1   OOS developments timeline [3]

successful business model for the orbital servicing regime. The Defence Advanced Research Projects Agency's (DARPA's) Robotic Servicing of Geosynchronous Satellites (RSGS) program intends to demonstrate versatile dexterous robotic operational capability in Geostationary Earth Orbit (GEO) that can extend satellite life spans, enhance resilience, and improve reliability, and it targets launch in 2023 [5]. A review of space robotics technologies for OOS of spacecraft is provided in [6] and an overview of the OOS missions along with the recent engineering developments in the related subfields is presented in [7].

Throughout the history of space missions, nearly all spacecraft have been assembled on the ground and delivered in orbit through a launch vehicle. The size of the launch vehicle imposes significant restrictions on the payload design, size, and volume to fit inside a launch vehicle fairings. For example, the James Webb Space Telescope (JWST) with a 25 m² aperture that is protected by a 5-layer, tennis court-sized sun-shield must fold up into the largest current launch fairings and deploy in orbit. Besides, the orbital deployment of JWST involves 400 individual sequences that each one must work perfectly, thus further increasing the risk to the mission. OOA offers the potential opportunity to construct significantly larger structures, such as high-definition space telescopes (HDSTs) through a series of separate launches and assemble them in orbit, thus breaking the limitations imposed by launch vehicle fairing dimensions. Several ongoing research and mission programs aim to demonstrate OOA in the next two to three years. The NASA OSAM-1 (On-orbit Servicing, Assembly, and Manufacturing 1), previously known as Restore-L mission [8], as OOS mission where a robotic spacecraft will demonstrate satellite servicing to extend satellites' lifespans – even if the satellites were not designed to be serviced on-orbit. Also, as a part of the mission, the SPace Infrastructure DExterous Robot (SPIDER) payload will demonstrate OOA with an in-space assembly of individual antenna elements to form a functional 9 ft (3 m) communications antenna.

Space debris is a problem in both Low Earth Orbit (LEO) and GEO. Three orbital regions in LEO have been identified as critical where most catastrophic collision is expected to occur because of the large concentration of mass with high chances of triggering a cascading effect, termed *Kessler syndrome* [9]. ADR involve removal of defunct satellite or spacecraft components to provide stability and reduce the risk of collision of space debris with operational satellites. ADR technologies enable the orbital manipulation of existing objects and help to carry out the removal of the debris while also allowing the reuse of unique orbital slots. A brief review of space debris capturing and removal methods along with comparison of the existing technologies is presented in [10].

The rest of this chapter is arranged as follows. Section 5.2 provides a brief overview of the equation of motion for relative navigation, followed by the camera pose estimation, spacecraft relative pose estimation, and emerging trends in the deep-learning-based spacecraft pose estimation. Section 5.3 discusses the different deep-learning approaches and the components in their respective spacecraft pose estimation framework. Section 5.4 discusses the simulations tools for generating realistic orbital images to train machine learning algorithms and followed by the technologies that are essential to developing the robotic ground-based testbeds for simulating close-proximity operation scenarios. Section 5.5 provides a brief overview of results and the analytical validation results comparison for the two deep-learning approaches for spacecraft pose estimation discussed in the chapter. The final section presents the summary and future trends in using the deep-learning methods in spacecraft Guidance, Navigation, and Control (GNC) to achieve autonomous navigation.

5.2   Theoretical foundation

In all the orbital robotic missions, the mission phases remain common in the early stages from launch to close-range operations with the target. The common mission phases includes Launch and Early Orbit Phase (LEOP), far-range rendezvous, mid-range rendezvous and close-range rendezvous phase. A far-range rendezvous denotes the phase where the approximate distance between servicer and target is in the range of 10s of km to 100s of m, and the observables in this phase are the bearing and range. During the mid-range rendezvous phase, the target will be in the range of 100s of m to 5 m and the observables are the position and the attitude. In the close-range rendezvous the target will be in the range between 5 and 1 m. Both mid- and close-range rendezvous involves estimation and tracking of the full 6-DoF pose of the target spacecraft to effectively align the servicer before performing any mission operation.

5.2.1   The equations of relative motion

Terminal rendezvous deals with the problem of relative motion, i.e. the motion of one satellite with respect to the other satellite. The two satellites in motion around a central body are named as the target and the chaser. When the chaser is close to the target, it becomes more convenient to keep one of the spacecraft as a fixed point, i.e. a reference frame is attached to the target's centre of mass and then the chaser motion is analysed with respect to the target coordinate frame. The target coordinate system is oriented such that the x-axis lying along the radius of the orbit from the Earth's centre, the y-axis is in the direction of the local horizon, and the z-axis is normal to the orbital plane completing the right-hand triad, as shown in Figure 5.2. The orbital position of the chaser and the target around the Earth is given by $\mathbf{r}_C$ and $\mathbf{r}_T$. The orbital angular rate of the target is given by ω and the position of the chaser with respect to the target is denoted by ρ. The position of the chaser in the inertial reference frame is denoted by

$$\mathbf{r}_C = \mathbf{r}_T + \rho \tag{5.1}$$

Differentiating the above equation with respect to the inertial coordinate frame results in

$$\ddot{\mathbf{r}}_C = \ddot{\mathbf{r}}_T + \ddot{\rho} + 2(\omega \times \dot{\rho}) + \dot{\omega} \times \rho + \omega \times (\omega \times \rho) \tag{5.2}$$

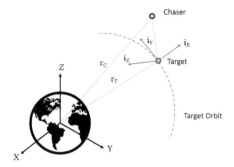

Figure 5.2   Reference frames for relative navigation

where $\ddot{\mathbf{r}}_C$ denotes the inertial acceleration of the chaser satellite, $\ddot{\mathbf{r}}_T$ is the inertial acceleration of the target satellite, $\ddot{\rho}$ is the relative acceleration of the chaser with respect to the target, $2(\omega \times \dot{\rho})$ denotes the Coriolis acceleration, $\omega \times (\omega \times \rho)$ is the centripetal acceleration, and $\dot{\omega} \times \rho$ is the Euler acceleration. When the target in circular orbit $\dot{\omega} = 0$, and resolving the above equation in x, y, and z components provide the proximity relative motion of the chaser in the target's orbital coordinate system and is given by

$$\ddot{x} - 2n\dot{y} - 3n^2 x = 0$$
$$\ddot{y} + 2n\dot{x} = 0 \tag{5.3}$$
$$\ddot{z} + n^2 z = 0$$

where n is the mean motion of the satellite. The above set of equations are known as the Clohessy-Wiltshire (CW) equations with no external disturbances and/or control accelerations. The non-homogeneous form of (5.3) is given by

$$\ddot{x} - 2n\dot{y} - 3n^2 x = d_x + u_x$$
$$\ddot{y} + 2n\dot{x} = d_y + u_y \tag{5.4}$$
$$\ddot{z} + n^2 z = d_z + u_z$$

where $u_{x,y,z}$ are the control acceleration and $d_{x,y,z}$ are the environmental disturbance. One of the key assumptions is that the distance between chaser and target vehicles is very small compared with the distance to the centre of the Earth. Because of the linearisation, the accuracy of the CW equations decreases with the distance between the two satellites.

The linear differential equations in (5.3) can be conveniently written in state space form, where the state vector $\mathbf{x}$ is given by $\begin{bmatrix} x & y & z & v_x & v_y & v_z \end{bmatrix}^T$.

$$\dot{\mathbf{x}}(t) = A\mathbf{x}(t) \tag{5.5}$$

where A is the system matrix. Given the initial position and velocity $\mathbf{x}(t_0)$, the position and velocity at some future time $\mathbf{x}(t)$, the solutions to CW equations can be formulated in terms of the state transition matrix as below:

$$\mathbf{x}(t) = \Phi(t)\mathbf{x}(t_0) \tag{5.6}$$

and the state transition matrix $\Phi(t)$ is given by

$$\Phi(t) = \begin{bmatrix} 4 - 3\cos nt & 0 & 0 & \frac{1}{n}\sin nt & \frac{2}{n}(1 - \cos nt) & 0 \\ 6(\sin nt - nt) & 1 & 0 & \frac{2}{n}(\cos nt - 1) & \frac{1}{n}(4\sin nt - 3nt) & 0 \\ 0 & 0 & \cos nt & 0 & 0 & \frac{1}{n}(\sin nt) \\ 3n\sin nt & 0 & 0 & \cos nt & 2\sin nt & 0 \\ 6n(\cos nt - 1) & 0 & 0 & -2\sin nt & 4\cos nt - 3 & 0 \\ 0 & 0 & -n\sin nt & 0 & 0 & \cos nt \end{bmatrix}. \tag{5.7}$$

The above equations are for the circular orbit, the linear relative dynamics equation set for elliptical orbits was derived by Tschauner and Hempel [11], named after them as the T–H equation set. Many attempts [12, 13] were made to solve the differential equations of relative motion for the elliptical orbit of arbitrary eccentricity. The state transition matrix forms describing the relative motion on an arbitrary elliptical orbit provided by Yamanaka and Ankerson [14] are given below:

$$\begin{bmatrix} \tilde{x}_0 \\ \tilde{y}_0 \\ \tilde{z}_0 \\ \tilde{v}_{x0} \\ \tilde{v}_{y0} \\ \tilde{v}_{z0} \end{bmatrix} = \frac{1}{1-e^2}\begin{bmatrix} 1 - e^2 & 0 & 3es(1/\rho + 1/\rho^2) & -es(1 + 1/\rho) & 0 & -ec + 2 \\ 0 & 1 - e^2 & 0 & 0 & 0 & 0 \\ 0 & 0 & -3s(1/\rho + e^2/\rho^2) & s(1 + 1/\rho) & 0 & c - 2e \\ 0 & 0 & -3(c/\rho + e) & c(1 + 1/\rho) + e & 0 & -s \\ 0 & 0 & 0 & 0 & 1 - e^2 & 0 \\ 0 & 0 & 3\rho + e^2 - 1 & -\rho^2 & 0 & es \end{bmatrix}\begin{bmatrix} \tilde{x}_0 \\ \tilde{y}_0 \\ \tilde{z}_0 \\ \tilde{v}_{x0} \\ \tilde{v}_{y0} \\ \tilde{v}_{z0} \end{bmatrix} \tag{5.8}$$

$$\begin{bmatrix} \tilde{x}_t \\ \tilde{y}_t \\ \tilde{z}_t \\ \tilde{v}_{xt} \\ \tilde{v}_{yt} \\ \tilde{v}_{zt} \end{bmatrix} = \begin{bmatrix} 1 & 0 & -c(1 + 1/\rho) & s(1 + 1/\rho) & 0 & 3\rho^2 J \\ 0 & c/\rho_{\theta-\theta_0} & 0 & 0 & s/\rho_{\theta-\theta_0} & 0 \\ 0 & 0 & s & c & 0 & (2 - 3esJ) \\ 0 & 0 & 2s & 2c - e & 0 & 3(1 - 2esJ) \\ 0 & -s/\rho_{\theta-\theta_0} & 0 & 0 & c/\rho_{\theta-\theta_0} & 0 \\ 0 & 0 & s' & c' & 0 & -3e(s'J + s/\rho^2) \end{bmatrix}\begin{bmatrix} \tilde{x}_0 \\ \tilde{y}_0 \\ \tilde{z}_0 \\ \tilde{v}_{x0} \\ \tilde{v}_{y0} \\ \tilde{v}_{z0} \end{bmatrix} \tag{5.9}$$

where e is the eccentricity, θ is the true anomaly, $\rho = 1 + e\cos\theta$, $s = \rho\sin\theta$, $c = \rho\cos\theta$, $s' = \cos\theta + e\cos 2\theta$, $c' = -\sin\theta - e\sin$, and $\tilde{}$ represents the pseudo-initial values. The transformed variables are calculated from the true state variablesand are given by

$$\tilde{\mathbf{r}} = \rho\mathbf{r}$$
$$\tilde{\mathbf{v}} = -e\sin\theta\mathbf{r} + (1/k^2\rho)\mathbf{v} \tag{5.10}$$

and the inverse transformation is given by

$$\mathbf{r} = \frac{1}{\rho}\tilde{\mathbf{r}}$$

$$\mathbf{v} = k^2(e\sin\theta\tilde{\mathbf{r}} + \rho\tilde{\mathbf{v}}).$$

(5.11)

5.2.2   Camera pose estimation

Recovering the pose of a camera from a single image is a fundamental computer vision problem and estimating the camera's pose, i.e. the position and orientation, is a key to the many robotic applications including localisation and navigation. Given an image I and the camera parameter C, an absolute pose estimator predicts the 3D pose, i.e. translation and rotation of the camera in the world coordinates. The location of the camera, i.e. translation, with respect to the origin of reference (world coordinates) is specified by $t \in \mathbb{R}^3$ and the orientation of the camera is represented as quaternion $q \in \mathbb{R}^4$.

Camera pose estimation is a central step in 3D computer vision approaches such as Structure-from-Motion (SfM) [15, 16], Simultaneous Localisation and Mapping (SLAM) [17], and visual odometry [18]. Traditional pose estimation methods employed local keypoints, keypoint detectors, and descriptors to extract keypoints. They use feature-matching techniques to compute the correspondences between multiple images and then construct the relation between object's 3D model and 2D pixel locations via perspective-n-point (PnP) projection to compute the 6-DoF camera pose parameters. Feature descriptors are used to encode distinct features in the images and have the ability to detect the same physical interest points under different viewing conditions. The local descriptors must be invariant for the image transformation such as scale, rotation, illumination, and viewpoint changes. A few feature descriptors offer faster and better performance in a particular environment but often fail to detect the similar features in a rapidly changing viewpoint or illumination scenarios.

In recent years, machine learning algorithms are rapidly developing especially Deep Neural Networks and Convolution Neural Networks (CNNs) for image processing. CNNs have emerged as the method of choice for a variety of problems and demonstrated impressive success in image classification, semantic segmentation, and object detection tasks. PoseNet [19] was the first learning-based architecture that implemented the idea of regressing the absolute pose with deep architecture based on modified GoogLeNet and it was used to relocate the camera on Structure-from-Motion models. The PoseNet architecture replaced the soft classification layer with a fully connected layer to directly regress the absolute pose from a single image. The orientation regression is done by minimising the L2 loss between quaternions. PoseNet offered several advantages over traditional approaches including short inference times low memory, and no feature engineering requirement. However, the PoseNet over-fitted into its training data and does not generalise well to the unfamiliar scenes. When compared to the feature-based state-of-the-art methods, the localisation error on different datasets (indoor and outdoor) was an order of magnitude larger [20].

Following the PoseNet, several deep-learning frameworks evolved over the years, and a brief review of Camera Pose Estimation with Deep Learning was provided in [20]. There are two main approaches for estimating the pose of rigid bodies using deep learning. The first method is to predict the camera viewpoint with respect to the object either through a direct regression or through casting the problem as classification into set discrete views. The idea of pose regression with end-to-end learning offered several advantages compared to the structure-based methods. Deep absolute pose regression does not require any feature engineering and relies on CNN-based encodings that were found to be more robust towards changes in the scene, such as lighting conditions and viewpoint. On the other hand, the classification approaches for pose estimation are found to be less suitable because of the fine-grained information necessary for describing the 3D pose parameters. However, the accuracy of the end-to-end learning is far behind the accuracy achieved via geometry-based solutions.

The second method is a keypoint-based or feature-based approach which predicts the 2D pixel location of distinct keypoints in a projected image and then use a PnP algorithm to compute the pose. PnP algorithms can estimate the 6-DoF pose of an object, provided the 3D location of keypoints, corresponding 2D locations in the projected image, and the intrinsic parameters of the camera. CNNs are considerable alternate to replace the feature descriptors and matching process, thus avoiding the complex feature engineering. CNN-based keypoint prediction is found to produce better results in facial keypoints and human pose estimation. Their success lies in the high discriminative capacity of the network, and their ability to aggregate information over a wide field of view allows for the resolution of ambiguities (e.g. symmetry) and for localising occluding joints [21].

5.2.3   Relative pose estimation

GNC technologies for autonomous rendezvous and docking, formation flying, or proximity operations of spacecraft or other vehicles require accurate, up-to-date measurements and estimates of relative range and attitude while in close formation and during rendezvous or proximity operations. The term relative-pose denotes the relative position and orientation of a target spacecraft with respect to the chaser/ service spacecraft and represents key information on the relative state for the navigation system. The task of a pose estimation algorithm is to predict the rigid-body transformation from the object coordinate system to the camera coordinate system.

Vision sensors can be divided into active and passive devices, where active techniques project light into the scene, i.e. LIght Detection And Ranging (LIDAR) sensors and Time-of-Flight (ToF) cameras, in contrast to passive techniques, which measure the visible radiation that is already present in the scene, i.e. monocular and stereo cameras. Passive sensors are preferred for space applications because of their low power and mass requirements. Within the passive sensors, monocular cameras are best suited with high flexibility and convenient in terms of operational range, mass, power consumption, and the required processing power. Inexpensive image sensors, coupled with advanced image processing and tracking algorithms, can provide a cost-effective and accurate sensing capability to obtain full 6-DoF relative pose information during proximity operations [22].

The prominent algorithms that are used today for pose estimation and tracking utilise fiducial markings on the tracked objects. The fiducial marking is more suitable for close-range applications and repetitive tasks; however, it is impractical to use fiducials in all cases. Model-based techniques rely on the a priori knowledge of the object whose pose and motion are to be estimated and avoid the need for using fiducial markings. Traditional monocular pose estimation solutions [22–24] rely on model-based approach, where it takes an input image, such as a spacecraft on-orbit, and detect the spacecraft features such as edges and corners using feature extractors. At the same time, an initial pose is generated to project the model onto the image. The 2D correspondences between the detected features and the projected features are calculated and the measured translation and rotation error between correspondences is then minimised to effectively align the features. This process will provide a refined pose estimate and the result is used to adjust the initial pose in the next frame, thus enabling a more efficient correspondence search and resulting in fine pose estimate. However, non-model-based techniques do not assume a priori knowledge of the target object's geometry, texture, or other visual identifications. These methods solve for the optimal camera pose estimation or motion through recovering identical features in the images and under epipolar or motion field constraints. Some of the challenges faced by the feature point extraction methods in feature detection and matching are sharp change in shadows/appearance change due to rotation/tumbling motion, the low Signal-To-Noise Ratio (SNR), and the high contrast which characterise the space images. Many of the algorithms may have difficulties with image-to-model feature correlation, feature persistence for tracking over more than a few frames, foreground-background segmentation, and detection and correction of correlation and tracking errors [22].

5.2.4   Emerging trends

With recent advancements in deep learning and the popularity of the European Space Agency's (ESA)'s Spacecraft Pose Estimation Challenge enabled new developments with the state-of-the-art performance in the visual pose estimation algorithms. Several research works and their results are published based on this competition including Spacecraft Pose Network (SPN) [25], Pose Estimation with Deep Landmark Regression [26], Pose Estimation with soft classification [27], and segmentation-driven approach [28].

SPN [25] used a combination of classification and regression approaches to compute the relative pose. It predicts the bounding box of the satellite in the image with an object detection network and the bounded sub-image is processed through a sub-network to perform classification on the 3D pose. During classification, the special orthogonal rotation group SO3 is discretised into m uniformly distributed fine-bins representing the base rotations and the network retrieves the n-most relevant rotations from the feature map of the detected object. The translation of the pose is estimated via the constraints from the bounding box dimensions to fit the entire object with the predicted rotation. Chen *et al.* [26] presented a keypoint-based approach to estimate the pose of the satellite. They regress the bounding box around the satellite using an object detection CNN and crop the image. The cropped image is then fed into the keypoint regression CNN to obtain the 2D locations of the landmarks. Finally, the 2D-3D landmark correspondences and non-linear optimisation were used to compute the pose estimates. Gerard [28] presented a segmentation-driven approach where the segmentation stream is used to identify the interested target object and the regression stream is used to predict 2D keypoint position. Finally, the iterative PnP with RANdom SAmple Consensus (RANSAC)-based version of the EPnP is used to compute the pose. One of the limitations is that the bad 2D keypoint prediction will lead to inaccurate PnP pose estimate. Proenca and Gao [27] proposed a hybrid approach which involves a simple regression branch for location estimation and soft-classification-based approach to predict the orientations. Dhamani *et al.* [29] presented a CNN-based approach to estimate the relative bearing (azimuth and elevation). The algorithm was developed for and deployed on the Seeker-1 mission, a CubeSat class technology demonstrator mission, intended to provide relative bearing estimates of the non-cooperative Cygnus vehicle from ranges of 5 to 40 m in real time (>1 Hz). Harvard *et al.* [30] proposed an architecture that uses existing keypoint localisation algorithms to identify robust keypoints and then train a CNN on this limited set of keypoints (with feature descriptor components) to create specialised descriptors. For each landmark, a visibility map is also generated through ray tracing. PnP-RANSAC was used to estimate the pose and non-linear filter to track the pose.

5.3   Deep-learning-based spacecraft pose estimation

The recent success of the deep-learning frameworks in computer vision suggests that they possess the potential to address wide-range of complex and challenging problems in computer vision applications. Since the focus of this chapter is on pose estimation, this section discusses the two major approaches for deep-learning-based spacecraft pose estimation, the feature/keypoint-based approach and the direct regression approach.

5.3.1   *Keypoint-based pose estimation*

A general framework of keypoint-based pose estimation using deep learning is shown in Figure 5.3.

5.3.1.1   Object detection

The first set of tasks in a keypoint-based pose estimation framework is to detect the object in the image. The object recognition problem denotes the more general problem of identifying/localising all the objects in a given image, subsuming the

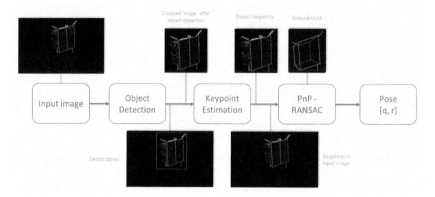

Figure 5.3   Keypoint-based framework for the spacecraft pose estimation [31]

problems of object detection and classification. Object detection methods based on deep learning falls into two categories: region-based frameworks and unified frameworks. Region-based frameworks, otherwise known as Region Proposal Networks (RPNs) [32–34], generate category-independent region proposals from an image, uses CNN to extract features from these regions, and then category-specific classifiers are used to determine the category labels of the proposals. On the other hand, the unified framework uses a single feed forward CNN to predict location information and orientation class probabilities of objects from the whole image, without involving region proposal generation or post-classification and feature resampling, encapsulating all computation in a single network. Though the RPNs have higher detection accuracy, the process is time-consuming; on the other hand, the unified pipeline framework is simple and can detect objects more quickly. For more information on object detection, a detailed review is provided in [35].

You Only Look Once v3 (YOLOv3) is one of the widely used deep-learning-based object detection methods. YOLOv3 [36] considers object detection as a regression problem and employs a grid-cell-based approach to detect the object. Figure 5.4 shows the network architecture of YOLOv3. The input image is divided into S × S grid and if the centre of the object falls into a grid cell, then it is responsible for detecting the object. Each grid cell is validated with a classifier which creates a set of bounding box propositions and then each proposition is filtered out when the probability is lower than a selected threshold. Each grid cell predicts C class probabilities, B bounding box locations, and confidence scores. YOLOv3 is less likely to make false predictions because it considers the entire image when making predictions, thus it implicitly encodes contextual information about object classes.

YOLOv3 borrows the idea of residual networks and uses successive 3 × 3 and 1 × 1 convolutional layers to perform feature extraction. There are five residual blocks and each residual block is composed of multiple residual units. YOLOv3 downsamples the input image by five times and predicts the targets at three different scales by combining the features from the last three downsampled layers. At predict one, the 32× downsampled feature map is used to detect big targets. At predict

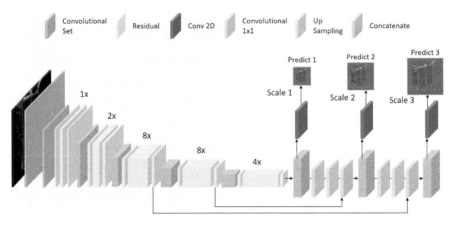

Figure 5.4   Network details of YOLOv3

two, the 16× downsampled feature map is used to detect medium-sized targets. At predict three, the 8× downsampled feature map is used to detect small targets. Feature fusion enables merging small feature maps with deep semantic information to the large feature maps with finer-grained information. To perform feature fusion, YOLOv3 resizes the deeper layers via upsampling and this enables the feature map sizes to match the previous layers. Through concatenation, the features from the earlier layer are merged with the scaled features from the deeper layer, thus enabling good performance to detect both large and small targets.

5.3.1.2   Landmark regression

After finding the bounding boxes of the object, the image is cropped to the specific size of the bounding box and then feed into the landmark regression network. Each training image is marked with the set of the ground truth (pixel) locations of landmarks z_i. These labels are then used in the training of a regression model to predict locations of the 2D landmarks in the test images. Several well-established regression algorithms are used in facial keypoint detection or human pose estimation can be used in the landmark regression of the satellite landmarks. MobileNetV2 in combination with YOLOV2 Lite was used in [37] to predict the 2D location of keypoints. HighResolution Net (HRNet) architecture with 32 channels in the highest resolution feature maps was used to regress the 2D landmark locations and the predicted output was a tensor of heatmaps corresponding to each landmark [26]. The rest of this section describes the HRNet architecture used to perform keypoint regression.

HRNet can maintain high-resolution representations throughout the process, hence the name HighResolution Net. Figure 5.5 shows the network architecture of HRNet and it starts from a high-resolution sub-network as the first stage, gradually adds high-to-low resolution sub-networks one by one to form more stages, and connects the multi-resolution sub-networks in parallel [38]. While connecting high-to-low resolution sub-networks in parallel HRNet is able to maintain high resolution

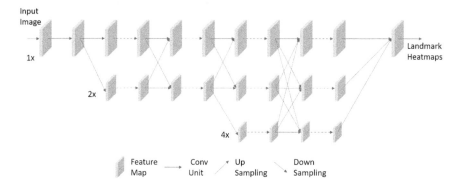

Figure 5.5   The HRNet architeture

throughout the process without any need of recovering the resolution through the low-to-high process; thus, it helps to predict the heatmap more precisely.

To improve the performance of the algorithm, the bounding box from the object detection phase is used to crop the input images to specific dimensions and then resize to fit the input dimensions. The output of the model is a tensor of 11 heatmaps each representing the location of 2D keypoint. To train the model, the loss function l is set to minimise the mean squared error between the predicted heatmaps $h(\mathbf{z}_i)$ and the ground truth heatmaps $h(\mathbf{z}_i^*)$. The ground truth heatmaps are generated as 2D normal distributions around the location of the keypoints with the specified standard deviations. Because HRNet maintains a high-resolution representation, it can produce high-resolution heatmaps with superior spatial accuracy [26]:

$$l = \frac{1}{N} \sum_{i=1}^{N} v_i (h(\mathbf{z}_i) - h(\mathbf{z}_i^*))^2 \qquad (5.12)$$

5.3.1.3   PnP + RANSAC

With the known 2D and 3D correspondences, PnP algorithm can be applied to compute the camera pose. However, one should note that not all found correspondences are correct, there are possibilities of false correspondences among the derived keypoints. To eliminate the outliers and estimate the camera pose with a higher probability, an outlier rejection algorithm is applied. The RANSAC algorithm [39] was proposed as a general parameter estimation approach to cope with a greater number of outliers in the input data. Unlike conventional sampling techniques that use as much of the data as possible to obtain an initial solution, RANSAC generates solution by using the minimum number of observations with the smallest set possible to estimate the underlying model parameters and then carry on to grow this set with consistent data inputs.

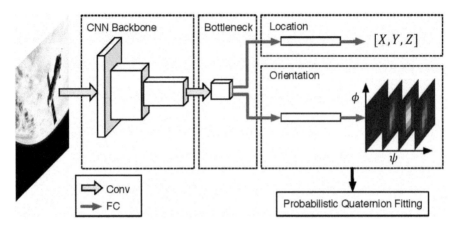

Figure 5.6   Pose estimator framework used in [27]

5.3.2   Non-keypoint-based pose estimation

The above section discussed the pose estimation problem addressed via a keypoint-based approach where the location of keypoints was predicted and used as an input to the PnP algorithm to estimate the pose. The alternate approach estimates the pose directly from the image without processing the keypoints. This can be done via direct regression of the 3D position and unit quaternion vectors; however, previous studies indicate that the orientation regression, i.e. a norm-based loss of unit quaternions, does not provide good estimates and resulted in larger error margin. This is mainly attributed to the loss function and its inability to represent the actual angular distance for any orientation representation. A non-keypoint-based approach for spacecraft pose estimation was presented in [27], where a ResNet-based architecture was used as the network backbone followed by two separate branches for the estimation of the location and orientation, as shown in Figure 5.6. This gives the framework an advantage of the low number of pooling layers as well as a good trade-off between accuracy and complexity. The last fully connected layer and the global average pooling layer of the original architecture were removed to keep spatial feature resolution. The global pooling layer was replaced by one extra 3×3 convolution with a stride of 2 (bottleneck layer) to compress the CNN features since the task branches are fully connected to the input tensor. The location estimation was carried out through a simple regression branch with two fully connected layers while minimising the relative error in the loss function. The location loss function was given by (5.13).

$$L_{loc} = \sum_i^m \frac{||t^i - t^i_{gt}||_2}{||t^i_{gt}||_2} \qquad (5.13)$$

For orientation estimation, a continuous orientation estimation via classification with soft-assignment coding was proposed. Each label (q_{gt}) is encoded as a Gaussian random variable in the orientation discrete output space. The network will be trained to output the probability mass function that corresponds to the true orientation. During training, each bin is encoded with the soft-assignment function f and given by

$$f(b_i, q_{gt}) = \frac{K(b_i, q_{gt})}{\sum_j^N K(b_i, q_{gt})} \tag{5.14}$$

where $K(x,y)$ is the kernel function and it uses the normalised angular difference between two quaternions.

$$K(x,y) = e^{-\frac{\left(\frac{2\cos^{-1}(|x^T y|)}{\pi}\right)^2}{2\sigma^2}} \quad \text{and} \quad \sigma^2 = \frac{\left(\frac{\Delta}{M}\right)^2}{12}$$

where the variance σ^2 denotes the quantisation error approximation, Δ/M represents the quantisation step, Δ is the smoothing factor that controls the Gaussian width, and M is the number of bins per dimension (i.e. Euler angle). Though the framework offered similar accuracy to the keypoint-based methods, the number of parameters used for training is quite high and also the architecture is found to be not scalable in terms of image and orientation resolution.

5.4 Advancements in simulators and experimental testbeds

5.4.1 Digital simulators

For every machine learning approach, training the model using a good dataset and fine-tuning the training parameters are the two important keys to its success. The training dataset for spacecraft pose estimation contains thousands of images of target object in orbit and to generate such dataset using real space imagery is an expensive process. The alternative approach is to prepare a synthetic dataset using photo-realistic visual simulators. These simulators take advantage of the state-of-the-art rendering engines and it can simulate the close to realistic visual of the target model under the given lighting conditions. Realistic image simulation tools were used in previous missions to aid Vision-Based Navigation (VBN) in space/planetary environments (such as Lunar environment, Asteroid surface) and it includes the ESA's PANGU (Planet and Asteroid Natural scene Generation Utility) [40] and the SurRender (Airbus internal simulator) [41]. PANGU is a simulator tool designed to aid the testing, verification, and validation of VBN used in the autonomous spacecraft missions (including the planetary landers and rovers). PANGU provides realistic simulations of onboard vision and LIDAR guidance sensors with high fidelity, real-time simulation if necessary. The simulator can be used in wide-range of test setups including off-line open-loop, closed-loop, and hardware-in-the-loop. Airbus's SurRender can be used in two different modes of image rendering, ray-tracing, and OpenGL. It can produce physically accurate images providing the known irradiance (each pixel contains an irradiance value expressed in W/m²). It was optimised to efficiently handle a wide range of resolutions and also to simulate sparse scenes in a space environment. It can generate images in real time in OpenGL mode for real-time hardware-in-loop software testing.

Another photo-realistic simulator tool, Orbit Visual Simulator (OrViS) (previously named as URSO, Unreal Rendered Spacecraft on Orbit), was developed based

Figure 5.7   Examples of rendered images of a Soyuz model simulated using OrViS

on Unreal Engine 4 (UE4) [27]. The simulator takes advantage of state-of-the-art rendering technologies implemented in UE4. Within the simulator, Earth was modelled as a high polygonal sphere with the high-resolution image applied as texture over the sphere's surface. The high-resolution image (21 600×10 800) of Earth from the Blue Marble Next Generation collection (https://visibleearth.nasa.gov) was used for the textures and it is further masked to obtain specular reflections from the ocean surface. An additional plugin was used to simulate atmospheric scattering to achieve close to a realistic Earth background. Lighting in the simulator environment is simply made of directional light and a spotlight to simulate the sunlight and the Earth albedo, respectively. Ambient lighting was disabled and to simulate the Sun, a body of emissive material with UE4 bloom scatter convolution was used. Further, lens flare was added to create realistic lighting conditions for the camera as exposed in the orbital environment. Target models are simulated with high-resolution textures and UE4 materials to achieve a realistic appearance, and the sample images of Soyuz spacecraft from OrViS is shown in Figure 5.7. UnrealCV plugin is used to communicate with the tool via server/client TCP connection. A random sampling of the parameters such as Earth yaw, orbital height, camera, operation range, and target orientation combined with the target to be within the camera viewing frustum are used to generate the necessary dataset of the target object for proximity operations.

One of the prominent datasets for spacecraft pose estimation available (at the time of writing) is the SPEED dataset [42]. This dataset was part of the Satellite Pose Estimation Challenge organised by the Space Rendezvous Laboratory (SLAB) at Stanford University and the Advanced Concepts Team (ACT) of the ESA. The dataset contains images of the Tango spacecraft from the PRISMA mission generated from two different sources using the same camera parameters, referred to as synthetic and real images. The synthetic images were generated using the OpenGL-based image rendering pipeline to generate photo-realistic images of the Tango spacecraft. The real images of the Tango spacecraft are captured using the Testbed for Rendezvous and Optical Navigation (TRON) facility of SLAB. Table 5.1 provides a brief comparison of the different visual simulators currently in use suitable for the purpose of generating photo-realistic images for relative navigation.

Table 5.1   A comparison of the orbital visual simulators

Visual simulator	Purpose	Rendering pipeline	Testing and validation	Sensor support	Requirement / interfaces	Outputs
ESA's PANGU [40]	Orbit and Planetary	OpenGL	Supports Hardware-in-loop (HiL) Testing	Optical (B & W, RGB)	64-bit PCs running Windows, Linux or on MacOS on commercial virtual machines	• Import/export planetary Digital Elevation Model (DEM) in PDS format • Asteroid shape models in ICQ or OBJ • Popular CAD formats. • JPG, TIF, PNG
AirBus Surrender [41]	Orbit and Planetary	Raytracing (CPU): physically accurate OpenGL (GPU): real time	Supports HiL Testing	• Active sensors: LIDAR, time-of-flight cameras • Optical (B & W, RGB, multispectral) or infrared	• Linux or Windows • Client-server protocol (TCP/IP) • Interfaces: Python 3, MATLAB / Simulink, C++, Lua	• DEM, Textures, albedo maps • JPG, TIF, PNG • PDS format • 3D meshes (OBJ, 3DS, PLY, Collada) • Images rendered in physical units (W/m2) • Slope maps • Depth maps
University of Surrey/ STAR Lab - OrViS [27]	Orbital (Relative Navigation)	Unreal Engine 4	No realtime HiL support	• Optical (B & W, RGB) • Possible to generate depth maps	• Online: Web app (All OS) • Offline: Linux • Interface: Python 3	• 3D meshes (FBX, OBJ) • Texture maps • JPG, TIF, PNG • Image rendered based on the lightning setup • Depth maps
Simulators based on Blender	Orbital (Relative Navigation)	Blender rendering Engines (cycles, Eevee)	-	Optical (B & W, RGB)	Offline	• Import/export all blender supported CAD formats • JPG, TIF, PNG • Rendered image
Stanford/ TRON Lab Simulator [43]	Orbital (Relative Navigation)	OpenGL	No real-time HiL support	Optical (B & W, RGB)	Offline	• 3D meshes, Texture maps • JPG, TIF, PNG • Rendered image

Dhamani *et al.* [29] used UE4 to simulate the open-source model of Cygnus spacecraft with a small set of LEO-like environments to generate the images and used the Microsoft's AirSim plugin for Unreal Engine to capture virtual images. Custom Python scripts were used to localise the camera around the target and capture the pictures with random backdrops, lighting conditions, and orientations. The dataset generated contains the images with bounding boxes of the target and centroid location. Another well-known open-source rendering tool is the Blender (https://www.blender.org). Oestreich *et al.* [44] used the Blender to generate the dataset, with the light source modified to simulate the local orbital lighting conditions as well as Earth albedo. Later, Gaussian noise is added to each image using MATLAB®'s Image Processing Toolbox™.

5.4.2   Ground-based physical testbeds

Ground-based hardware-in-the-loop experiments are essential for testing and validating autonomous spacecraft GNC systems. With growing interests in autonomous on-orbit space missions, such as space rendezvous and docking and on-orbit services, various ground-based testbeds have been developed worldwide.

Modern ground-based testbeds for on-orbit robotics have two main features: (1) simulating relative motion and micro-gravity dynamics of one or multiple orbiting objects in space; (2) simulating the orbital lighting condition to evaluate the sensors and guidance algorithms, which are realised by using the following methodologies:

- Experiment objects are levitated by using air-bearing systems or robotic arms to compensate the gravitational forces and moved as they are operating in orbits.
- Staged lighting setup is used in a reflection controlled room with light-absorbing materials to cover the experiment area.

Air-bearing tables were first introduced to simulate up to 3-DoF frictionless motion in a flat surface [45]. Figure 5.8a shows an air-bearing testbed. The key components of an air-bearing platform are a controlled jet and gas tank. The platform levitates the experiment object to create a contactless and frictionless motion. The supporting platform usually needs to use extra weight to relocate the mass centre to keep the air-bearing system stable. The benefit of this system is that it can reflect real dynamics to simulate 3-DoF micro-gravity dynamics. The drawback is the complexity in the gas-jetting system; the air-bearing platform's running time is limited by the gas supply and can only simulate up to 3-DoF motion. The latest air-bearing system for space robotics research is the Navy Postgraduate School's POSEIDYN (Proximity Operation of Spacecraft: Experimental hardware-In-the-loop Dynamic Simulator) system. It has been successfully used for the research of close-approximate spacecraft navigation and control [47, 48].

The state-of-the-art testbeds are capable of simulating high-DoF dynamics by using robotic arms. The robotic arms compensate the objects' gravitational forces and move them to follow the desired trajectories as they are 'flying' in the micro-gravity environment. This approach is first used for spacecraft rendezvous and

(a) Air-bearing friction-less table at University of Surrey
Credit: Surrey Space Centre

(b) Air-bearing spinning platform for spin-stablized spacecraft attitude control experiments
Credit: Surrey Space Centre- STAR LAB

Figure 5.8   Two examples of ground-based testing facilities by using air-bearing systems, (b) system is retrieved from [46]

docking testings [49]. Benefiting from the high-DoF motion simulation, the popularity of the robotic ground-based testbeds are increasing. GMV's Platform-Arts and DLR's OOS-SIM (On-Orbit Servicing Simulator) (shown in Figure 5.8) [50, 51] are the two most recently space robotic testbeds developed by space contractor and space agency, respectively. Both testbeds use commercial high-accuracy industrial robotic arms to simulate the relative motions of the experimenting targets in space.

Universities and research institutes have also been developing the testbeds but by pursuing more budget solutions. For example, the STAR LAB developed a similar high-DoF testbed [52]. Instead of using industrial robotic arms, this testbed uses a much smaller laboratory robotic arm and an open-sourced operating system – ROS (Robot Operating System).

Figure 5.9 shows two examples of the ground-based space robotics testbed for comparison. Both testbeds are developed based on robotic technologies and have a

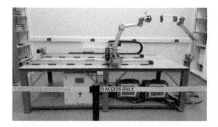

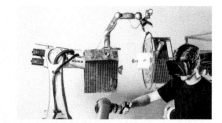

(a) Re-configurable orbital robotics testbed.
Credit: Surrey Space Centre – STAR LAB

(b) DLR's OOS-SIM.
Credit: DLR (CC-BY 3.0)

Figure 5.9   Two examples of ground-based testing facilities using high-DoF robotic arms for space robotics research

similar setup – two robotic arms simulate relative motions between chaser and target spacecraft and visual sensors are implemented around the testbed.

- Target Arm: The robotic arm that carries and compensates the experimental object's weight and simulates the full 6-DoF free-motion of rigid body dynamics in the microgravity environment. The arm has a force-torque sensor installed to measure the contact forces/torques fed back into the controller to simulate the corresponding motion of the experimental object as in space.
- Service Arm: The service arm simulates the motion of the chaser spacecraft or the robotic arm on the chaser spacecraft if the testbed is developed for on-orbit assembly.
- Visual Sensors: This part includes the choice and placement of sensors as well as staging the lighting conditions and other visual features in the laboratory environment to retain high-fidelity.

Table 5.2 shows a technical comparison chart between the examples of the ground-based testbeds for autonomous spacecraft GNC systems. Note that for the testbeds using robotic arms, the payload's capacity depends on the load capacity of the robotic arm. The accuracy of simulating dynamic motion depends on the robotic arm control system. Therefore, high-fidelity dynamic simulation requires high-end industrial robotic arms. The testbeds usually include a high-fidelity orbital propagation model to simulate the relative motion between the service arm and the target arm even when the experiments are taking place in the close-proximity range.

For example, the STAR LAB's testbed uses two identical and relatively low-cost 6-DoF robotic arms as the target arm and service arm. This robot arm has a maximum payload limit, 5 kg, which is enough to off-load a standard 2U CubeSat. The model shown in the figure is a modular telescope segment for an on-orbit assembly experiment.

Figure 5.10 shows the STAR LAB testbed's customised setup for testing on-orbit assembly research work. The main hardware components include two 6-DoF robotic arms, two force-torque sensors, and an RGB-D camera. In this figure, the experiment object is mounted on the target arm simulating attitude dynamics in the micro-gravity environment.

Compared with the other testbeds listed above, the STAR LAB testbed has compact size and potentially costs less than the testbeds using larger industrial robotic arms. Moreover, the STAR LAB testbed implements a high-fidelity orbital propagation model to simulate the relative motion between the service arm and the target arm in the close-proximity range. The limitation of the testbed is the UR5 robot arm has a payload capacity of 5 kg. Therefore, the experiment targets need to be prepared within the capacity but the actual dynamic properties can be scaled in simulation.

The target arm has three operating modes which represent the existing OOA and OOS paradigms:

- Fix-position: The target arm joints can be configured and locked to set the target to a fixed location within the workspace. In this mode, the target arm simulates

Table 5.2   A comparison chart of the ground-based testbeds for close-approximate space experiments. Recreated from [52].

	STAR LAB's orbital Robotics Testbed	NPS's POSEIDYN	GMV's platform-art	DLR's OOS-SIM
DoF of the target	6	3	6 + 1	6
Simulation range	Proximity	Proximity	Rendezvousandproximity	Proximity
Orbital mechanics	Included	Partly included	Included	Not included
Max. target mass (kg)	5	N/A	150	120
Service/chaser spacecraft	Simulated	Up to 9.5 kg	Scaled or up to 150kg	Scaled or up to 120kg
Operating range	$3 \times 2 \times 2.5$ m^3	4×4 m^2	15 m	23.28 m^3
Perception sensor	RGB/RGB-D camera	Multi-camera system	LIDAR	Stereo camera
Nominal tracking error/ground truth error	0.05 m and 1 deg	Various (N/A)	0.007 m and 0.8 deg	Less than 0.005 m
Nominal dynamic simulation latency	Desired 300–400 ms	N/A	100 ms	16 ms (closed-loop system)
ROS supported	Yes	Part	No	No
Customised connector/end-effector	Yes	Yes	Yes	No

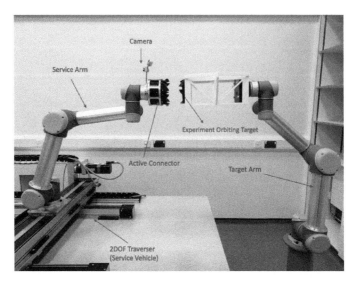

Figure 5.10   *The STAR LAB's Orbital Robotics Testbed has a typical ground-*
based testbed configuration: service arm (left) and target arm
(right). The end-effector of the service arm is exchangeable. This
setup uses a pair of form-fit mechanical connectors

a fixed locaiton on the same spacecraft to deliver building blocks to the service
arm.

• Free-floating: The target arm uses the force-torque sensor's feedback to simulate spacecraft attitude in micro-gravity environment. In this mode, the experiment is assumed to be carried out in close-proximity range without considering
orbital mechanics.

• Free-orbiting: The target arm moves to simulate user-defined rendezvous trajectory (constrained by orbital mechanics) and this mode is for evaluating the
relative navigation algorithms.

Section 5.4.3 describes the design and concept of the methodology of simulating the relative motion between the service spacecraft and the target.

5.4.3   *The methodology of simulating relative motion*

The flowchart in Figure 5.11 shows the algorithm of micro-gravity dynamics simulation delivered by the target arm. The simulation starts performing from an initial
trajectory of the target. This trajectory is relative to a fixed reference frame, which
is predefined. For the present setup, the origin of the fixed reference frame is at the
centre of the first joint of the service arm. The relative trajectory of the target is prop-
agated by the main simulation computer and executed by the controller interface in
the ROS environment. The target arm has a force-torque sensor, which measures any
external force and torque from the dynamic contacts during the assembly process. A

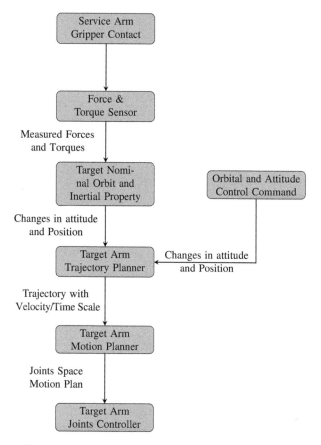

Figure 5.11   *Flowchart for simulating the relative motion between the service
and target arms*

new motion and relative trajectory are then calculated and updated to the robotic arm
motion planner to simulate the dynamic response of the modular telescope segment
in the micro-gravity environment.

One of the key features of the testbed is the high-fidelity model of orbital
mechanics for the close-proximity range, which is usually omitted in other space
robotics testbeds to reduce the simulation complexity. However, for smaller free-
flying orbiting objects, the attitude and orbit changes are nontrivial. Figure 5.12
illustrates a general case for on-orbit assembly/service where the trajectory of the
target has shifted after the service arm has exerted a contact force opposite to its
orbiting direction. For example, if the initial orbit of the in-orbit assembly is a circular orbit, the new orbit of the target after the contact with the service arm shifts
to an elliptical orbit. d_{r1} is the relative displacement of the target between the initial
and the new position. As the position of the service spacecraft is considered to be
fixed and is the reference frame for the testbed, the change of the trajectory of the

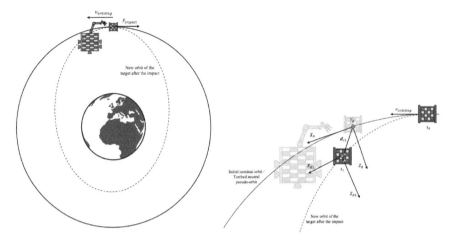

Figure 5.12   The relative motion (d_{t1} in the local reference frame) of the target after a dynamic contact. The impact force direction is opposite to the nominal target orbiting velocity

end-effector of the target arm can intuitively simulate the displacement between the service spacecraft and the target. In addition, the relative attitude of the target represented in the original orbital reference frame is also updated and simulated by the target arm.

The solution for the relative motion, d_{t1}, can be calculated by using the CW equation shown in (5.7) with Yamanaka-Ankerson's approach shown in (5.8) and (5.9). This approach has the added benefit that it retains nonlinear high fidelity for the testbed. High-fidelity simulation of the relative motion can be computationally expensive in more challenging situations. The process can be simplified depends on the experiment scenarios. For example, in the experiment scenarios of on-orbit assembly of large space infrastructures, the service spacecraft with the service arm attached is usually heavier than the targets. The STAR LAB testbed is developed for smaller orbiting target whose mass is around 2–5 kg without using scaled models. In the scenarios that the service arm sits on the top of a much larger orbiting platform, e.g., a space station, the dynamic response of the platform can be negligible after the dynamic contact.

The next step is the physical realisation of the relative motion of the target. This is done by using the commercial robotic arm controller interface and the MoveIt framework in ROS, as shown in Figure 5.13. The control interface between ROS and the robotic arm offers trajectory and velocity following controllers, which ensure the simulated trajectory to be executed and followed by the joints movements of the robotic arm. Besides, modern planners enable motion planning of the arm for trajectory tracking. The available pacakges also help to address the common robotic control and planning problems such as obstacle avoidance and mitigating of singularity. Several different planning algorithms with optimisation can ensure the robot arm moves in a smooth and continuous pattern.

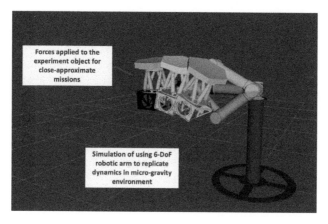

Figure 5.13   *ROS MoveIt framework to control the simulated relative motion of the target*

Figure 5.14 shows an example of the relative motion simulation on the target arm of the STAR LAB's Orbital Robotics Testbed. In this simulation, an arbitrary simulated impact force was applied to the corner of the target, a free-floating modular space telescope segment, for 1 s. The observing reference frame is the service spacecraft reference frame fixed relative to the camera reference frame, assuming the camera is mounted on the service spacecraft.

Due to the limited workspace of the robot arm, the simulation window is too short to observe the final orbit of the target in this example. The robotic arms that have a larger workspace can extend the simulation duration of the relative motion for the experiments with larger momentum exchanges or beyond close-proximity range. In addition, other options to extend the simulation range include using traversers to move the target arm.

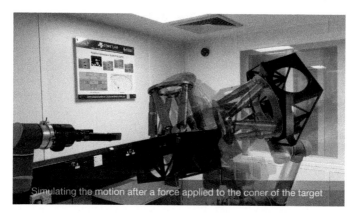

Figure 5.14   *An example demonstrating the relative motion of an orbiting target. (Courtesy to CAS-CIOMP for providing a sample of orbiting target)*

The accuracy of the robotic arm tracking the relative motions depends on both the robot control interfaces and the low-level controllers. The control interfaces translate the trajectory information from the computer to the robot arm. In the situation that the robot arm does not support velocity and acceleration tracking, the position points need to be modified with special attention to the time sequence. As for the low-level controller, the modern industrial robot arms usually have factory calibrated and tuned software to retain high precision to retain high precision. Another useful feature for on-orbit robotics testbeds is impedance control. To realise a low-impedance controller, torques shall be measured at joint level. This is usually available on more advanced robotic arms or from adding external force-torque sensors. In conclusion, the fidelity and capability of the orbital robotics testbed mostly depend on the hardware performance and capability of the robotic arms used in the testbed setup.

5.5   Analytical results and comparison

A comparative analysis of keypoint-based and non-keypoint-based deep-learning approaches for spacecraft pose estimation was presented in [31]. The Soyuz dataset from OrViS simulator was used to train and test the deep-learning models. The Soyuz dataset contains 5 000 images, of which 10 per cent is reserved for testing and another 10 per cent for validation.

For the keypoints-based pose estimation approach, the framework shown in Figure 5.3 was used. To perform object detection, ResNet-based Faster R-CNN [53] was used. ResNet-50 was used as a backbone and transfer learning approach was used to train the model. During the training, the models are loaded with the pre-trained weights from COCO (Common Objects in Context) dataset [54] and further tuned to identify the Soyuz spacecraft using the bounding box coordinates. The training parameters include the input image size of 320 with a batch size of 8, learning rate of $5e^{-3}$, Stochastic Gradient Descent (SGD) with a momentum of 0.9, and a weight decay regularisation of $5e^{-4}$. Input image augmentations such as random rotation, random translation, coarse dropouts, Gaussian noise, random brightness, and contrast were randomly added to make the training model more robust. The training converges quickly around 30 epochs for both the models.

For keypoint regression, HigherHRNet [55] is used, which, in turn, uses a HRNet mentioned in Section 5.3 as backbone, to regress the 2D landmark locations. The architecture used has 32- channels in the highest resolution feature maps and provides an output at two different scales: 1/4 and 1/2. To achieve good visibility of keypoints, 21 keypoints were selected all around the Soyuz spacecraft. For training the model to regress the locations of the keypoints, the input image size is set to 640 with a batch size of 4 and an adam optimiser is used with $1e^{-3}$ learning rate and 0.9 momentum. Image augmentation performed on the input image includes random rotation $(-30°, 30°)$, random translation up to $(-30\%, 30\%)$, coarse dropouts, Gaussian noise, random brightness, and contrast. For keypoint regression, the training time is quite extensive and the image augmentation plays a

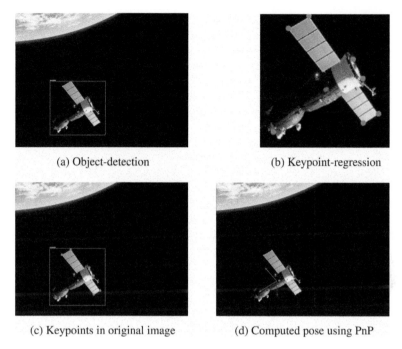

(a) Object-detection   (b) Keypoint-regression

(c) Keypoints in original image   (d) Computed pose using PnP

Figure 5.15   Sample from keypoint-based approach

key role in identifying the right keypoints. The sample images from the keypoint-based approach are shown in Figure 5.15 with Figure 5.15a highlighting the detected object in the bounding box and Figure 5.15b showing the location of detected keypoints above the set threshold, and the locations of the points merged in the original image are shown in Figure 5.15c whereas Figure 5.15d shows the computed pose along with the ground truth (which was overlapped).

For non-keypoint-based approach, the network uses ResNet-50 with a bottleneck width of 32 filters and the training starts with the weights from the backbone of Mask R-CNN trained on COCO dataset. To perform orientation soft classification, the network uses 16 bins per Euler angle. Camera rotation perturbations with a maximum magnitude of $10°$ were included to augment the dataset. Networks were trained on images with the input images resized to half their original size to provide a batch size of 4, SGD with a moment of 0.9, and a weight decay regularisation of $1e^{-4}$. The learning rate was scheduled using step decay depending on the model convergence; however, it was found to depend highly on the orientation estimation method, the number of orientation bins, augmentation pipeline, and the dataset.

The results were recorded as the mean absolute location error and the mean angular error. The results from the keypoint-based and non-keypoint-based deep-learning approaches for the OrViS dataset are shown in Table 5.3 and the errors presented are mean absolute location error and the mean angular error. The results suggest that the keypoint approach performs better compared to the non-keypoint

Table 5.3   Results for different approaches for Soyuz dataset [31]

Method	Location error	Angular error
Non-keypoint-based pose estimation [27]	0.8 m	7.4°
Keypoint-based pose estimation [31]	0.3 m	4.9°

approach. Similar results were seen in the Spacecraft Pose Estimation Challenge [56], where a keypoints method offered better estimates compared to the non-keypoint-based approach. The advantage and disadvantage of both approaches are discussed further. The keypoint-based approach is more scalable, i.e. it is possible to increase the number of keypoints in the target and trained accordingly. In the keypoint-based approach, it is possible to upgrade the model and improve the model performance by replacing the key components (i.e for object detection or keypoint estimation) to state-of-the-art algorithms. For a given image with partly occluded target object, if the keypoint-detection algorithm is able to properly detect the minimum number of keypoints required to compute the pose, the estimated pose will not differ more in accuracy. One of the limitations of the keypoint approach is that the good geometric model or the knowledge of keypoint locations is essential to train the model and compute the pose. The non-keypoint-based approach has a few limitations, the orientation resolution achievable with the non-keypoint-based approach depends on the number of bins used to encode the quaternion space during probabilistic soft classification. With the increase in the number of bins, the number of parameters in the trained model increases. For example, with $64 \times 64 \times 64$ bins, a more accurate result for the SPEED dataset was achieved and it requires around 500 M parameters [27]. In comparison, the number of parameters for both object detection and the keypoint-based methods is around 50–80 M based on the configurations. The advantage of the non-keypoint-based approach is that it does not rely on the PnP solvers to estimate the pose and hence it averts the pitfall of wrong keypoint detection leading the PnP estimate to stick with local minima. One of the solutions to this problem is to use keypoints confidence values to prune bad predictions, thus reducing the chances of getting stuck with the local minima. In both cases, the models trained on synthetic images are found to be transferable to detect objects and keypoints on real images.

5.6   Recommendations and future trends

This chapter presented the background of the spacecraft pose estimation and the state-of-the-art algorithms that are using deep-learning techniques. The two current major approaches for deep-learning-based spacecraft pose estimation, the keypoint-based and the non-keypoint-based approachs were presented. The keypoint approach provides many advantages compared to the non-keypoint approach, including a lower number of training parameters, upgradable sub-components, and better overall performance. The keypoint framework has two sub-components that can be easily swapped at any time to the state-of-the-art algorithms to get better performance and can be trained independently. This chapter summarised the simulators that are used to create the realistic orbital image of the target and the technology behind it. Finally, a brief overview of the robotic ground-based testbeds that are currently used for close-proximity operation scenarios simulation was presented.

The application of deep-learning in spacecraft pose estimation is gaining momentum; however, the overall development is in the early stages and currently being considered for the technology demonstration missions to prove its robustness. This highlights the performance and ease of use of the new algorithms compared to the traditional approach and the future looks more promising than ever before. Adding further improvements to deep-learning algorithms and increasing its technology readiness levels will attract more space missions to include deep-learning-based pose estimation in their GNC framework and can fine-tune the algorithms suitable for their specific mission application. Performing further sensor data fusion along with the estimated target pose will enable the spacecraft to execute complex autonomous tasks and thus increasing GNC autonomy in orbit.

The development of robust deep-learning models depends on the quality of the dataset and the training parameters optimisation; hence it is equally important to have better ground-based simulators that can render realistic target images. The computing hardware is becoming easily accessible and it provides the simulators (such as UE4) the necessary resources (like GPU) to render highly sophisticated photo-realistic orbital images. A future version of UE will bring more control through micro-polygon geometry and this offers the possibility to create fine geometric details as the eye can see and can simulate close to realistic dynamic lighting conditions. Generating such simulations in real time, one can simulate and validate the operations of the entire mission in simulation. In the ground-based testbed, it is not viable to simulate the entire mission operation in a single testbed setup and also it is expensive to have such a dedicated testbed. The re-configurable smaller testbed will provide an advantage to academic researchers to define and test the algorithms in a limited setup while improving the technology readiness levels of these algorithms.

With the orbital space becoming congested and sustainable operations in space becoming a priority, private space companies are expecting a change in the way how the spacecraft is built and utilised. This change in the idea will pave for more OOS missions and thus bring the autonomous operations into focus on future missions. The vision-based navigation will be a critical component for future OOS missions. Apart from the algorithms, there are numerous developments in the hardware, for example, advanced optical camera systems and LIDARs for visual navigation. The computing power needed to process the LIDAR point clouds in real time onboard the spacecraft is considered a drawback. LIDARs can be used in combination with monocular cameras to support the critical decision-making process. The VBN algorithms shall be able to run on the low-power computing platform and/or COTS systems, and this will enable more small satellite missions in the future. The autonomous visual navigation offers several challenges and the pose estimation using deep-learning addressed in this chapter presents a pathway for future missions to achieve full 6-DoF pose estimation of non-cooperative spacecraft using monocular cameras.

References

[1]   Sachdev S.S., Fuller B.R. 'The shuttle remote manipulator system and its use in orbital operations'. The Space Congress® Proceedings – Space: The Next Twenty Years; 1983. pp. 1–21.

[2]   Nanjangud A., Blacker P.C., Bandyopadhyay S., Gao Y., *et al.* 'Robotics and AI-Enabled On-Orbit operations with future generation of small satellites'. *Proceedings of the IEEE.* 2018;**106**(3):429–39.

[3]   Matos de Carvalho T.H., Kingston J. 'Establishing a framework to explore the servicer-client relationship in on-orbit servicing'. *Acta Astronautica.* 2018;**153**:109–21.

[4]   IET-E&T. *One satellite services another in orbit for the first time. IET engineering and technology [online].* 2020. Available from https://eandt.theiet.org/content/articles/2020/04/one-satellite-has-serviced-another-in-orbit-for-the-first-time/ [Accessed 1 Nov 2020].

[5]   Roesler G. *Robotic servicing of geosynchronous satellites (RSGS). Defense Advanced Research Projects Agency [online].* 2017. p. 1. Available from http://wwwdarpamil/program/robotic-servicing-of-geosynchronous-satellites [Accessed 14 Jul 2017].

[6]   Flores-Abad A., Ma O., Pham K., Ulrich S., *et al.* 'A review of space robotics technologies for on-orbit servicing'. *Progress in Aerospace Sciences.* 2014;**68**(5):1–26.

[7]   WJ L., Cheng D.Y., Liu X.G., *et al.* 'On-Orbit Service (OOS) of spacecraft: a review of engineering developments'. Progress in Aerospace Sciences; 2019 May. p. S0376042118301210.

[8]   GSFC N. 'Restore-L - proving satellite servicing'. *Nasa – Gsfc.* 2016.

[9]   Kessler D.J., Cour-Palais B.G. 'Collision frequency of artificial satellites: the creation of a debris belt'. *Journal of Geophysical Research.* 1978;**83**(A6):2637–46.

[10] Shan M., Guo J., Gill E. 'Review and comparison of active space debris capturing and removal methods'. *Progress in Aerospace Sciences.* 2016;**80**(A6):18–32.

[11] Tschauner J., Hempel P. 'Rendezvous with a target in an elliptical orbit'. *Astronautica Acta.* 1965;**11**(2):104–9.

[12] Carter T.E. 'State transition matrices for terminal rendezvous studies: brief survey and new example'. *Journal of Guidance, Control, and Dynamics.* 1998;**21**(1):148–55.

[13] Wolfsberger W., Weiß J., Rangnitt D. 'Strategies and schemes for rendezvous on geostationary transfer orbit'. *Acta Astronautica.* 1983;**10**(8):527–38.

[14] Yamanaka K., Ankersen F. 'New state transition matrix for relative motion on an arbitrary elliptical orbit'. *Journal of Guidance, Control, and Dynamics.* 2002;**25**(1):60–6.

[15] Agarwal S., Furukawa Y., Snavely N., *et al.* 'Building Rome in a day'. *Communications of the ACM.* 2011;**54**(10):105–12.

[16] Schonberger J.L., Frahm J.M. 'Structure-from-motion revisited'. The IEEE Conference on Computer Vision and Pattern Recognition (CVPR); 2016. pp. 4104–13.

[17] Durrant-Whyte H., Bailey T. 'Simultaneous localization and mapping: Part I'. *IEEE Robotics & Automation Magazine.* 2006;**13**(2):99–110.

[18] Nister D., Naroditsky O., Bergen J. 'Visual odometry'. Proceedings of the 2004 IEEE Computer Society Conference on Computer Vision and Pattern Recognition, 2004; 2004. pp. 652–9.

[19] Kendall A., Grimes M., Cipolla R. 'PoseNet: a convolutional network for real-time 6-DOF camera relocalization'. 2015 IEEE International Conference on Computer Vision (ICCV); 2015. pp. 2938–46.

[20] Shavit Y., Ferens R. 'Introduction to camera pose estimation with deep learning. arXiv; 2019'. *arXiv preprint arXiv.* 1907;**05272**.

[21] Pavlakos G., Zhou X., Chan A., *et al.* '6-DoF object pose from semantic keypoints'. 2017 IEEE International Conference on Robotics and Automation (ICRA); 2017. pp. 2011–8.

[22] Kelsey J.M., Byrne J., Cosgrove M. 'Vision-based relative pose estimation for autonomous rendezvous and docking'. *2006 IEEE Aerospace Conference Big Sky, MT, USA; IEEE*; 2006. pp. 1–20.

[23] Capuano V., Alimo S.R., AQ H., *et al.* 'Robust features extraction for on-board monocular-based spacecraft pose acquisition'. AIAA Scitech 2019 Forum; 2019. p. 2005.

[24] Petit A., Marchand E., Kanani K. 'A robust model-based tracker combining geometrical and color edge information'. 2013 IEEE/RSJ International Conference on Intelligent Robots and Systems; 2013. pp. 3719–24.

[25] Sharma S., D'Amico S. 'Pose estimation for non-cooperative rendezvous using neural networks'. AIAA/AAS Space Flight Mechanics Meeting; 2019.

[26] Chen B., Cao J., Parra A., *et al.* 'Satellite pose estimation with deep landmark regression and nonlinear pose refinement'. 2019 IEEE/CVF International Conference on Computer Vision Workshop (ICCVW); 2019. pp. 2816–24.

[27] Proença P.F., Gao Y. *Deep learning for spacecraft pose estimation from photorealistic rendering [online].* 2020. Available from https://arxiv.org/abs/1907.04298 [Accessed 01 Feb 2020].

[28] Gerard K. Segmentation-driven Satellite Pose Estimation. 2019. Available from https://indico.esa.int/event/319/attachments/3561/4754/pose_gerard_segmentation.pdf.

[29] Dhamani N., Martin G., Schubert C., *et al.* 'Applications of machine learning and monocular vision for autonomous on-orbit proximity operations'. AIAA Scitech 2020 Forum; 2020. pp. 1–18.

[30]   Harvard A., Capuano V., Shao E.Y., *et al.* 'Pose estimation of uncooperative spacecraft from monocular images using neural network based keypoints'. AIAA Scitech 2020 Forum; 2020. pp. 1–10.

[31]   Rathinam A., Gao Y. 'On-orbit relative navigation near a known target using monocular vision and convolutional neural networks for pose estimation'. International Symposium on Artificial Intelligence, Robotics and Automation in Space (iSAIRAS); 2020. pp. 1–6.

[32]   Ren S., He K., Girshick R., *et al.* 'Faster R-CNN: towards real-time object detection with region proposal networks'. Advances in Neural Information Processing Systems; 2015. pp. 91–9.

[33]   Girshick R., Donahue J., Darrell T., *et al.* 'Rich feature hierarchies for accurate object detection and semantic segmentation'. Proceedings of the IEEE Conference on Computer vision and Pattern Recognition; 2014. pp. 580–7.

[34]   Girshick R. 'Fast R-CNN'. 2015 IEEE International Conference on Computer Vision (ICCV); 2015. pp. 1440–8.

[35]   Liu L., Ouyang W., Wang X., *et al.* 'Deep learning for generic object detection: a survey'. *International Journal of Computer Vision.* 2020;**128**(2):261–318.

[36]   Redmon J., Farhadi A. 'YOLOv3: an incremental improvement'. *arXiv.* 2018.

[37]   Park T.H., D'Amico S. *ESA Pose Estimation Challenge 2019.* Stanford University; 2019. TN-19-01.

[38]   Sun K., Xiao B., Liu D., *et al.* 'Deep high-resolution representation learning for human pose estimation'. Proceedings of the IEEE Conference on Computer Vision and Pattern Recognition; 2019. pp. 5693–703.

[39]   Fischler M.A., Bolles R.C. 'Random sample consensus: a paradigm for model fitting with applications to image analysis and automated cartography'. *Communications of the ACM.* 1981;**24**(6):381–95.

[40]   Parkes S., Martin I., Dunstan M., *et al.* 'Planet surface simulation with PANGU'. Space OPS 2004 Conference; 2004. p. 389.

[41]   Brochard R., Lebreton J., Robin C., *et al.* 'Scientific image rendering for space scenes with the surrender software'. 69th International Astronautical Congress (IAC); 2018 Oct.

[42]   Sharma S., Park T.H., D'Amico S. *Spacecraft Pose Estimation Dataset (SPEED).* Stanford Digital Repository; 2019.

[43]   Sharma S., Beierle C., D'Amico S. 'Pose estimation for non-cooperative spacecraft rendezvous using Convolutional neural networks'. *IEEE Aerospace Conference Proceedings.* 2018;**2018**:1–12.

[44]   Oestreich C., Lim T.W., Broussard R. 'On-orbit relative pose initialization via convolutional neural networks'. AIAA Scitech 2020 Forum; 2020. pp. 1–22.

[45]   Schwartz J.L., Peck M.A., Hall C.D. 'Historical review of air-bearing spacecraft simulators'. *Journal of Guidance, Control, and Dynamics.* 2003;**26**(4):513–22.

[46]   Chanik A., Gao Y., Si J. 'Modular Testbed for spinning spacecraft'. *Journal of Spacecraft and Rockets.* 2017;**54**(1):90–100.

[47] Wilde M., Ciarcià M., Grompone A., Romano M., *et al.* 'Experimental characterization of inverse dynamics guidance in docking with a rotating target'. *Journal of Guidance, Control, and Dynamics.* 2016;**39**(6):1173–87.

[48] Virgili-Llop J., Romano M. 'Simultaneous capture and detumble of a resident space object by a free-flying spacecraft-manipulator system'. *Frontiers in Robotics and AI.* 2019;**6**(March):14.

[49] Bell R., Morphopoulos T., Pollack J., *et al.* 'Hardware-in-the loop tests of an autonomous GN&C system for on-orbit servicing'. AIAA Space 2003 Conference and Exposition; 2003 Sep. pp. 1–18.

[50] Colmenarejo P., Barrena V., Graziano M. 'Ground validation of active debris removal technologies and GNC systems'. *Proceedings of the International Astronautical Congress, IAC.* 2013;**3**:2278–87.

[51] Artigas J., De Stefano M., Rackl W., *et al.* 'The OOS-SIM: An on-ground simulation facility for on-orbit servicing robotic operations'. Proceedings – IEEE International Conference on Robotics and Automation; 2015 Jun. pp. 2854–60.

[52] Hao Z., Mavrakis N., Proenca P., *et al.* 'Ground-based high-DOF al and robotics demonstrator for In-orbits space optical telescope assembly'. Proceedings of the International Astronautical Congress; 2019. pp. 1–11.

[53] He K., Zhang X., Ren S., *et al.* 'Deep residual learning for image recognition'. Proceedings of the IEEE Conference on Computer Vision and Pattern Recognition; 2016. pp. 770–8.

[54] Lin T., Maire M., Belongie S.J., *et al. Microsoft COCO: common objects in context. CoRR.* 2014. Available from http://arxiv.org/abs/1405.0312.

[55] Cheng B., Xiao B., Wang J., *et al.* 'HigherHRNet: scale-aware representation learning for bottom-up human pose estimation'. CVPR; 2020. pp. 1–10.

[56] Kisantal M., Sharma S., Park T.H., *et al.* 'Satellite pose estimation challenge: dataset, competition design and results'. IEEE Transactions on Aerospace and Electronic Systems; 2020.

Chapter 6

Inertial parameter identification, reactionless path planning and control for orbital robotic capturing of unknown objects

Chu Zhongyi[1], Hai Xiao[1], and Ma Ye[1]

With the exploration of space, the number of orbiting spacecrafts has been accumulating. A considerable part of them are non-cooperative targets such as the rocket end-stages or disposed satellites, which seriously threaten active spacecrafts. So the reasonable disposal of space non-cooperative targets is very urgent. On-orbit capture is the premise of most on-orbit operations. As non-cooperative objects are unable to supply any prior information and its inertial parameters are not available, the capture process owns the feature of complex time varying, strong coupling and nonlinearity, and thereby the space operation via space robotic system requires better adaptive and real-time property. This chapter focuses on the inertial parameter identification of a non-cooperative object, the adaptive reactionless path planning strategy for the robot arm during the identification and stability control strategy for the whole system. The relative content is organized as follows:

The first section establishes the basic dynamic model of the space robotic system that operates in the post-capture stage. According to the system's kinematics and dynamics equations, the two basic equations of identification for the non-cooperative object are constructed: the equation of Momentum Conservation as well as the equation of Newton-Euler, to obtain the basis of two-step identification and the error mechanism analysis.

The second section proposes the theory of error mechanism analysis and designing an improved inertial parameter identification method via the contact force measure. To deal with the strong coupling existing in the conventional identification equation, the two-step scheme is proposed, and the sufficient condition for identification as well as the error mechanism analysis is deduced for the improved identification method, which employs the contact force information to deal with the error accumulation and thus can improve the accuracy.

The third section designs an adaptive reactionless path planning method for the manipulator to deal with the motion disturbance and proposes a robust adaptive

[1]School of Instrumentation and Optoelectronics Engineering, Beihang University, Beijing, China

control strategy for joints' controllers and feedforward control strategy for the space-craft's controller to ensure the stability of the whole system. The Slide-windowed Recursive Least Square (RLS) algorithm is employed to identify and compensate the momentum coefficient matrix that updates online, and thus the Adaptive Reaction Null Space (ARNS) algorithm is constructed and the dynamic path planning is completed for the manipulator. The robust adaptive control strategy for the manipulator is proposed to track the planned path, and the feedforward control strategy via coupling torque compensation for the spacecraft ensures the attitude stabilization in the process of capturing and parameter identification.

6.1   Introduction

The continuous advancement of world science and technology has promoted the rapid development of space science and technology [1, 2]. A considerable number of non-cooperative objects, such as rocket end stages and disposed satellites [3], threaten active space vehicles. So the reasonable disposal of non-cooperative objects is very urgent. On-orbit capture is the premise of most on-orbit operations. The accurate capture is a prerequisite to ensure effective operation in orbit and the normal operation of the spacecraft, which has become a hot spot in the research of space technology in various countries. Research on space capture operations among various countries mainly include research on capturing objects and development of capture operation tools. Capture tools mainly include space rope nets, rope systems [4, 5], and space robotic arms [6, 7] systems. The objects of the capture operation are mainly oriented to space cooperative targets whose inertia characteristics are accurately known and space non-cooperative targets whose inertia characteristics are unknown.

On-orbit capture technology based on a space multi-degree-of-freedom robotic arm system, as shown in Figure 6.1, is one of the earliest developed on-orbit capture methods. Since the 1980s, different countries have begun the development of space robotic arm systems. For example, the space shuttle Columbia in 1981 became a major event in the history of human space as it carried a robotic arm for the first time. Canadian SMRS manipulators [9] and German ROTEX [10] intelligent robots are world-renowned for their highly equipped sensing systems. Japan-produced ETS-VII [11] spacecraft has conducted experiments in autonomous orbital capture technology, although the capture uses cooperative targets. The US Defence Advanced Research Projects Agency's recent 'Phoenix' program [12] focuses on the recovery and utilization of space dumped stars. China has launched Tiangong-2 Space Laboratory [13] in 2016, which was equipped with a 6-Degree of Freedom (DOF) lightweight arm for space assembly mission verifications. As Chinese space development blueprint [13], a 90-ton space will be established by 2025. It will have a space robotic arm system to carry out cabins movement and assembly. As the complexity of space operations increases, more lightweight and flexible operating tools are required, such as the development of space tethering systems, which are light enough to be negligible relative to the spacecraft itself and can extend the distance

Figure 6.1   Space robotic manipulator system in International Space Station (ISS) [8]

for space operations. However, the operation of the tether system and the net is more suitable for the operation of towing the target star, and its operating characteristics require the tether under tension. Once the tether is in a relaxed state, its operation will fail. Therefore, the current use of rigid robotic arms still maintains huge advantages in ensuring the accuracy, stability, safety, and flexibility of space operations and occupies an important dominant position in the intelligent operating system.

Today on-orbit capture missions are still geared towards cooperative targets in actual space operations, while capture missions for non-cooperative targets, such as diposed satellites and aerolites in Figure 6.2, are still at the theoretical and ground test stages. At present, the transformation from space non-cooperative targets to cooperative targets has become a feasible operation method [16]. Accurate inertia parameter identification of non-cooperative targets and compensation will improve capture performance. The theoretical research in this area has attracted the attention of a wide range of scholars and has achieved preliminary development. It is believed

Figure 6.2   Non-cooperative targets (disposed satellites [14], aerolite [15], etc.)

that in the near future, space capture missions based on non-cooperative targets will become practical.

The inertial parameter identification for non-cooperative targets involves the system's sensor accuracy, identification algorithms, the motion of the joints of the robotic arm, and the control strategies of the joints. Fast, accurate, and stable operation for non-cooperative targets is necessary, which demands more effective parameter identification methods, non-reactive path planning method, and coordinated control strategy.

6.1.1   Relative work and development status

The space operation based on capturing non-cooperative targets in space mainly involves three aspects: one is the method for identifying inertia parameters of non-cooperative targets in space; the other is the dynamic path planning problem of the space manipulator without reaction; and the third is the coordinated control strategy of the space manipulator system. The status and development trends of these three aspects at home and abroad will be described in detail in the following sections.

6.1.1.1   Method for identifying inertial parameters of space non-cooperative targets

Space non-cooperative targets are called space unknowns [17], space unknown targets [18], space passive bodies [19], etc., in domestic and foreign literature, including non-cooperative spacecraft, failure satellites, space debris, and space Garbage, etc., referring to objects or targets whose inertia attributes exist in the space environment are completely or partially unknown and cannot be cooperatively captured. Since the input of the space manipulator control system is based on the dynamic parameters of each part of the system, obtaining accurate system dynamic parameters is a prerequisite to ensure that the controller performs the control task effectively and the stable operation of the system. Space non-cooperative targets due to their inertia uncertainty of parameters (mass, centre of mass position, and three-dimensional inertia in space) lead to inaccurate and abrupt changes in the dynamic parameters of the system after capture, which, in turn, affects the control accuracy of the entire system. In severe cases, the spacecraft may be unstable or even complete failure. Therefore, it is necessary to obtain accurate non-cooperative target inertia parameters.

The current research is mainly for offline [17] or online identification of inertial parameters of non-cooperative targets [20]. Identification principles include system dynamics equations [21] and system momentum conservation equations [17]. Methods include parameter identification in the post-capture phase and pre-capture phase [17]. Identification methods are written and solved to include non-cooperative targets linear identification equations for inertia parameters. Identification algorithms for linear equations mainly involve the conventional least-squares algorithm and the least mean square algorithm.

In terms of identification principles, currently the main consideration is that the spacecraft body-space manipulator-noncooperative target system is free-floating.

Based on this, the system's momentum conservation equation and dynamic equation at the capture stage are constructed. Among them, the momentum conservation equation mainly considers that the system receives little external force (space microgravity, etc.) in a short period of time, which can ensure that the system's momentum remains constant for a certain period of time. As for the momentum equation, the inertia parameter of the noncooperative target is also included. The dynamic equations include the system's Newton-Euler equation and Lagrange equation, which mainly involve the internal force of the system, that is, by analysing the force of the non-cooperative target to carry out effective identification of inertia parameters. Lagrangian method [22] needs to write the generalized dynamic equations of the entire system and solve the partial differential equations, which simplifies the system dynamic relationship and is suitable for offline processing. The elements in the formula have no specific physical meaning. The mechanical method [17] presents recursive calculations involving specific force and moment information and is suitable for online processing.

In terms of identification methods, it includes the identification strategy in the pre-capture stage and the post-capture stage. The pre-capture phase identification is mainly to estimate the inertia parameters of the end manipulators of the space manipulator as they approach the non-cooperative target. The estimated parameters can be collected and measured (visual tracking) by the motion information of the non-cooperative target and further processed. Identification in the post-capture phase is mainly performed by the end manipulator on the non-cooperative target after the capture is stable. The operation process needs to assist the continuous and effective movement of the robotic arm to ensure the accuracy and stability of the identification.

The identification method currently implements the linear separation of inertia parameters of non-cooperative targets based on the obtained system's momentum conservation equations and dynamic equations. The precise known dynamic parameters are separated from the unknown parameters (non-cooperative target inertia parameters) to solve the linear equation. The current methods based on linear equations mainly involve the least-squares algorithm with its derivative algorithm, and the least mean square algorithm with its derivative algorithm. The least-squares algorithm is fast and accurate, and can effectively reduce the correlation of data. However, the calculation complexity is high, and as the number of calculation increases, the performance of the algorithm decreases due to the accumulation of white noise. The minimum mean square algorithm is simple and feasible. The amount of calculation is small, and the problem of white noise accumulation can be effectively overcome. However, the convergence speed is slow and it is easy to fall into a local optimal value.

6.1.1.2 Reactionless path planning for non-cooperative objects capture

The capture operation of non-cooperative objects by the space robotic arm in the free-floating state [23] due to the coupling effect [24] will disturb the attitude of the base, resulting in attitude deviation and serious problems such as failure of the spacecraft and abnormal antenna communication. To ensure the stability of the spacecraft's base body attitude, the control system can quickly maintain the base body's attitude without disturbance through its reaction wheel device or jet power device. However, this method increases the risk of saturation of the reaction wheel and the exhaustion of jet fuel. At present it is more inclined to use the coupling motion relationship of the robotic arm to eliminate its interference on the spacecraft base, as the robotic arm consumes the power generated by solar sails, which greatly reduces the energy requirements of the spacecraft system.

At present, there are two main methods with the robotic arm itself to eliminate its disturbance to the spacecraft body: the first is to use the reactionless robotic arm actuator [25, 26] and the specially designed mechanical arm configuration. The advantage is that the performance is solid and stable. However, the requirements for the design accuracy of the mechanism [27, 28] are extremely high, and it is difficult to achieve on-orbit maintenance. Once the structure is damaged this method will permanently fail. The second is to use the non-reactive path planning method of the robotic arm, that is, to obtain an optimal trajectory through theoretical calculations and meet the minimum coupling between the spacecraft body and the robotic arm. This method is flexible and practical. Trajectory planning strategy can be adjusted according to the coupling effect. It has become a method of suppressing disturbances for sustainable research in the future. Dubowsky [29] invented Euclidean Distance Map (EDM) to estimate the optimal trajectory of the robotic arm with the least disturbance. But this method demands a huge amount of calculations and is not suitable for on-orbit dynamic systems. Quinn [30] proposed an optimal strategy for reducing the reaction force of the robotic arm by utilizing the redundancy of the robotic arm. The redundancy of the robotic arm determines the characteristics of the trajectory planning. RNS algorithm proposed in the 1990s has also been used to eliminate the coupling between the robotic arm and the spacecraft body. Because of its simple algorithm, it has been widely used so far. Yoshida [31] first proposed the concept of RNS in 1992. This concept was used to solve the dynamic interaction of space floating manipulators, and then applied to track the end effector of the robotic arm and suppress vibration [32]. The space engineering test satellite ETS-VII achieved non-reactive capture of space cooperation targets through the RNS algorithm [32]. The RNS algorithm is also used in the reactionless control phase before and after a collision when capturing a target. However, the original RNS algorithm cannot be applied to a system containing non-cooperative targets, because this algorithm requires accurate dynamic parameters of the system. If the precise dynamic parameters cannot be obtained, the trajectory planning without reaction of the space manipulator will fail. Thai Chau Nguyen-Huynh [20] proposed an adaptive reactionless path planning method in 2013. Its purpose is to modify the original RNS algorithm to make it an adaptive algorithm that can change with the existence of external uncertainties. This method takes into account the time-varying characteristics of the system. However, conventional least squares will follow the gain matrix, which results in the saturation of the algorithm itself. Therefore, the development of a dynamic trajectory planning algorithm based on online parameter identification has become an urgent problem.

6.1.1.3 Attitude stable control method of spacecraft-manipulator-target system

The path planning of a robotic arm system that includes uncertain dynamic parameters needs to consider the adaptive process of the robotic arm joint path, and the effective tracking of the planned path by the joint controller. Considering the complex disturbance factors inside the spacecraft, such as the uncertainty of the system model and the sudden changes in the dynamics of the system, the joint controller needs sufficient robustness and fast adaptability. Single robust control [33–35] or adaptive control [36–38] cannot meet the needs of capturing non-cooperative target control. In the currently developed hybrid robust adaptive control strategies [39–42], the adaptive control part (neural network [43], fuzzy theory, etc.) is mainly aimed at the uncertain part of the system and the external disturbance part. The robust control part mainly performs secondary compensation for the residual error of the adaptive process. However, the traditional robust adaptive controller focuses on the control accuracy and lacks in-depth research on the control efficiency and tracking speed. This is very important to overcome the sudden changes in system dynamic parameters caused by non-cooperative targets. Some scholars have proposed adaptive controllers that use more efficient optimization algorithms, such as extreme learning machines [44], which show extremely fast learning and recognition speed for the uncertain part of the system compared to traditional neural networks, and so they are widely used for identification and control of systems with large uncertainties [45, 46]. However, the random generation of the network input weight elements of the extreme learning machine often results in poor performance of the algorithm. Therefore, some improved extreme learning machine algorithms [47–49] have emerged, which mainly improved the input weight elements through particle swarm optimization, making them updated towards optimal direction. At present, there are still a few related researches on the path planning in the process of identification. Due to the large uncertain factors in the manipulator system caused by the non-cooperative targets after capturing space, the adaptive reactionless trajectory planning based on online update combined with joint control tracking needs further research.

6.2 Joint kinetic model of spacecraft and unknown object

The identification of inertia parameters of non-cooperative targets in the post-capture requires complete system kinematics information, such as relative position and velocity, and dynamic information of the system and the estimation of the coupling moment. In addition, when the system is in a free-floating state, the linear angular momentum of the entire system and the Jacobian matrix are used for path planning of the robotic arm and parameter identification of non-cooperative targets. This section is based on the basic physical theories such as kinematics and dynamics involved in the identification. Relative physical quantities are deduced and explained in detail as the basic for the later identification, reactionless path planning, and control strategy.

6.2.1   System kinematic analysis

6.2.1.1   System position vector analysis

The displacement vector $\mathbf{R}_p$ of any point P relative to the coordinate system on a space multi-degree-of-freedom multi-arm robot can be expressed in different coordinate systems (inertial system or base system). For the solution of the space dynamics equations and the linear velocity, linear momentum, and Jacobian matrix, it is necessary to obtain the space position vectors of the mass centres and joints of each part of the system relative to the inertial system or the body system. The kinematic relationship of the system is solved based on the direct path method. First, the displacement vector of the spacecraft base's centre of mass with respect to the inertial system is set as $\mathbf{R}_B$. The displacement vector of the centre of mass C_k of the link k (k = 1, 2,..., N) on the robot arm m (m = 1, 2, 3,..., M) relative to the spacecraft base system is $\mathbf{r}_{mk}$. The displacement vector of the joint k relative to the spacecraft base system is $\mathbf{p}_{mk}$. The displacement of centre of mass of the link k relative to the base frame $\mathbf{R}_{mk}$ can be expressed as

$$\mathbf{R}_{mk} = \mathbf{R}_B + \mathbf{r}_{mk} \tag{6.1}$$

where $\mathbf{r}_{mk}$ can be expressed as:

$$\mathbf{r}_{mk} = \mathbf{b}_{m0} + \sum_{j=1}^{k-1} \left(\mathbf{a}_{mj} + \mathbf{b}_{mj} \right) + \mathbf{a}_{mk} \tag{6.2}$$

where $\mathbf{a}_{mj}$ and $\mathbf{b}_{mj}$ (j = 1, 2, ..., N) describe the displacement of joint j and $j+1$ relative to the mass centre C_j.

6.2.1.2   System velocity vector analysis

Based on the above analysis, velocity of C_k can be obtained by differentiating (6.1) and (6.2) and expressed as

$$\mathbf{v}_{mk} = \dot{\mathbf{r}}_{mk} = \mathbf{v}_B + \boldsymbol{\omega}_{mk} \times \mathbf{r}_{mk} \tag{6.3}$$

where $\mathbf{v}_B$ is the velocity of the centre of mass.

The angular velocity of kth link on the mth $^{mk}\boldsymbol{\omega}_{mk}$ arm can be expressed in the kth joint fixed frame as

$$^{mk}\boldsymbol{\omega}_{mk} = \mathbf{T}_{mk} \cdot \boldsymbol{\omega}_B + \sum_{j=1}^{k} \dot{\theta}_{mj}, \ (m = 1, 2, 3, ..., M), (k = 1, 2, ..., N) \tag{6.4}$$

Equation (6.4) is calculated in the inertial coordinate, where $\boldsymbol{\omega}_B$ is the angular velocity of the spacecraft. $\dot{\theta}_{mk}$ is the angular velocity of the kth joint on the mth arm described under the kth joint fixed frame. $\mathbf{T}_{mk}$ is the transformation matrix from inertial coordinate to the kth joint fixed frame on the mth arm.

Based on the link recursive relationship, the acceleration of kth joint on the mth arm can be expressed, respectively, in joint fixed frame and inertia coordinate as

$$\mathbf{a}_{cmk} = T_{1\_12} \cdot \left(^{m(k-1)}\omega_{m(k-1)} \times \left(^{m(k-1)}\omega_{m(k-1)} \times b_{m(k-1)} \right) + ^{m(k-1)}\omega_{m(k-1)} \times b_{m(k-1)} + \mathbf{a}_{cm(k-1)} \right)$$
$$+ ^{mk}\dot{\omega}_{mk} \times a_{mk} + ^{mk}\omega_{mk} \times \left(^{mk}\omega_{mk} \times a_{mk} \right)$$

$$\tag{6.5}$$

$$A_{cmk} = T_{mk}^{T} \cdot a_{cmk} \tag{6.6}$$

6.2.1.3   Velocity Jacobian matrix

To obtain velocity Jacobian matrix, the linear and angular velocity are expressed as

$$\left\{ \begin{array}{c} v_{mk} \\ \omega_{mk} \end{array} \right\} = J_{km} \cdot v \tag{6.7}$$

where v is the generalized velocity:

$$v = (v_B^T, \ \omega_B^T, \ \dot{\theta}_{11}, \ ..., \dot{\theta}_{1N}, \ \dot{\theta}_{21}, \ ..., \ \dot{\theta}_{2N}, \ ....., \ \dot{\theta}_{MN})^T \tag{6.8}$$

J_{km} is Jacobian matrix and can be obtained through (6.5) and (6.6) as

$$J_{km} = \begin{bmatrix} I_{3\times3} & J_1 & J_2 \\ 0_{3\times3} & I_{3\times3} & J_3 \end{bmatrix} \tag{6.9}$$

where

$$J_1 = -[T_0 \cdot^0 r_0^{(m)} + \sum_{i=1}^{k} [T_{mi}(a_{mi} + b_{mi})]]^\times \tag{6.10}$$

$$J_2 = - \left(\sum_{i=1}^{k-1} \left[T_{mk} \cdot \left(a_{mi} + b_{mi} \right) \right]^\times \cdot E_{mi} \right) - T_{mk} \cdot a_{mi}^\times E_{mk} \tag{6.11}$$

$$J_3 = E_{mk} \tag{6.12}$$

where T_0 is the transformation matrix from inertial coordinate to the base frame and T_{mk} is the transformation matrix from inertial coordinate to the kth joint fixed frame on the mth arm. $[\bullet]^\times$ is defined as

$$[r]^\times = \begin{bmatrix} 0 & -r_z & r_y \\ r_z & 0 & -r_x \\ -r_y & r_x & 0 \end{bmatrix} \tag{6.13}$$

E_{mk} is defined as

$$E_{mk} = \begin{bmatrix} 0_{3\times b} & T_{m1} \cdot z & ... & T_{mk} \cdot z & 0 \end{bmatrix}_{3\times K} \tag{6.14}$$

where

$$b = \sum_{l=1}^{m-1} N_l \tag{6.15}$$

$$z = [0 \ \ 0 \ \ 1]^T \tag{6.16}$$

6.2.1.4   System linear and angular momentum calculation

In the free-floating space system (zero initial condition), its linear angular momentum is conserved, satisfying the following relationship:

$$P = m_B \cdot v_B + \sum m_{mk} \cdot v_{mk} = 0 \tag{6.17}$$

$$L = I_B \cdot \omega_B + r_B \times m_B \cdot v_B + \sum I_{mk} \cdot \omega_{mk} + \sum r_{mk} \times m_{mk} \cdot v_{mk} = 0 \tag{6.18}$$

Equations (6.9)–(6.12) are taken into account and the following can be obtained:

$$\begin{pmatrix} P \\ L \end{pmatrix} = \begin{pmatrix} JP1 & JP2 \\ JL1 & JL2 \end{pmatrix} \begin{pmatrix} v_B \\ \omega_B \end{pmatrix} + \begin{pmatrix} JS1 \\ JS2 \end{pmatrix} \dot{\theta} \tag{6.19}$$

where

$$JP1 = M \cdot I_{3\times3} \tag{6.20}$$

$$JP2 = \sum_{m-1}^{M} \sum_{k-1}^{N} J_1(m, k) \tag{6.21}$$

$$JL1 = m_B \cdot r_B{}^\times + \sum_{m=1}^{M} \sum_{k=1}^{N} m_{mk} r_{mk}{}^\times \tag{6.22}$$

$$JL2 = I_B + \sum_{m=1}^{M} \sum_{k=1}^{N} m_{mk} r_{mk}{}^\times J_1(m, k) + \sum_{i=1}^{m} \sum_{j=1}^{k} T_{ij}{}^T \cdot I_{ij} \cdot T_{ij} \tag{6.23}$$

$$JS1 = \sum_{m=1}^{M} \sum_{k=1}^{N} m_{mk} J_2(m, k) \tag{6.24}$$

$$JS2 = \sum_{m=1}^{M} \sum_{k=1}^{N} m_{mk} r_{mk}{}^\times J_2(m, k) + \sum_{j=1}^{m} \sum_{j=1}^{k} T_{ij}{}^T \cdot I_{ij} \cdot T_{ij} \cdot J_3(i, j) \tag{6.25}$$

$$\dot{\theta} = (\dot{\theta}_{11}, ..., \dot{\theta}_{1N}, \dot{\theta}_{21}, ..., \dot{\theta}_{2N}, ..., \dot{\theta}_{MN})^T \tag{6.26}$$

where M is the total mass of the system, m_B is the base mass, I_B is the moment of inertia, m_{mk} and I_{mk} are the mass and moment of inertia of the kth link on the mth arm.

6.2.2 System kinetic analysis

Each link's kinetic equation can be obtained through Newton-Euler formula as

$$\begin{cases} F_{mk} = m_{mk} A_{cmk} + F_{m(k+1)} \\ N_{cmk} = I_{mk} \cdot{}^{mk} \dot{\omega}_{mk} + {}^{mk} \omega_{mk} \times I_{mk} \cdot{}^{mk} \omega_{mk} \\ N_{mk} = a_{mk} \times {}^{mk} F_{mk} + b_{mk} \times {}^{mk} F_{m(k+1)} + N_{cmk} \\ n_{mk} = T_{mn}{}^T \cdot N_{mk} \end{cases} \tag{6.27}$$

where ${}^{mk}\omega_{mk}$ and ${}^{mk}\dot{\omega}_{mk}$ are the angular velocity and angular acceleration of the kth link on the mth arm, A_{cmk} is the linear acceleration of the link mass, F_{mk} and ${}^{mk}F_{mk}$ are the force at the mass centre under inertial coordinate and joint fixed frame, and N_{mk} and N_{cmk} are the torque of the link under inertial coordinate and joint fixed frame.

The standard kinetic equation can be obtained from (6.27) as

$$H(q) \ddot{q} + C(q, \dot{q}) \dot{q} + d = \tau \tag{6.28}$$

The matrix $H(q)$ represents the system dynamics inertia matrix. $C(q, \dot{q})$ is the coupling term including the system centrifugal force and the Coriolis force. d represents the disturbance existing inside the system. τ is the defined generalized force, that is, the driving torque of each joint of the robot arm. q is the defined generalized coordinate and satisfies (6.29).

$$q = \left(\theta_{11}, ..., \theta_{1N}, \theta_{21}, ..., \theta_{2N}, ..., \theta_{MN}\right)^{T} \tag{6.29}$$

Now consider the standard equations of dynamics of the actual system after capturing non-cooperative targets:

$$\hat{H}(q)\ddot{q} + \hat{C}(q,\dot{q})\dot{q} + d = \tau \tag{6.30}$$

$\hat{H}(q)$ and $\hat{C}(q,\dot{q})$ both contain unknown attributes for non-cooperative targets. Therefore, the matrix uncertainty can be defined as follows

$$\begin{cases} \tilde{H}(q) = \hat{H}(q) - H(q) \\ \tilde{C}(q,\dot{q}) = \hat{C}(q,\dot{q}) - C(q,\dot{q}) \end{cases} \tag{6.31}$$

According to the basic physical relationships such as kinematics and dynamics involved in the identification method of inertia parameters of space non-cooperative targets, this section explains the kincmatics analysis and generalized dynamics model of the system as the research basis and premise of the subsequent parameter identification theory and method.

6.3 Unknown object inertial parameter identification

6.3.1 Basic theory of identification

A rigid robotic system composed of a spacecraft, manipulator (*n*-degree-of-freedom), and space unknown object in the post-capture phase is derived in this section, as shown in Figure 6.3.

The condition of conservation of momentum can be fulfilled with a space robotic system, considering a spacecraft in a free-floating state and neglecting the effect of microgravity in a short time. From Figure 6.3, Σ_I, Σ_B, Σ_i, and Σ_U are the inertial frame and the body-fixed frames of the spacecraft, link *i*, and unknown object, respectively; $p_B(p_0)$, p_i, $p_U(p_{n+1})$ are the position vectors of the reference point of the body frame on the spacecraft, joint *i*, and unknown object, respectively, expressed

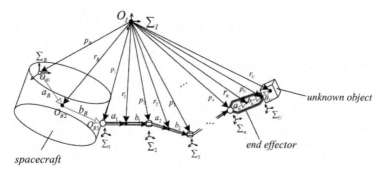

Figure 6.3   *Spacecraft-manipulator-unknown object system (space robotic system)[50]*

in Σ_I; r_B (r_0), r_i, and r_U are the position vectors of the centroids of the spacecraft, link i, and unknown object, respectively, expressed in Σ_I; $\omega_B(\omega_0)$, ω_i, and ω_U are the angular velocity vectors of the spacecraft, link i, and unknown object, respectively, expressed in Σ_I; and θ_i is the angle vector formed by the output of the encoder installed in joint i, expressed in Σ_I. As described in Section 6.2

$$a_i = r_i - p_i \tag{6.32}$$

$$b_i = p_{i+1} - r_i \tag{6.33}$$

$$a_U = r_U - p_U \tag{6.34}$$

$$\omega_i = \omega_{i-1} + \dot{\theta}_i \tag{6.35}$$

$$\dot{r}_i = \dot{p}_i + \omega_i \times a_i \tag{6.36}$$

$$\dot{p}_i = \dot{r}_{i-1} + \omega_{i-1} \times b_{i-1} \tag{6.37}$$

$$\omega_U = \omega_n \tag{6.38}$$

$$\dot{P}_U = \dot{r}_n + \omega_n \times b_n \tag{6.39}$$

$$\dot{r}_U = \dot{P}_U + \omega_U \times a_U \tag{6.40}$$

The kinematics of the space manipulator system in the post-capture phase is considered here and the system is assumed to be a rigid mechanism. The use of reaction wheels and other momentum exchange devices is not considered in the study. The captured target at the end is fixed to the end manipulator. The linear and angular momentum of the entire space robot system is set to zero. Therefore, (6.41)–(6.43) can be illustrated as follows:

$$P = m_B\dot{r}_B + \sum_{i=1}^{n} m_i\dot{r}_i + m_U\dot{r}_U \tag{6.41}$$

$$L = I_B\omega_B + \sum_{i=1}^{n} I_i\omega_i + I_U\omega_U + r_B \times m_B\dot{r}_B + \sum_{i=1}^{n} r_i \times m_i\dot{r}_i + r_U \times m_U\dot{r}_U \tag{6.42}$$

$$\left\{ \begin{array}{c} -{}^U\dot{p}_U \\ {}^U L_K \end{array} \right\} = \left\{ \begin{array}{cc} {}^U P_K[{}^U\omega_U^\times] & 0 \\ 0 & [{}^U P_K^\times][{}^{\#U}\omega_U] \end{array} \right\} \left\{ \begin{array}{c} \frac{1}{m_U} \\ {}^U a_U \\ {}^U I_U^\# \end{array} \right\} \tag{6.43}$$

From (6.41) and (6.43), P and L denote the linear momentum and angular momentum of the whole space robotic system, respectively, expressed in the inertial frame; m_B (m_0), m_i, and m_U are the masses of the spacecraft, link i, and unknown object, respectively. $I_B(I_0)$, I_i, and I_U are the inertial tensors relative to the individual centroids of the spacecraft, link i, and unknown object, respectively. From the linear identification (6.43), m_U, a_U, and I_U are the inertial parameters of the unknown object to be identified, and ${}^U P_K$, ${}^U L_K$, ${}^U\dot{p}_U$, ${}^U\omega_U$, ${}^U a_U$, and ${}^U I_U$ are matrices (vectors) of P_K, L_K, $\dot{p}_U$, ω_U, a_U, and I_U in the inertial frame, expressed in Σ_U. From [17], P_K, L_K, $\omega_U^\times$, $\#\omega_U$, and $I_U^\#$ can be described as

$$P_K = m_B\dot{r}_B + \sum_{i=1}^{n} m_i\dot{r}_i \tag{6.44}$$

$$\boldsymbol{L_K} = -\boldsymbol{I_B}\boldsymbol{\omega_B} - \sum_{i=1}^{n} \boldsymbol{I_i}\boldsymbol{\omega_i} - m_B \dot{\boldsymbol{r}}_B \times \left(\boldsymbol{b_B} + \sum_{i=1}^{n}(\boldsymbol{a_i} + \boldsymbol{b_i})\right) -$$
$$\sum_{i=1}^{n} m_i \dot{\boldsymbol{r}}_i \times (\boldsymbol{L_i}) \tag{6.45}$$
$$\boldsymbol{L_i} = \boldsymbol{b_i} + \sum_{j=i+1}^{n}(\boldsymbol{a_j} + \boldsymbol{b_j})$$

$$\boldsymbol{\omega_U}^{\times} = \begin{bmatrix} 0 & -\omega_{Uz} & \omega_{Uy} \\ \omega_{Uz} & 0 & -\omega_{Ux} \\ -\omega_{Uy} & \omega_{Ux} & 0 \end{bmatrix} \tag{6.46}$$

$$\boldsymbol{I_U^{\#}} = \begin{bmatrix} I_{Uxx}, & I_{Uxy}, & I_{Uxz}, & I_{Uyy}, & I_{Uyz}, & I_{Uzz} \end{bmatrix}^T \tag{6.47}$$

Equation (6.43) can be decomposed as

$$\begin{cases} -\dot{p}_{Ux} = P_{Kx} \cdot \dfrac{1}{m_U} - \omega_{Uz} \cdot a_{Uy} + \omega_{Uy} \cdot a_{Uz} \\[2mm] -\dot{p}_{Uy} = P_{Ky} \cdot \dfrac{1}{m_U} + \omega_{Uz} \cdot a_{Ux} - \omega_{Ux} \cdot a_{Uz} \\[2mm] -\dot{p}_{Uz} = P_{Kz} \cdot \dfrac{1}{m_U} - \omega_{Uy} \cdot a_{Ux} + \omega_{Ux} \cdot a_{Uy} \end{cases} \tag{6.48}$$

$$\begin{cases} L_{Kx} = -P_{Kz} \cdot a_{Uy} + P_{Ky} \cdot a_{Uz} + \omega_{Ux} \cdot I_{Uxx} + \omega_{Uy} \cdot I_{Uxy} + \omega_{Uz} \cdot I_{Uxz} \\ L_{Ky} = P_{Kz} \cdot a_{Ux} - P_{Kx} \cdot a_{Uz} + \omega_{Ux} \cdot I_{Uxy} + \omega_{Uy} \cdot I_{Uyy} + \omega_{Uz} \cdot I_{Uyz} \\ L_{Kz} = -P_{Ky} \cdot a_{Ux} + P_{Kx} \cdot a_{Uy} + \omega_{Ux} \cdot I_{Uxz} + \omega_{Uy} \cdot I_{Uyz} + \omega_{Uz} \cdot I_{Uzz} \end{cases} \tag{6.49}$$

Two crucial conclusions can be illustrated.

First, to ensure that the coefficient matrix of the linear identification equation (6.43) is in a non-singular state, there must exist kinematic information on all the translational and rotational motions from the three orthogonal axes in the inertial frame to maintain the integrity of the identification for all inertial parameters. On account of that the motions of the three orthogonal axes in the inertial frame can be decomposed into the motions of three axes from an arbitrary body frame in the space robotic system, also incorporating Σ_U, the necessary condition of intact identification can be stated using existing translation and rotation motions in all three orthogonal directions from the body frame of the unknown object, which can be ensured by actuating joints of various directions in the post-capture phase or by the collision motivation in the capture-collision phase for the identification of the unknown object.

Second, linear (6.43) contains ten inertial parameters ($m_U, a_{Ux}, a_{Uy}, a_{Uz}, I_U^{\#}$) to be identified, with strong coupling features of unknown parameters. The ten inertial parameters can be ideally obtained by the perfect coefficient matrix (or vector) of linear (6.43), which determines the accuracy of the identified parameters. However, as they are inevitably impacted by measurement or estimation errors, inaccurate coefficients cause poor accuracy of the coupling of the inertial parameters (6.43). To address the strong coupling of linear (6.43), a practical decoupling procedure is essential, and the linear identification (6.43) can thus be decomposed into (6.50) and (6.51). Therefore, a two-step procedure of identification can be implemented to improve the accuracy of the inertial parameter identification.

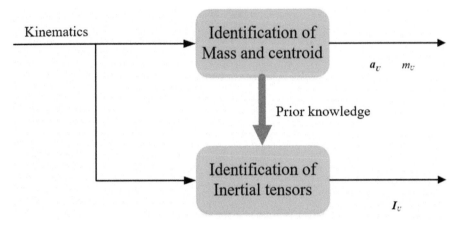

Figure 6.4   Two-step identification [50]

$$-\dot{p}_U = P_K \cdot \frac{1}{m_U} + [\omega_U{}^\times] \cdot a_U \qquad (6.50)$$

$$L_K - [P_K{}^\times] \cdot a_U = [\#\omega_U] \cdot [I_U^{\#}] \qquad (6.51)$$

6.3.2   Identification scheme incorporating information of contact force together with force/torque of end-effector

In fact, there exist estimation errors of the spacecraft and manipulator that can be accumulated to the coefficient matrix or vectors of the identification equation in a similar way as the accumulation by measurement errors. Therefore, compared with acquiring coefficient information through indirect deduction from the spacecraft and manipulator, with the error accumulation process introducing more uncertainty, a straightaway mode of measurement without too much accumulated calculation is essential and can improve the identification accuracy. According to [50], ΔP_K and ΔL_K cause the greatest distortions to the identified inertial parameters, containing accumulated errors of both the kinematic and inertial properties of the spacecraft and manipulator. To reduce the errors of ΔP_K and ΔL_K caused by the accumulated calculation from measurement or estimation errors, a new identification scheme incorporating the contact force together with the force/torque of the end-effector is employed to improve the identification precision, as shown in Figure 6.4.

Figure 6.5 shows the force of the end manipulator based on the post-capture stage, where f_U and n_U represent the external force and moment acting on the contact point of the unknown object; f_n/n_n represent the force/torque acting on the endjoint of the end effector. In practice, f_U, f_n and n_n can be measured by tactile (Figure 6.3) sensors and force/torque sensors. The derivations of the rest of the mechanical information are indicated in (6.52)–(6.55).

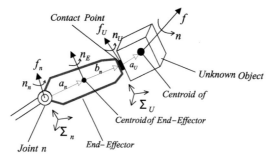

Figure 6.5 Force analysis of end-effector [50]

$$\tilde{\boldsymbol{f}} = \tilde{\boldsymbol{f}}_U \tag{6.52}$$

$$\tilde{\boldsymbol{n}} = \tilde{\boldsymbol{n}}_U - \boldsymbol{a}_U \times \tilde{\boldsymbol{f}}_U \tag{6.53}$$

$$\tilde{\boldsymbol{n}}_U = \tilde{\boldsymbol{n}}_n - \tilde{\boldsymbol{n}}_E - \boldsymbol{a}_n \times \tilde{\boldsymbol{f}}_n - \boldsymbol{b}_n \times \tilde{\boldsymbol{f}}_U \tag{6.54}$$

$$\begin{cases} \tilde{\boldsymbol{n}}_E = I_E \tilde{\dot{\boldsymbol{\omega}}}_E + \tilde{\boldsymbol{\omega}}_E \times (I_E \tilde{\boldsymbol{\omega}}_E) \\ \tilde{\boldsymbol{\omega}}_E = \tilde{\boldsymbol{\omega}}_U = \tilde{\boldsymbol{\omega}}_n \end{cases} \tag{6.55}$$

Through (6.52)–(6.55), an improved parameter identification equation can be obtained.

From (6.41)–(6.42) and (6.44)–(6.45), the momentum of the system can be decomposed into (2.56) and (2.57):

$$\begin{cases} \hat{\boldsymbol{P}} = \hat{\boldsymbol{P}}_K + \hat{\boldsymbol{P}}_U \\ \hat{\boldsymbol{P}}_U = m_U \tilde{\dot{\boldsymbol{r}}}_U \end{cases} \tag{6.56}$$

$$\begin{cases} \tilde{\boldsymbol{L}} = -\tilde{\boldsymbol{L}}_K + \tilde{\boldsymbol{L}}_U + \tilde{\boldsymbol{P}}_K \times \boldsymbol{a}_U \\ \tilde{\boldsymbol{L}}_U = I_U \tilde{\boldsymbol{\omega}}_U \end{cases} \tag{6.57}$$

Differentiate (6.56) and (6.57) and calculate the increment in a short time to get the final identification equations as:

$$\begin{cases} d\tilde{\boldsymbol{P}}_K = -\tilde{\boldsymbol{f}} \cdot dt \\ d\tilde{\boldsymbol{L}}_K - d\left(\tilde{\boldsymbol{P}}_K \times \boldsymbol{a}_U\right) = d\tilde{\boldsymbol{L}}_K - d\left(\tilde{\boldsymbol{P}}_K^{\times} \cdot \boldsymbol{a}_U\right) = \tilde{\boldsymbol{n}} \cdot dt \end{cases} \tag{6.58}$$

$$\begin{cases} \delta\tilde{\boldsymbol{P}}_K = -\int_0^{\delta t} \tilde{\boldsymbol{f}} \cdot dt \\ \delta\tilde{\boldsymbol{L}}_K - \delta\left(\tilde{\boldsymbol{P}}_K \times \boldsymbol{a}_U\right) = \delta\tilde{\boldsymbol{L}}_K - \delta\left(\tilde{\boldsymbol{P}}_K^{\times} \cdot \boldsymbol{a}_U\right) = \int_0^{\delta t} \tilde{\boldsymbol{n}} \cdot dt \end{cases} \tag{6.59}$$

6.3.3   Solution of the modified identification equation using the hybrid RLS-APSA algorithm

Equation (6.59) can be used to construct linear equations and their simplified forms (6.62). Equation (6.62) represents the non-cooperative target inertia parameter vector. A and b represent the corresponding coefficient matrix and vector.

$$A_1 \Omega_1 = b_1$$

$$A_1 = \left[-\int_0^{\delta t} f_U \cdot dt \ \left[(\delta\omega_U)^\times \right] \right] \quad b_1 = -\delta \dot{p}_U \quad \Omega_1 = \begin{bmatrix} \frac{1}{m_U} \\ a_U \end{bmatrix} \tag{6.60}$$

$$A_2 \Omega_2 b_2$$
$$A_2 = [\#(\delta\omega_U)] \quad b_2 = \int_0^{\delta t} (n_n - a_n \times f_n - b_n \times f_U - a_U \times f_U) dt - I_E \omega_U \tag{6.61}$$
$$\Omega_2 = [I_U^\#]$$

$$A\Omega = b \tag{6.62}$$

This section proposes a hybrid algorithm [51–53], which includes two parts: RLS and APSA (affine projection symbol algorithm). The main purpose of RLS [54] is to maintain convergence speed and reduce the correlation of measurement data, and to quickly obtain approximate results of identification parameters, whereas APSA [55] has the function of immune to coloured noise and impulsive noise. At the same time, it has low computational complexity but can only maintain a slow convergence rate. RLS and APSA standard regression forms are shown as

$$\Omega_{k+1} = \hat{\Omega}_k + K_{(k+1)}[b_{k+1} - A_{(k+1)}\hat{\Omega}_{(k)}]$$
$$K_{k+1} = \frac{P_{(k)} A_{(k+1)}}{\lambda I + A_{(k+1)}^T P_{(k)} A_{(k+1)}}$$
$$P_{(k+1)} = \frac{1}{\lambda}[I - K_{(k+1)} A_{(k+1)}^T] P_{(k)} \tag{6.63}$$
$$\hat{\Omega}_{(k+1)} = \hat{\Omega}_{(k)} + \mu \frac{X_{(k)}\mathrm{sgn}(E_{(k)})}{\sqrt{\mathrm{sgn}(E^T_{(k)}) \cdot X_{(k)} \cdot X_{(k)}^T \cdot \mathrm{sgn}(E_{(k)}) + \varepsilon}}$$
$$X_{(k)} = [A_{(k)}; A_{(k-1)}; A_{(k-2)} ... A_{(k-M+1)}]$$

$$E_{(k)} = b_{(k)} - X_{(k)}^T \cdot \Omega_{(k)} \tag{6.64}$$

In (6.63)–(6.64), $A_{(k)}$ and $b_{(k)}$ are the k-time state of the matrix/vector A *and* b. $\Omega_{(k)}$ is the estimated value of k-time of Ω. In (6.63), the forgetting factor λ adjusts the convergence rate and weakens the influence of historical data on the current data. In (6.64), the step factor μ and the adjustment factor ε (take a small positive number) adjust the convergence speed of APSA. M represents the $(k-M)$th measurement time.

To merge the two algorithms, a switching mechanism is needed to switch the two algorithms. The switching mechanism is designed as:

$$\left| \frac{d}{dt} \|e_{(k)}\|_2^2 \right| \underset{APSA}{\overset{RLS}{\gtrless}} \rho_0 \tag{6.65}$$

$$e_{(k)} = b_{(k+1)} - A_{(k+1)} \hat{\Omega}_{(k)} \qquad (6.66)$$

6.4 Adaptive reactionless control strategy during manipulation of unknown object

Section 6.3 mainly analysed the basic theory of parameter identification. Through theoretical analysis, it was found that for the complete identification of the inertia parameters, the space manipulator in the post-capture stage needs to have different orthogonal directions of movement. However, due to the coupling effect of the system, the movement of the manipulator will inevitably cause the attitude of the spacecraft base body unstable. To ensure that the motion of the robotic arm minimizes the attitude disturbance of the spacecraft base, this section focuses on the control strategy of the robotic arm based on adaptive reactionless trajectory planning to achieve its own disturbance-free motion. At the same time, considering the system's attitude stabilization during the identification, this section has conducted a study on the system's coordinated control method.

To realize the reactionless motion of the robotic arm, the integrated adaptive reactionless control strategy involves adaptive path planning and the corresponding control strategy. The path planning would be adaptively updated by the SW(Slide Window)-RLS(Recursive Least Squares) because of the existing time-variable elements in the coefficient matrix of the ARNS (Adaptive Reaction Null Space) algorithm. In addition, the robust adaptive control strategy via PSO(Particle swarm optimization)-ELM(Extreme Learning Machine) algorithm is proposed to fast-track the planned adaptive path of the robotic arm to ensure enough robustness and the adaptive nature of the system control performance. The schematic of the proposed adaptive reactionless control strategy is illustrated in Figure 6.6 and Figure 6.7.

6.4.1 Adaptive reactionless path planning via SW-RLS

The space manipulator system follows the principle of conservation of linear angular momentum. According to the theory in Section 6.2, the expression of linear and angular momentum of the system can be expressed as

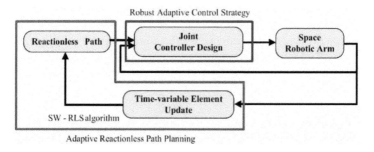

Figure 6.6   Adaptive reactionless control strategy [35]

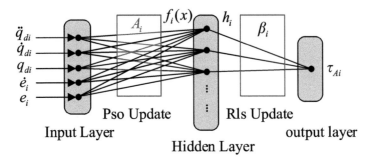

Figure 6.7   *ELM network diagram [57]*

$$\begin{bmatrix} P \\ L \end{bmatrix} = \begin{bmatrix} mI & m\tilde{r}_{0g}^T \\ 0 & H_\omega \end{bmatrix} \begin{bmatrix} v_0 \\ \omega_0 \end{bmatrix} + \begin{bmatrix} J_{T\omega} \\ H_{\omega\phi} \end{bmatrix} \dot{q} + \begin{bmatrix} 0 \\ r_g \times P \end{bmatrix} \tag{6.67}$$

In the above expression, v_0 and ω_0 represent the linear velocity and angular velocity of the spacecraft, respectively, expressed in the inertial frame; $\dot{q}$ represents the derivation of the generalized variable q, i.e., joint rate; m denotes the total mass of the robotic system; and I is the identity matrix. The coefficient matrix of angular velocity q and linear velocity $\dot{q}$ are based on the inertial matrix and coupling inertial matrix, accordingly.

Given the angular momentum of the robotic system from (6.67), with initial momentum is expressed as

$$L = H_\omega \omega_0 + H_{\omega\phi}\dot{q} + r_g \times P_0 = L_0 \tag{6.68}$$

i.e.,

$$H_\omega + \omega_0 + H_{\omega\phi}\dot{q} = L_0 - r_g \times P_0 \tag{6.69}$$

Based on the RNS algorithm [56], the desired path planning of the robotic arm to realize reactionless motion for the spacecraft can be obtained by (6.70),

$$\dot{q}_{d|RNS} = H_{\omega\phi}^+(L_0 - r_g \times P_0) + (I - H_{\omega\phi}^+ H_{\omega\phi})\dot{\zeta} \tag{6.70}$$

where $\dot{\zeta}$ is the arbitrary vector and follows $\dot{\zeta} \in \mathbf{R}^n$.

Then, considering the captured unknown object with unknown properties, the coefficient matrix (base inertial matrix, coupling inertial matrix) incorporating the uncertain dynamic parameter cannot be obtained precisely, which affects the desired path to realize the absolutely reactionless motion. Therefore, the ARNS algorithm is employed to realize the adjustment online for the coefficient matrix to obtain the precise property of the post-capture system.

Equation (6.69) is rewritten, illustrated by (6.71), in which the signal ∧ represents the current estimate value that incorporates the dynamic parameter of the captured unknown object.

$$\dot{q} = \hat{H}_{w\phi}^+(L_0 - r_g \times P_0) - \hat{H}_{w\phi}^+ \hat{H}_\omega \omega_0 + (I - \hat{H}_{w\phi}^+ \hat{H}_{w\phi})\dot{\zeta} \tag{6.71}$$

Then, (6.70) and (6.71) are combined and expressed as

$$
\dot{q}_{d/RNS} - \dot{q} = (H_{w\phi}^{+} - \hat{H}_{w\phi}^{+})(L_0 - r_g \times P_0) + \hat{H}_{w\phi}^{+}\hat{H}_{w\phi} - H_{w\phi}^{+}H_{w\phi}\dot{\zeta}
$$

$$
= [K_1 \ K_2 \ K_3] \begin{bmatrix} 1 \\ \omega_0 \\ \dot{\zeta} \end{bmatrix}
\tag{6.72}
$$

where

$$
\begin{cases}
K_1 = \left(H_{w\phi}^{+} - \hat{H}_{w\phi}^{+}\right)\left(L_0 - r_g \times P_0\right) \\
\qquad K_2 = \hat{H}_{w\phi}^{+}\hat{H}_\omega \\
K_3 = \left(\hat{H}_{w\phi}^{+}\hat{H}_{w\phi} - H_{w\phi}^{+}H_{w\phi}\right)
\end{cases}
\tag{6.73}
$$

The identification algorithm SW-RLS is used to obtain the coefficient matrix online to deal with the time-variable parameters from K_1, K_2, and K_3. The simplified linear regression form of (6.72) is defined as follows:

$$
\psi \Omega = \phi
\tag{6.74}
$$

where Ω denotes the coefficient matrix $[K_1 \ K_2 \ K_3]^\mathrm{T}$; Ψ, Φ denote the terms $[1 \ \omega_0 \ \dot{\zeta}]^T$ and $[\dot{q}_{d/RNS} - \dot{q}]^\mathrm{T}$, respectively.

The specific form of SW-RLS is expressed in (6.75), where k represents the kth solution of the corresponding element; u represents the length of the sliding window; E_1 and E_2 represent the prior and posterior errors. PN is the inverse of the autocorrelation matrix, and its initial conditions satisfy $\mathbf{P}_{(0)} = \delta \mathbf{I}$ and $\delta > 1$.

$$
\begin{aligned}
\varepsilon_1 &= \Phi_{(k+1)} - \Psi_{(k+1)}\hat{\Omega}_{(k)} \\
\Delta_{1(k+1)} &= \frac{PN_{(k)}\Psi_{(k+1)}}{1+\Psi_{(k+1)}{}^T PN_{(k)}\Psi_{(k+1)}} \\
PN_{1(k+1)} &= [I - \Delta_{1(k+1)}\Psi_{(k+1)}{}^T]PN_{(k)} \\
\hat{\Omega}_{1(k+1)} &= \hat{\Omega}_{(k)} + \Delta_{1(k+1)}\varepsilon_1 \\
\varepsilon_2 &= \Phi_{(k-u+1)} - \Psi_{(k-u+1)}\hat{\Omega}_{1(k+1)} \\
\Delta_{(k+1)} &= \frac{PN_{1(k+1)}\Psi_{(k-u+1)}}{-1+\Psi_{(k-u+1)}{}^T PN_{1(k+1)}\Psi_{(k-u+1)}} \\
PN_{(k+1)} &= [I - \Delta_{(k+1)}\Psi_{(k-u+1)}{}^T]PN_{1(k+1)} \\
\hat{\Omega}_{(k+1)} &= \hat{\Omega}_{1(k+1)} + \Delta_{(k+1)}\varepsilon_2
\end{aligned}
\tag{6.75}
$$

6.4.2 Robust adaptive control strategy via the PSO-ELM algorithm

6.4.2.1 Adaptive control term via PSO-ELM algorithm

In this section, the robust adaptive control strategy using the PSO-ELM algorithm is proposed. To compensate for the uncertain term $D'(i)$ for the ith subsystem and to realize the desired tracking for each joint, the proposed control strategy for each joint is illustrated in Figure 6.5.

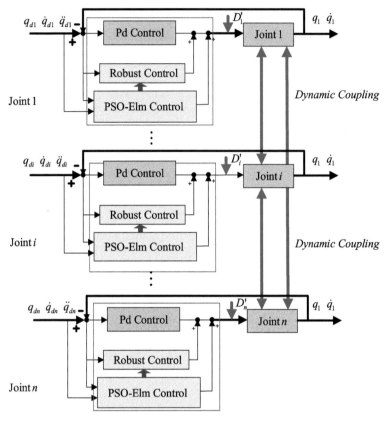

Figure 6.8   Proposed control strategy for the joint controller

The ELM algorithm is employed to construct an adaptive control strategy to quickly compensate for the abrupt properties of the control system. The PSO algorithm is employed to optimize the input feature of the ELM network to address any input feature variation from the optimal value. The schematic of the PSO-ELM algorithm is illustrated in Figure 6.8.

The input vector x_i is defined as

$$x_i = \begin{bmatrix} \ddot{q}_{di} & \dot{q}_{di} & q_{di} & e_i & \dot{e}_i \end{bmatrix} \tag{6.76}$$

Therefore, the input and output for the hidden layer can be obtained as

$$f_i(x_i) = A_i x_i \tag{6.77}$$

$$h_i = g_i(f_i(x_i) + b_i) = g_i(A_i x_i + b_i) \tag{6.78}$$

where A_i is input weight matrix, b_i is hidden bias layer, and g_i is activation function, normally sigmoid function.

The output of the network can be obtained via (6.79), with the output weight matrix βi

$$\tau_{Ai} = \boldsymbol{h}_i \boldsymbol{\beta}_i \tag{6.79}$$

The proposed adaptive law for $\boldsymbol{\beta}_i$ is given as

$$\begin{aligned}
\dot{\hat{\boldsymbol{\beta}}}_i &= [\dot{\hat{\boldsymbol{\beta}}}_{i1} \; \dot{\hat{\boldsymbol{\beta}}}_{i2} ... \dot{\hat{\boldsymbol{\beta}}}_{ij} ...] \\
\dot{\hat{\beta}}_{ij} &= \gamma_{ij} \boldsymbol{h}_i r_i \; (j = 1, 2, ...l) \\
\hat{\tau} &= \boldsymbol{h}_i \hat{\boldsymbol{\beta}}_i
\end{aligned} \tag{6.80}$$

where $\dot{\hat{\boldsymbol{\beta}}}_i$ denotes the estimated growth rate of the estimated weight matrix $\hat{\boldsymbol{\beta}}_i$; l is the number of hidden layer nodes; γ denotes the step-length regulatory factor; and $\hat{\tau}_{Ai}$ represents the estimated output torque, i.e., the output of the adaptive controller.

To avoid the input weight A_i and the hidden bias layer b_i of the ELM network varying from the optimal value, A_i and b_i elements are set as the particles requiring optimization, and these elements are reformed into the set S, with amount m as following

$$S = \left(s_1, s_2, ..., s_i, ...s_m\right)^T \tag{6.81}$$

Then, the PSO algorithm is used to optimize the particles S, according to the principle of obtaining the least value of the fitness function $F(S)$, which is chosen by the control tracking error e in this chapter. Thus, if the tracking error e exhibits a downtrend, the obtained particles move towards the optimal value. Then, the update via (6.82) is adopted; however, if the tracking error is on the rise, the original particle values are maintained.

$$\begin{aligned}
v_i^{k+1} &= w v_i^k + c_1 R_1 (p_i^k - S_i^k) + c_2 R_2 (g_i^k - S_i^k) \\
S_i^{k+1} &= S_i^k + v_i^{k+1}
\end{aligned} \tag{6.82}$$

From (6.82), elements s_i and v_i represent the position and velocity of particle i, respectively; p_i and g_i denote the best partial position and the best global position for particle i, respectively, obtained by the empirical data from the experimental tests; w denotes the inertial weight serving as a trade-off between the global and local exploration capabilities of the swarm; c_1 and c_2 are the weights of the stochastic acceleration terms, determined in advance and pulling each particle towards p_i and g_i; R_1 and R_2 are the random variables from the range of [0,1]; k denotes the kth iteration for the corresponding elements. To avoid overrun, the limitations for the position and velocity are set as $s_i \in \left[S_{min}, S_{max}\right]$ and $v_i \in \left[V_{min}, V_{max}\right]$, respectively.

6.4.2.2   Robust control strategy

The main purpose of the robust control term in this chapter is to reject model uncertainties. The residual errors are produced by the adaptive control process, shown as follows:

$$E_i = \tau_{Ai} - \tau_{Ai}^* = \boldsymbol{h}_i \boldsymbol{\beta}_i - \boldsymbol{h}_i \boldsymbol{\beta}_i^* \tag{6.83}$$

where τ_{Ai}^* denotes the optimal approximation of τ_{Ai}. Noting that there exists an upper bound of E_i with $|E_i| \leq \varepsilon_i$, the basic form of the robust control term can be illustrated, where τ_{Ri} represents the estimated output of the robust control term and r_i represents the ith element of system error r.

Therefore, the estimated output of the proposed control strategy $\tau\_$ for the robotic arm can be obtained by (6.84), in which the positive definite matrix $\mathbf{K}$ represents the gain of the PD control strategy.

$$\hat{\tau} = \begin{bmatrix} \hat{\tau}_1 & \hat{\tau}_2 \dots & \hat{\tau}_i \dots & \hat{\tau}_n \end{bmatrix}$$
$$\hat{\tau}_i = \hat{\tau}_{PDi} + \hat{\tau}_{Ai} + \hat{\tau}_{Ri} = \mathbf{K}r_i + \mathbf{h}_i\hat{\boldsymbol{\beta}}_i + \varepsilon \cdot sign(r_i) \tag{6.84}$$

6.4.2.3   Stability analysis of the proposed control strategy

Considering the following Lyapunov function

$$L = \tfrac{1}{2}\mathbf{r}^T \mathbf{M}\mathbf{r} + \tfrac{1}{2}\sum_{i=1}^{n}\sum_{j=1}^{l}\tfrac{1}{\gamma_{ij}}\tilde{\beta}_{ij}\tilde{\beta}_{ij}$$
$$\tilde{\beta}_{ij} = \hat{\beta}_{ij} - \beta_{ij}^* \tag{6.85}$$

The first derivation of (6.85) is derived through (6.84).

$$\dot{L} = \mathbf{r}^T \mathbf{M}\dot{\mathbf{r}} + \tfrac{1}{2}\mathbf{r}^T \dot{\mathbf{M}}\mathbf{r} + \sum_{i=1}^{n}\sum_{j=1}^{l}\tfrac{1}{\gamma_{ij}}\tilde{\beta}_{ij}\dot{\hat{\beta}}_{ij}$$
$$= -\sum_{i=1}^{n}\mathbf{h}_i\tilde{\beta}_i r_i - \mathbf{r}^T \mathbf{K}\mathbf{r} - \boldsymbol{\varepsilon}\cdot \mathbf{r}^T sign(\mathbf{r}) + \mathbf{r}^T\sum_{i=1}^{n} E_i + \sum_{i=1}^{n}\sum_{j=1}^{l}\tfrac{1}{\gamma_{ij}}\tilde{\beta}_{ij}\dot{\hat{\beta}}_{ij} \tag{6.86}$$

Combining (6.86) and (6.80), the relationship of (6.87) can be obtained as follows:

$$\dot{L} = -\mathbf{r}^T\mathbf{K}\mathbf{r}-\varepsilon\mathbf{r}^T sign(\mathbf{r}) + \mathbf{r}^T\sum_{i=1}^{n} E_i$$
$$= -\mathbf{r}^T\mathbf{K}\mathbf{r} - \mathbf{r}^T\left(\boldsymbol{\varepsilon}\cdot sign(\mathbf{r}) - \sum_{i=1}^{n} E_i\right) \le -\mathbf{r}^T\mathbf{K}\mathbf{r} - \sum_{i=1}^{n}(\varepsilon_i - E_i)|r_i| \le 0 \tag{6.87}$$

Therefore, the stability condition is sufficient according to (6.87), which proves the stability of the proposed control law.

6.5   Numerical simulation

6.5.1   Inertial parameter identification simulation

To verify the validity of the proposed identification method in this chapter, a simplified dynamic model of space robotic system with a manipulator (3-DOF) is established by ADAMS, with orthogonal three joints installed on manipulator, as depicted in Figure 6.9. And the geometric and inertial parameters of space robot model are displayed in Table 6.1, expressed in body frame of all the parts from robotic system. The robotic system keeps a zero-momentum initial condition, with the initial linear or angular velocity keeps zero of all the parts from robotic system. To ensure the adequate motion condition for identification, the orthogonal joints installed on the manipulator were simultaneously driven by cosine signals of acceleration, by

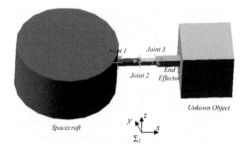

Figure 6.9   Dynamic model of space robotic system [50]

maintaining the motion of robotic system in a period of 100 s in the post-capture phase, as displayed in Table 6.2. The identification process is implemented by constructed ADAMS-MATLAB co-simulation platform in an off-line state, with setting data sample period τ of 0.005 s and integration period δt of 0.1 s, i.e., sampling rate γ of 0.05.

Considering not just Gaussian white noise existing, one adds the measurement errors in the simulation by Gaussian white noise v1, Gaussian coloured noise v2, and impulsive noise v3 in all measured data, i.e., $(\tilde{\varphi}, \tilde{\psi}, \tilde{\eta})$, $\tilde{\dot{p}}_B$, $\tilde{\omega}_B$ $\tilde{f}_U$, $\tilde{f}_n$, $\tilde{n}_n$ ($\Delta\dot{\theta}_i$ was ignored by the digital output of $\tilde{\theta}_i$), in which the Gaussian white noise and Gaussian coloured noise are presented by (6.88) and (6.89), respectively, and the impulsive noise is presented in Table 6.3. From (6.88) and (6.89), the random signal $x_1(t)$, $x_2(t)$ owns zero mean value and standard deviation by 1 per cent of signal magnitude.

And considering aerospace operation by fuel consumption, the estimation errors by the prior knowledge of spacecraft's mass and centroid are 5 per cent of nominal values in the process of identification.

$$v_1 = x_1(t) \tag{6.88}$$
$$v_2 = x_2(t) + 0.02x_2(t - 0.01) \tag{6.89}$$

A optimal parameter setting by algorithm of conventional RLS or RLS-APSA is essential and distinct with identifying m_U, a_U, and I_U, for the organization form of coefficient matrix and vector incorporating various kinematic and inertial information. Therefore, a suitable setting is employed in the simulation and presented in Table 6.4.

The superior accuracy of modified identification equation by (6.58) and (6.59) is verified by incorporating contact force as well as force/torque of end-effector. The inertial parameters of unknown object is identified via identical algorithm of APSA by parallel parameter setting from Table 6.4. A comparison process is implemented using the modified identification equations (6.58) and (6.59) and also conventional identification equation. The simulation results are presented in Figure 6.10 and Table 6.5.

To confirm the stability of identification process emerged by the proposed RLS-APSA, another comparison process is implemented by identifying the inertial parameters using hybrid RLS-APSA and also conventional RLS, via identical

Table 6.1 *Geometric and inertial parameters of space robot model with 3-DOF manipulator*

Link i	Mass (kg)	Length (m)		Inertia tensor (kg m²)					
		a_i	b_i	$I_i(1,1)$	$I_i(2,2)$	$I_i(3,3)$	$I_i(1,2)$	$I_i(1,3)$	$I_i(2,3)$
0	1 000	0	1	1 000	1 000	500	0	0	0
1	5	0.2	0.2	0.1	0.1	0.05	0	0	0
2	10	0.2	0.2	0.1	0.2	0.2	0	0	0
3	5	0.2	0.2	0.1	0.1	0.05	0	0	0
Unknown object	100	0.5	-	10	20	10	0	0	0

Table 6.2   Actuating signals of each joint

| Time (s) | Joint 1/ | | Joint 2/ | | Joint 3/ | |
	rad/s	rad/s²	rad/s	rad/s²	rad/s	rad/s²
0	0	$2\pi\cos(\pi t)$	0	$2\pi\cos(\pi t)$	0	$2\pi\cos(\pi t)$
~0 to 100	-	$2\pi\cos(\pi t)$	-	$2\pi\cos(\pi t)$	-	$2\pi\cos(\pi t)$

Table 6.3   Impulsive noises v3

Time (s)	0–30	30	30–35	35	35–40	40	40–55	55	55–100
Impact strength	0	1	0	−0.5	0	1	0	−0.5	0

scheme of modified identification equation. The simulation results are presented in Figure 6.11 and Table 6.6.

With the parallel parameter setting by RLS-APSA, the identification scheme incorporating contact force and force/torque of end-effector presents notably superior results of identification compared with conventional scheme, except a_{Uy}, a_{Uz} with no obvious distinction. And there are fluctuations of identified values in the initial phase of identification process for both identification schemes and a break point from fluctuations to stability process, which are determined by the detection threshold ρ_1, switching the algorithm from RLS to APSA to keep the stability of identification process. From Table 6.5, the identification scheme incorporating contact force and force/torque of end-effector owns the major deviations of 6.57 and 4.50 per cent from m_U and I_{Uyy} accordingly, compared with conventional identification scheme owning the major deviations of 10.50 and 17.25 per cent from m_U and I_{Uyy}, respectively. Therefore, there is a remarkable performance improvement of accuracy by introducing measured information of contact force and force/torque of end-effector into conventional identification scheme.

Under the various measurement noises, the identification process by the conventional RLS presents the fluctuation and drifting; however, the RLS-APSA still keeps a stable state with tiny fluctuation compared with conventional RLS. And the two algorithms also own parallel parameter setting in the initial phase of identification by RLS algorithm to quickly estimate the approximate values of identified inertial parameters. Therefore, the identification process based on RLS-APSA algorithm is significantly improved in performance of immunity and stability by various measurement noises relative to conventional RLS.

In conclusion, the identification scheme incorporating measured information contact force and force/torque of end-effector improves the precision of the identified inertial parameters significantly, and the proposed RLS-APSA hybrid algorithm ensures the stability of identification process effectively.

Table 6.4   Parameter setting

Parameter	λ			μ			ε			M			ρ_1		
	m_U, a_U		I_U	m_U, a_U		I_U	m_U, a_U		I_U	m_U, a_U		I_U	m_U, a_U		I_U
Value	0.999		0.9995	0.00001		0.01	1		1	2		2	0.0005		0.01

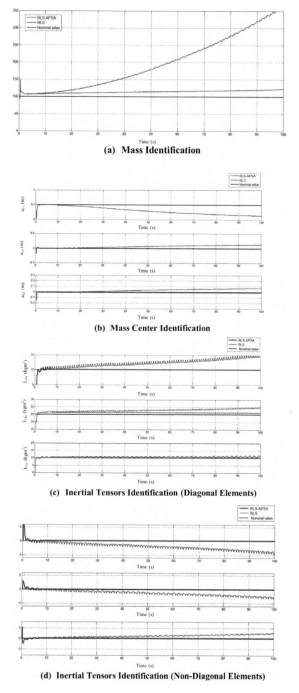

(a) **Mass Identification**

(b) **Mass Center Identification**

(c) **Inertial Tensors Identification (Diagonal Elements)**

(d) **Inertial Tensors Identification (Non-Diagonal Elements)**

Figure 6.10   Comparison results of conventional RLS algorithm and hybrid RLS-APSA algorithm (modified identification equation) [50]

Table 6.5   *Identification results of modified and conventional identification equations with RLS-APSA algorithm*

Inertial parameter	Nominal value	Modified method mean value (70–80 s)/Error With Noise A With Noise B With Noise C	Conventional method mean value (70–80 s)/ Error
m_U/kg	100	100.94/0.94% 101.23/1.23% 102.34/2.34%	107.98/7.98%
a_{Ux}/m	0.5	0.498/−0.4% 0.497/−0.6% 0.494/−1.2%	0.482/−3.6%
a_{Uy}/m	0	0.0017/- 0.0018/- 0.002/-	0.006/-
a_{Uz}/m	0	0.00005/- 0.00007/- 0.00009/-	0.0035/-
I_{Uxx}/kg m²	10	10.02/0.2% 10.03/0.3% 10.06/0.6%	11.8/18%
I_{Uxy}/kg m²	0	0.12/- 0.13/- 0.15/-	0.2/-
I_{Uxz}/kg m²	0	0.04/- 0.6/- 0.9/-	−0.18/-
I_{Uyy}/kg m²	20	20.2/1% 20.3/1.5% 20.7/3.5%	23.8/16%
I_{Uyz}/kg m²	0	0.21/- 0.27/- 0.72/-	1.3/-
I_{Uzz}/kg m²	10	10.1/1% 10.15/1.5% 10.35/3.5%	12.1/21%

6.5.2   Path planning and control simulation

To verify the validity of the proposed control strategy, a planar dynamic model of space robot with 4-DOF robotic arm operating in the free-floating mode is established by the MATLAB®/Simulink software, as shown in Figure 6.12. The sampling period is 0.001 s for the established simulation platform. The whole system operating in the post-capture stage owns initial linear/angular motion with the initial angular velocity of 0.25 rad/s for captured object during pre-captured stage, attached with precise knowledge about the spacecraft as well as the robotic arm and unknown dynamic parameter about captured object in the end-effector. The

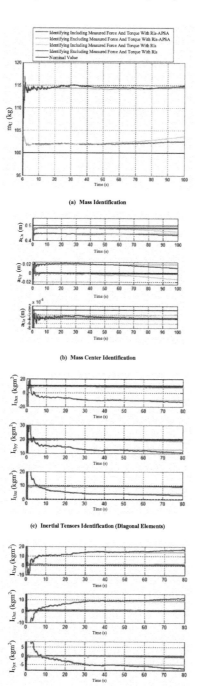

Figure 6.11   Comparison results of conventional identification equation and modified identification equation (RLS-APSA) [50]

Table 6.6   Identification results of two algorithms

Inertial parameter	Nominal value	Modified method mean value (90–100 s)/Error		Conventional method mean value (90–100 s)/Error	
		RLS	RLS-APSA	RLS	RLS-APSA
m_U/kg	500	-/-	510.12/2.02%	-/-	519.72/3.94%
a_{Ux}/m	0	-/-	−0.002/-	-/-	0.026/-
a_{Uy}/m	−0.25	-/-	−0.252/0.8%	-/-	−0.29/16%
a_{Uz}/m	0	-/-	0.005/-	-/-	0.023/-
I_{Uxx}/kg m²	100	-/-	101.12/1.12%	-/-	161.81/62.81%
I_{Uxy}/kg m²	0	-/-	−0.52/-	-/-	−24.58/-
I_{Uxz}/kg m²	0	-/-	−5.71/-	-/-	−26.9/-
I_{Uyy}/kg m²	200	-/-	201.33/0.67%	-/-	222.12/11.06%
I_{Uyz}/kg m²	0	-/-	3.28/-	-/-	6.45/-
I_{Uzz}/kg m²	200	-/-	200.94/0.47%	-/-	145.37/−27.32%

geometric and inertial parameters of the system employed in the model are shown in Table 6.7, where a_i and b_i denote the position vector from the joint i to the centroid of link i and also from the centre of link i to the joint $(i + 1)$, respectively; the I_i denotes the inertial tensor along z-axis; the link 0 represents the spacecraft, and the attitude angle and joint angle can be expressed by α and θ_i, respectively, in this chapter. The parameter setting and initial condition of the model are illustrated in Table 6.8, and the interior disturbance term d in the robotic system is set by the function $d_i = (0.8*\text{sign}(\dot{q}_i)+5*\sin(10t)+2*\textbf{\textit{rand}}(0,1))$ for each joint, where the signal $\textbf{\textit{rand}}(0,1)$ means the value is randomly produced from the range of $(0,1)$.

The total linear/angular momentum of the system in the post-capture stage is shown in Figure 6.13, where we can conclude that the whole system is subject to the law of linear/angular momentum conservation. In the simulation process, consider the adaptive reactionless path planning via SW-RLS algorithm and conventional RLS algorithm, respectively; the planned path via the two various algorithms can be

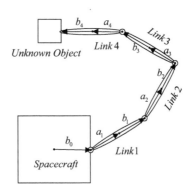

Figure 6.12   Space robot with 4-DOF robotic arm

Table 6.7   Geometric and inertial parameters of the space robot

Link i	Mass (kg)	a_i (m)	b_i (m)	I_i (kg m^2)
0	500	-	0.4	500
1	10	0.25	0.25	0.5
2	10	0.25	0.25	0.5
3	10	0.25	0.25	0.5
4	10	0.25	0.25	0.5
Captured object	20	0.5	-	5

obtained with parameter setting in Table 6.8 and is illustrated in Figure 6.14, where we can conclude that the planned path for robotic arm via the SW-RLS algorithm exhibits time-variable feature that keeps pace; yet the planned path via the conventional RLS exhibits saturation phenomenon that cannot track the time-variable property of the coefficient matrix.

Then we employ several various control schemes to track the planned path produced by the SW-RLS algorithm, expressed as follows:

Scheme 1. Adaptive control strategy via RBF (Radial Basis Function) algorithm: 60 hidden layer nodes (l=60); the centre f_{iC}, the width σ, and the bias of hidden nodes affiliated to $rand(-1,1)$, $rand(0,1)$, and $rand(0,1)$, respectively; the activation function selected by the RBF $h_i = exp(-f_i(x) - f_{iC})^2/(2\sigma)^2)$; the output of controller $\hat{\tau}_i = \hat{\tau}_{PDi} + \hat{\tau}_{Ai}$ with corresponding K in Table 6.8.

Table 6.8   Parameter setting of dynamic simulation

Parameter	Value	Parameter	Value
K	Diag(500,500,500)	$\alpha(0)$	0 rad
Λ	Diag(2,2,2)	$\dot{\alpha}(0)/\omega_0$	0.1 rad/s
$c_1 = c_2$	0.2	$\theta_1(0)$	0 rad
ω	2	$\theta_2(0)$	$\pi/2$ rad
$S_{max} = -S_{min}$	1	$\theta_3(0)$	0 rad
$V_{max} = -V_{min}$	1	$\theta_4(0)$	$-\pi/2$ rad
γ	1	$\dot{\theta}_i(0), i = 1,2,3,4$	0 rad
m	20	v_0	$[0.5\ 0.2]^T$
l	60		
$\dot{\zeta}$	$[1, 1, 1, 1]^T$		
δ	1.2		
ε_i	2		

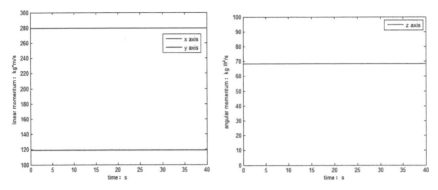

Figure 6.13   Total linear/angular momentum measurement

Scheme 2. Adaptive control strategy via ELM algorithm: all the initial conditions in accord with scheme 1 except that the activation function selected by the Sigmoid function with $h_i = 1/(1 + exp(-f_i(x)))$.

Scheme 3. Robust adaptive control strategy via ELM algorithm: all the initial conditions in accord with scheme 2 except an appending robust controller with the output of $\hat{\tau}_i = \hat{\tau}_{PDi} + \hat{\tau}_{Ai} + \hat{\tau}_{Ri}$.

Scheme 4. Proposed control strategy: all the initial conditions in accord with scheme 3 except an optimization of input weight matrix for the ELM algorithm via the PSO.

The comparison of tracking results for the control Schemes 1–4 is shown in Figure 6.15, and Tables 6.9 and 6.10, with tracking performance and tracking error, respectively. And then we measure the angular velocity of the spacecraft shown in Figure 6.16, which exhibits the existing disturbance from the robotic arm's motion to the spacecraft as well as to the stability of attitude.

According to the tracking results from Figure 6.15 and Table 6.9, the adaptive control strategy via ELM exhibits better speed performance compared with adaptive control strategy via RBF, although the accuracy performances via steady-state error for these two schemes are inferior to the robust adaptive control strategy via ELM

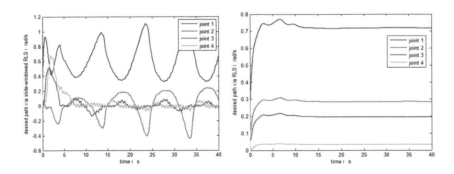

Figure 6.14   Adaptive reactionless path planning via various algorithm

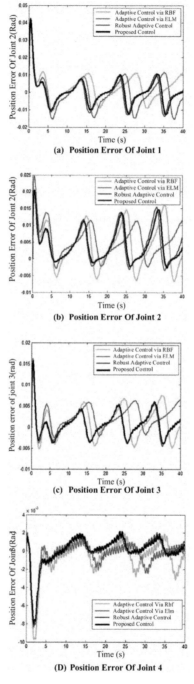

(a) **Position Error Of Joint 1**

(b) **Position Error Of Joint 2**

(c) **Position Error Of Joint 3**

(D) **Position Error Of Joint 4**

Figure 6.15   Position tracking effort of each joint controller via various control strategy

Table 6.9   Mean value of position tracking errors for each joint controller (5–40 s)

Error (°)	Adaptive control via RBF	Adaptive control via ELM	Adaptive control via ELM	Proposed method
Joint 1	0.268	−0.195	0.121	0.117
Joint 2	0.432	0.386	0.258	0.252
Joint 3	0.124	0.175	0.117	0.112
Joint 4	−0.091	−0.781	0.069	0.059

because of appending the robust control strategy. And the proposed control strategy via PSO-ELM algorithm exhibits slight improvement of the tracking accuracy and shorter transient time yet with extremely negligible delay of speed compared with the robust adaptive control strategy via ELM algorithm, which reflects the optimization process from the PSO algorithm to seek and update the optimal input weight matrix of network. In addition, Figure 6.16 and Tables 6.11 and 6.12 exhibit the stability of the spacecraft's attitude after capturing an unknown object, where the reactionless control strategy via the proposed scheme owns the faster and relatively less angular velocity of spacecraft as well as least disturbance, which verifies the validation of the proposed method in this chapter.

6.6   Experimental results

The previous chapters have conducted theoretical research on the inertial parameter identification methods of space non-cooperative targets and the system control strategies in the identification process. In this chapter, we will build a ground mechanical test system, design the manipulator operation experiment in the capture stage, and obtain the results of the inertia parameter identification experiment by collecting and processing the relevant experimental data to confirm the validity of the proposed theory.

The ground test system is mainly composed of KUKA GmbH Robot and control cabinet, end-effector including 6-dimensional force/torque sensor, non-cooperative

Table 6.10   Position tracking steady-state mean square errors for each joint controller (5–40 s)

Error (°)	Adaptive control via RBF	Adaptive control via ELM	Adaptive control via ELM	Proposed method
Joint 1	0.371	0.429	0.381	0.374
Joint 2	0.418	0.301	0.309	0.307
Joint 3	0.241	0.215	0.161	0.151
Joint 4	0.0201	0.187	0.182	0.181

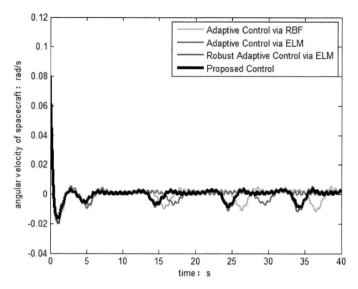

Figure 6.16   Angular velocity measurement of the spacecraft via various control strategy

target (cuboid, cylinder), PC, and other four parts. The experimental system is in the post-capture stage.

KUKA GmbH Robot (KR150/1 series) is a 6-DOF industrial robot arm, mainly composed of robot equipment (including arms, rocker arms, turntables, drives, base frames, hydropneumatic balancing systems), control cabinets, and connecting wires. The movement of each joint is restricted (including the limitation). The minimum rotation angle exceeds 180°.

The 6-dimensional force/torque sensor uses the GAMMA-SI-65-5 series of high-robust and high-precision *ATI* force/torque sensors. The highest sampling frequency is 28.5 kHz, the channel-to-noise ratio is 75; and the normal force measurement range does not exceed 200 N (resolution 1/20), the tangential force measurement range does not exceed 65 N (resolution 1/40), and the three-dimensional torque measurement range does not exceed 5 Nm (resolution 5/3 333).

Table 6.11   Mean value of spacecraft's angular velocity

Error (°)	Adaptive control via RBF	Adaptive control via ELM	Adaptive control via ELM	Proposed method
	−0.217	−0.191	−0.187	−0.172

Table 6.12   Position tracking steady-state mean square of spacecraft's angular velocity (5–40 s)

Error (°)	Adaptive control via RBF	Adaptive control via ELM	Adaptive control via ELM	Proposed method
	0.402	0.381	0.392	0.378

The non-cooperative targets are cylinders and cubes. The processing material is 6 061 aluminium alloy (hardness HB = 95, density ρ = 2 750 kg/m³). The whole system is shown as Figure 6.17.

The software part is divided into manipulator control software operation interface and 6-dimensional force/torque sensor software operation interface. The robot arm VW KUKA Control Panel (VKCP) can input the programmed software to control the motion of the robot arm, or directly drive the motion of the robot arm through the right joint drive switch. The 6-dimensional force/torque sensor uses the LabVIEW control interface to monitor.

The test system considers the identification method of non-cooperative target inertia parameters based on terminal force and moment information, and makes comparison with conventional identification methods. In the experiment, the end line angular momentum estimation based on the end touch measurement is considered to construct the coefficient matrix in the identification equation. To meet the complete identifiable condition of the inertia parameters, the two joints (different rotation axis directions) are simultaneously driven in the experiment to ensure the richness of non-cooperative target kinematics information and the conditions for complete identification of inertia parameters. In addition, considering the gravity effect of ground experiments, the data processing part uses gravity compensation to ensure the simulation of microgravity in space.

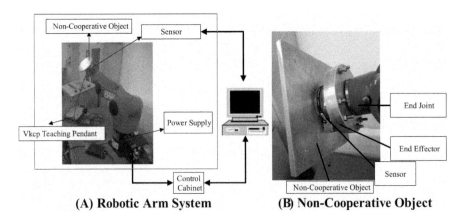

(A) Robotic Arm System   (B) Non-Cooperative Object

Figure 6.17   Experimental system

Figure 6.18   VKCP controller and sensor interface

$$-\delta\tilde{\boldsymbol{p}}_U + \int_0^{\delta t} g \cdot dt = \ -\int_0^{\delta t}\tilde{\boldsymbol{f}} \cdot dt \ \cdot \frac{1}{m_U} + \left[\left(\delta\tilde{\boldsymbol{\omega}}_U\right)^\times\right] \cdot \boldsymbol{a}_U \tag{6.90}$$

That is, to supplement the acceleration of gravity g received by the terminal fixture, such processing will include consideration of the identification situation under the action of gravity. Since gravity acts on the centre of mass of the terminal fixture, it does not affect the external moment of the terminal fixture.

In the experiment, the sampling frequency of the 6-dimensional force/torque sensor and the joint motion of the mechanical arm are both set to 100 Hz. Now consider the following two sets of experiments: The mechanical arm part drives joints 4 and 5 at the same time; the joint movement speed is set to 58 per cent of the maximum speed (250 mm/s) and the experiment running time is 15 s.

In Figure 6.19, f1, f2, and f3, respectively, represent the force information of the x, y, z axes relative to the inertial system during the movement of the non-cooperative target; f1, f2, and f3 (in the right part of Figure 6.19), respectively, represent the torque information of the x, y, z axes relative to the inertial system. Taking into account the measurement noise of the sensor and the tremor of the movement of the robot arm, the sensor's direct measurement data needs to be processed. This data processing uses Gaussian kernel model algorithm for denoising and smoothing.

Now the above experimental detection data is processed offline based on inertia parameter identification, and the identification results of non-cooperative target inertia parameters are given in Tables 6.13 and 6.14:

In terms of identification accuracy, the improved identification equation shows a clear advantage over the conventional identification equation. Although the terminal contact force and force/torque information contain measurement noise, it makes the direct use of the terminal touch force and torque information. The accumulation of measurement error is weakened and the identification accuracy is improved. In addition, considering the machining error of the robotic arms and the system modelling error, there is still a certain gap between the experimental identification result and the true value at this stage. Through the comparison of identification algorithms, the hybrid algorithm RLS-APSA proposed in this chapter can still ensure the stability of the identification process under the condition of system content measurement noise, and the identification results based on the traditional RLS algorithm show an unstable divergence state, which further proves the effectiveness of the proposed hybrid algorithm.

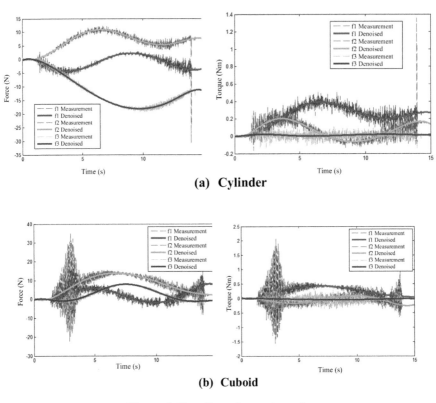

(a) Cylinder

(b) Cuboid

Figure 6.19   Experimental results

Table 6.13   Identification results – cylinder

Inertial parameter	Nominal value	Modified method/error RLS-APSA	Conventional method/error RLS-APSA
m_U/kg	1	1.57/0.57	2.72/1.72
a_{Ux}/m	0.01	0.018/0.008	0.293/0.293
a_{Uy}/m	0	0.003/0.003	0.052/0.052
a_{Uz}/m	0	−0.004/−0.004	−0.004/−0.004
I_{Uxx}/kg m^2	1	0.71/−0.29	−2.9/−3.9
I_{Uxy}/kg m^2	0	0.70/0.70	1.8/1.8
I_{Uxz}/kg m^2	0	0.13/0.13	5.7/5.7
I_{Uyy}/kg m^2	0.78	1.79/1.01	2.38/1.6
I_{Uyz}/kg m^2	0	0.95/0.95	−5.38/−5.38
I_{Uzz}/kg m^2	0.78	1.08/0.31	4.9/4.32

Table 6.14   *Identification results – cuboid*

Inertial parameter	Nominal value	Modified method/error RLS-APSA	Conventional method/error RLS-APSA
m_U/kg	1	1.36/0.36	2.48/1.48
a_{Ux}/m	0.01	0.012/0.002	0.452/0.442
a_{Uy}/m	0	0.02/0.02	0.021/0.021
a_{Uz}/m	0	−0.005/−0.005	−0.016/−0.016
I_{Uxx}/kg m^2	1	0.82/−0.18	−4.6/−5.6
I_{Uxy}/kg m^2	0	0.8/0.8	11.3/11.3
I_{Uxz}/kg m^2	0	0.15/0.15	0.8/0.8
I_{Uyy}/kg m^2	0.82	1.35/0.53	2.24/1.42
I_{Uyz}/kg m^2	0	0.32/0.32	−2.88/−2.88
I_{Uzz}/kg m^2	0.82	0.51/−0.31	2.13/1.31

6.7   Recommendations and future work

As the service level of space in orbit continues to improve, on-orbit capture operations based on non-cooperative targets with unknown inertial attributes have become the bottleneck restricting on-orbit operations. For the orbit capture of non-cooperative targets, the identification of non-cooperative target mass, rotational inertia, and other inertia parameters is the basic premise of the operation. The uncertainty of the system dynamics caused by the unknown inertia parameters will lead to the failure of the system. Therefore, this chapter takes the inertia parameter identification method of space non-cooperative targets as the research goal, focusing on the identification of inertia parameters and the path planning and control of the space manipulator and carries out research on the design of the system attitude stability control strategy, and verifies the theoretical method proposed in this chapter by building a joint simulation platform and a ground manipulator test platform. The main research results of this chapter are summarized as follows:

1. This chapter conducts research based on traditional inertia parameter identification methods, theoretically deduces the complete identification conditions of inertia parameters and the identification error mechanism of the parameters, and concludes that the cumulative process of measurement errors is the main reason affecting the accuracy of inertia parameter identification. This chapter reduces the accumulation process of measurement errors by incorporating the terminal touch information into the identification, thereby improving the accuracy of parameter identification. At the same time, considering the complexity of the measurement environment, this chapter proposes a hybrid identification algorithm based on RLS-APSA to ensure the stability of the parameter identification process.

2. Aiming at the posture of the system during the post-acquisition phase parameter identification process, the manoeuvre arm's motion is disturbed. In this

chapter, an adaptive reactionless trajectory planning based on sliding window least squares is proposed in consideration of the system's dynamic parameter mutations. Based on the design of a robust adaptive joint controller, the planned trajectory can be effectively tracked to achieve the minimum disturbance of the attitude of the spacecraft substrate by the movement of the robot arm. At the same time, this chapter uses the coordinated control of the spacecraft base body based on the coupled torque compensation feedforward control and the robust adaptive control of the robot arm to ensure the stability of the system attitude.

3. Through the construction of the ADAMS-MATLAB joint simulation platform, the theoretical verification of the spatial non-cooperative target inertia parameter identification method, the reactionless trajectory planning of the manipulator, and the system coordination control are simulated and verified. Finally, a ground test system is built to verify the identification method proposed in this chapter.

The research work of this chapter has made some progress, but there are still some deficiencies that need to be further studied in depth:

1. According to the identification error theory, the measurement error (accuracy) of the terminal force information mainly determines the accuracy of parameter identification, and how to use and improve the detection accuracy of the force sensor in the future experimental or on-orbit system, especially for the terminal tangential force and directional force detection, is the key to the identification of inertia parameters. In addition, the switching mechanism of the two hybrid algorithms in the hybrid identification algorithm can only be obtained through experience and cannot be adaptively obtained online. Therefore, the implementation of the hybrid algorithm still requires a lot of ground offline experiments to obtain the prerequisites for its on-orbit application.

2. To verify the trajectory planning and coordinated control of the space manipulator system in the free-floating state, a ground air floatation experimental platform that can simulate the space floating environment should be built; so the focus of future research should be shifted to design physical air floatation experimental platform and then carry out related theoretical verification.

References

[1]   Flores A., Ma O., Pham K., Ulrich S. 'A review of space robotics technologies for on-orbit servicing'. *Progress in Aerospace Sciences*. 2014;**68**:1–26.
[2]   Vasile M. 'Preface: Advances in asteroid and space debris science and technology-Part 2'. *Advances in Space Research*. 2016;**57**:1605–6.
[3]   Sullivan B.R., Akin D.L. 'A survey of serviceable spacecraft failures'. *Databases*. 2001;**1**(4540).

[4]   Williams P. 'Tether capture and momentum exchange from hyperbolic orbits'. *Journal of Spacecraft and Rockets*. 2010;**47**(1):205–10.

[5]   Lu Y., Huang P., Meng Z., Hu Y., Zhang F., Zhang Y. 'Finite time attitude takeover control for combination via tethered space robot'. *Acta Astronautica*. 2017;**136**:9–21.

[6]   Xu Y., Kanade T. *Space robotics: dynamics and control*. Springer Science & Business Media; 1992. pp. 25–7.

[7]   Yoshida K. 'Achievements in space robotics'. *IEEE Robotics & Automation Magazine*. 2009;**16**(4):20–8.

[8]   NASA Office of Communications. The International Space Station's Canadarm2 robotic arm and Dextre robotic hand [online]. 2019. Available from images-assets.nasa.gov/image/iss058e005961/iss058e005961~orig.jpg [Accessed 30 Oct 2020].

[9]   Zhang K., Zhou H., Wen Q., Sang R. 'Review of the development of robotic manipulator for International Space Station'. *Chinese Journal of Space Science*. 2010;**30**(6):612–19.

[10] Smith D., Martin C., Kassebom M., Petersen H., Shaw A., Skidmore B. 'A mission to preserve the geostationary region'. *Advances in Space Research*. 2004;**34**(5):1214–18.

[11] Oda M. 'Space robot experiments on NASDA's ETS-VII satellite – preliminary overview of the experiment results'. *Proceedings 1999 IEEE International Conference on Robotics and Automation*. Detroit, USA: IEEE; May 1999. pp. 1390–5.

[12] Shoemaker J., Wright M. 'Orbital express space operations architecture program'. *Space Systems Technology and Operations*. Florida, USA: International Society for Optics and Photonics. 2003; August 2003. pp. 1–9.

[13] Liu Y., Cui S., Liu H., *et al.* 'Robotic hand-arm system for on-orbit servicing missions in Tiangong-2 space laboratory'. *Assembly Automation*. 2019;**39**(5):999–1012.

[14] NASA Office of Communications. *RemoveDEBRIS satellite launch* [online]. 2018. Available from images-assets.nasa.gov/image/iss056e025331/iss056e025331~orig.jpg [Accessed 30 Oct 2020].

[15] NASA Office of Communications. *Rocky ring of debris around vega artist concept* [online]. 2013. Available from images-assets.nasa.gov/image/PIA16610/PIA16610~orig.jpg [Accessed 30 Oct 2020].

[16] Xu W., Liang B., Li C., Liu Y., Wang X. 'A modelling and simulation system of space robot for capturing non-cooperative target'. *Mathematical and Computer Modelling of Dynamical Systems*. 2009;**15**(4):371–93.

[17] Murotsu Y., Senda K., Ozaki M., Tsujio S. 'Parameter identification of unknown object handled by free-flying space robot'. *Journal of Guidance, Control, and Dynamics*. 1994;**17**(3):488–94.

[18] Tao Z., Li C., Yin Z., Hong C. 'TLE conversion of the non-cooperative space object using bearing-only measurements'. *Acta Photonica Sinica*. 2009;**38**(12):3230.

[19] Rekleitis G., Papadopoulos E. 'On on-orbit passive object handling by coop-
      erating space robotic servicers'. *2011 IEEE/RSJ International Conference on
      Intelligent Robots and Systems*. San Francisco, USA: IEEE; 2011; Sep 2011.
      pp. 595–600.
[20] Nguyen T., Sharf I. 'Adaptive reactionless motion and parameter identifi-
      cation in postcapture of space debris'. *Journal of Guidance, Control, and
      Dynamics*. 2013;**36**(2):404–14.
[21] Lampariello R., Hirzinger G. 'Modeling and experimental design for the on-
      orbit inertial parameter identification of free-flying space robots'. *ASME 2005
      International Design Engineering Technical Conferences and Computers
      and Information in Engineering Conference*. Long Beach, USA: American
      Society of Mechanical Engineers Digital Collection; 2005; Jan 2005. pp.
      881–90.
[22] Sun J., Zhang S., Chu Z. 'Adaline network identification for non-cooperative
      object inertial parameters'. *Journal of Aeronautic*. 2016;**37**(9):2799–808.
[23] Dubowsky S., Papadopoulos E. 'The kinematics, dynamics, and control of
      free-flying and free-floating space robotic systems'. *IEEE Transactions on
      Robotics and Automation*. 1993;**9**(5):531–43.
[24] Lu W., Geng Y., Chen X., Wang F. 'Relative position and attitude coupled
      control for butting spacecraft'. *Journal of Aeronautic*. 2015;**32**(5):857–65.
[25] Wu Y., Gosselin C. 'Synthesis of reactionless spatial 3-DoF and 6-DoF
      mechanisms without separate counter-rotations'. *The International Journal
      of Robotics Research*. 2004;**23**(6):625–42.
[26] Agrawal S., Fattah A. 'Reactionless robots: novels designs and concept stud-
      ies'. *7th International Conference on Control, Automation, Robotics and
      Vision*. Singapore, Singapore: IEEE; Dec 2002. pp. 809–14.
[27] Fattah A., Agrawal S. 'Design and modeling of classes of spatial reaction-
      less manipulators'. *2003 IEEE International Conference on Robotics and
      Automation*. IEEE; 2003. Taipei, Taiwan; Sep 2003. pp. 3225–30.
[28] Agrawal S., Fattah A. 'Reactionless space and ground robots: novel designs
      and concept studies'. *Mechanism and Machine theory*. 2004;**39**(1):25–40.
[29] Torres M., Dubowsky S. 'Minimizing spacecraft attitude disturbances in
      space manipulator systems'. *Journal of Guidance, Control, and Dynamics*.
      1992;**15**(4):1010–17.
[30] Quinn R.D., Chen J.L., Lawrence C. 'Base reaction control for space-based
      robots operating in microgravity environment'. *Journal of Guidance, Control,
      and Dynamics*. 1994;**17**(2):263–70.
[31] Nenchev D., Umetani Y., Yoshida K. 'Analysis of a redundant free-flying
      spacecraft/manipulator system'. *IEEE Transactions on Robotics and
      Automation*. 1992;**8**(1):1–6.
[32] Fukazu Y., Hara N., Kanamiya Y., Sato D. 'Reactionless resolved acceleration
      control with vibration suppression capability for JEMRMS/SFA'. *2008 IEEE
      International Conference on Robotics and Biomimetics*. Bangkok, Thailand:
      IEEE; 2009; Feb 2009. pp. 1359–64.

[33] Huang P., Yuan J., Liang B. 'Adaptive sliding-mode control of space robot during manipulating unknown objects'. *In 2007 IEEE International Conference on Control and Automation; Guangzhou.* China: IEEE; 2007; May 2007. pp. 2907–12.

[34] Zadeh S.M.H., Khorashadizadeh S., Fateh M.M., Hadadzarif M. 'Optimal sliding mode control of a robot manipulator under uncertainty using PSO'. *Nonlinear Dynamics.* 2016;**84**(4):2227–39.

[35] Chu Z., Li J., Lu S. 'The composite hierarchical control of multi-link multi-DOF space manipulator based on UDE and improved sliding mode control'. *Proceedings of the Institution of Mechanical Engineers, Part G: Journal of Aerospace Engineering.* 2015;**229**(14):2646–58.

[36] Faieghi M.R., Delavari H., Baleanu D. 'A novel adaptive controller for two-degree of freedom polar robot with unknown perturbations'. *Communications in Nonlinear Science and Numerical Simulation.* 2012;**17**(2):1021–30.

[37] Zhang W., Ye X., Jiang L., Zhu Y., Ji X., Hu X. 'Output feedback control for free-floating space robotic manipulators base on adaptive fuzzy neural network'. *Aerospace Science and Technology.* 2013;**29**(1):135–43.

[38] Smith A.M.C., Yang C., Ma H., Culverhouse P., Cangelosi A., Burdet E. 'Novel hybrid adaptive controller for manipulation in complex perturbation environments'. *Plos One.* 2015;**10**(6):e0129281.

[39] Chu Z., Cui J., Sun F. 'Fuzzy adaptive disturbance-observer-based robust tracking control of electrically driven free-floating space manipulator'. *IEEE Systems Journal.* 2012;**8**(2):343–52.

[40] Qin L., Liu F., Liang L., Gao J. 'Fuzzy adaptive robust control for space robot considering the effect of the gravity'. *Chinese Journal of Aeronautics.* 2014;**27**(6):1562–70.

[41] Zhang W., Qi N., Ma J., Xiao A.Yang. 'Neural integrated control for a free-floating space robot with suddenly changing parameters'. *Science China Information Sciences.* 2011;**54**(10):2091–9.

[42] Chen Z., Chen L. 'Robust adaptive composite control of space-based robot system with uncertain parameters and external disturbances'. *2009 IEEE/RSJ International Conference on Intelligent Robots and Systems.* St. Louis, USA: IEEE; 2009. pp. 2353–8.

[43] Widrow B., Lehr M.A. '30 years of adaptive neural networks: perceptron, madaline, and backpropagation'. *Proceedings of the IEEE.* 1990;**78**(9):1415–42.

[44] Huang G., Zhu Q., Siew C. 'Extreme learning machine: a new learning scheme of feedforward neural networks'. *2004 IEEE international joint conference on neural networks.* Budapest, Hungary: IEEE; 2004. pp. 985–90.

[45] Rong H., Wei J., Bai J., Zhao G., Liang Y. 'Adaptive neural control for a class of MIMO nonlinear systems with extreme learning machine'. *Neurocomputing.* 2015;**149**:405–14.

[46] Petković D., Danesh A.S., Dadkhah M., Misaghian N., Shamshirband S., Zalnezhad E. 'Adaptive control algorithm of flexible robotic gripper by extreme learning machine'. *Robotics and Computer-Integrated Manufacturing.* 2016;**37**:170–8.

[47] Zhu Q., Qin A., Suganthan P., Huang G. 'Evolutionary extreme learning machine'. *Pattern Recognition*. 2005;**38**(10):1759–63.

[48] Kennedy J. 'Particle swarm optimization'. *Proc. of 1995 IEEE Int. Conf. Neural Networks*. 2011;**4**(8):1942–8.

[49] Xu Y., Shu Y. 'Evolutionary extreme learning machine–based on particle swarm optimization'. *International Symposium on Neural Networks*. Dallas USA: Berlin: Springer; 2006; May 2006. pp. 644–52.

[50] Chu Z., Ma Y., Hou Y., Wang F. 'Inertial parameter identification using contact force information for an unknown object captured by a space manipulator'. *Acta Astronautica*. 2017;**131**(5):69–82.

[51] Tabandeh S., Melek W.W., Clark C.M. 'An adaptive niching genetic algorithm approach for generating multiple solutions of serial manipulator inverse kinematics with applications to modular robots'. *Robotica*. 2010;**28**(4):493–507.

[52] Franti P., Rahardja S., Huang H. 'Cascaded RLS–LMS prediction in MPEG-4 lossless audio coding'. *IEEE Transactions on Audio, Speech, and Language Processing*. 2008;**16**(3):554–62.

[53] Ysebaert G., Vanbleu K., Cuypers G., Moonen M., Pollet T. 'Combined RLS-LMS initialization for per tone equalizers in DMT-receivers'. *IEEE Transactions on Signal Processing*. 2003;**51**(7):1916–27.

[54] Venturi S.H.K., Panahi I. Hybrid RLS-NLMS algorithm for real-time remote active noise control using directional ultrasonic loudspeaker. *IECON 2014-40th Annual Conference of the IEEE Industrial Electronics Society*; Dallas, USA, IEEE; 2014. pp. 2418–24.

[55] Eweda E., Macchi O. 'Convergence of the RLS and LMS adaptive filters'. *IEEE Transactions on Circuits and Systems*. 1987;**34**(7):799–803.

[56] Nenchev D., Yoshida K., Vichitkulsawat P., Uchiyama M. 'Reaction null-space control of flexible structure mounted manipulator systems'. *IEEE Transactions on Robotics and Automation*. 1999;**15**(6):1011–23.

Chapter 7

Autonomous robotic grasping in orbital environment

Nikos Mavrakis[1] and Yang Gao[1]

Capturing a target will always be a crucial part of any orbital activity for space robots. This chapter aims to present an overview of the developed technologies and algorithms that enable a robot to grasp a target in microgravity. Studies on human grasping in microgravity are described both for providing a solid background on orbital grasping and as inspiration for the development of robotic systems to aid astronauts. The chapter also describes the most important past and future applications of target capturing with robotic arms. The core analysis of the chapter consists of a large number of studies and engineering milestones on orbital robotic capturing that are categorised based on the means of interaction with the target, as well as reporting the state-of-the-art grasping methods. A number of important missions that have grasping as their basic demonstrated technology are also presented. The chapter ends with outlining the most important physical, algorithmic, and operational challenges in orbital robotic grasping, and setting up the capabilities that future robotic systems need to possess.

7.1   Introduction

In 1997, the JAXA ETS-VII mission demonstrated the first use of an on-orbit autonomous robotic manipulator for docking with a target satellite. The manipulator had 6 degrees-of-freedom (DOF), a length of 2 m, and was equipped with a 2-fingered gripper. The robot captured a grapple on a target satellite at a distance of 20 cm, guided by visual markers. This experiment pioneered the use of autonomous robots in satellite capturing, paving the way for future space applications involving autonomous robots.

Since that day, a large number of robots have already been deployed for on-orbit operations. The increased interest in orbital activities both from space agencies and from the private sector has led to further development of such robotic systems. A common and persisting property for such systems is the need to make contact with

[1]STAR LAB, Surrey Space Centre, University of Surrey, UK

a target, capture it, and manipulate it according to the mission objectives. It is no trivial problem with various solutions have been studied and a small number also has been demonstrated in space. The aim of this chapter is to describe the problem of orbital grasping. Inspired from human grasping experiments in microgravity conditions, the chapter attempts to outline the underlying challenges of orbital robotic grasping, categorise the existing methods and hardware used for orbital grasping, and showcase a number of cutting-edge approaches.

The core aim of this chapter is to provide an outline of the scientific and engineering aspects and challenges in the field of grasping for orbital robotics. An analysis of experiments of human grasping in weightlessness is given to provide a background and showcase the difficulties faced by humans in this environment. The main applications of robotic grasping are then described to provide the applicability and potential of the field. A large number of studies and demonstrated technologies for orbital grasping are then analytically categorised and described with respect to their hardware and method of interaction with the target. The state of the art in the field is also described. The chapter continues by describing a number of past and future milestone missions that have performed orbital capturing of targets and ends with defining a number of challenges for the present and future of orbital grasping.

7.2   Human grasping in space

Humans are masters of dexterous and intelligent grasping, and human grasping has been extensively used as an inspiration for robotic applications. To develop robots that can grasp objects in microgravity environments, it is necessary to understand how the human grasping processes and mechanisms change when they are exposed to space environment. Research on human grasping in microgravity has been conducted on the MIR station, International Space Station (ISS), and parabolic flights.

The weight of a handled object plays an important role in human grasping. It affects a number of grasping choices, such as the grasp type and the location of the object, or the usage of one or two robotic arms. Object weight is the physical property most significantly altered in microgravity and it is the product of object mass and gravitational acceleration. In nature, mass can be realised as gravitational (static) or inertial (dynamic). Gravitational mass is felt when an object lies on a person's hand, and inertial mass is felt when a human alters the motion state of the object. The two masses are equivalent in physical sense. Humans perceive mass in one of the two ways, or as a combination [1]. In orbit, the gravitational acceleration is eliminated and humans experience mostly the inertial mass, as has been experimentally proven [2]. On Earth, the static and dynamic influences of mass induce force loads when an object is held or moved. When grasping the object, a human introduces grip forces that resist these loads and prevent slippage. The applied grip forces are barely strong enough to resist the static load when the object is not moving. Humans increase or decrease the grip force in phase with the variation of inertial loads when the object is moved, and they do

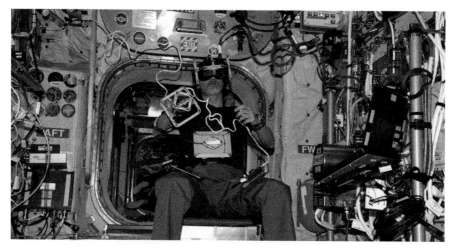

Figure 7.1   *The ESA GRASP experiment is one of the experiments that study how astronauts vary their gripping force with changes in the handled load [15]. The results of these experiments can be used to build better robotic controllers for orbital grasping. Image: ESA*

so through an anticipatory mechanism that predicts whether the object is going to slip during motion [3]. It has been shown that microgravity does not significantly affect this anticipatory mechanism [4], although until humans get accustomed to microgravity they tend to apply higher grip force than the necessary to stabilise the object [5]. This mechanism works under various loads induced by combinations of gravity, object and hand dynamics, and motion acceleration [6, 7]. The anticipatory mechanism is decoupled for static and dynamic loads [8]. Higher torque loads (i.e. the torque needed to be applied to stabilise the object) tend to slow down the mechanism's adaptation to microgravity [9]. When humans manage to adapt this mechanism in microgravity, they tend to adapt it quicker to higher gravity environments, such as those similar to Moon or Mars [10].

In addition, it has been shown that humans lose accuracy of arm reaching motion when experiencing brief periods of microgravity for the first time and that prolonged exposure corrects this behaviour [11]. The duration of reaching motion is also increased in microgravity [12]. However, Fitt's law, i.e., the modelling relationship between reaching velocity, reaching distance, and accuracy, is shown to be unaffected by microgravity [13].

Such studies have been conducted as part of major experiments designed to study how human grasping is altered in microgravity. A notable experiment is the GRIP experiment [14], flown over 20 parabolic flights. The ongoing ESA GRASP experiment on board the ISS will demonstrate how these grasping behaviours are altered after prolonged exposure in microgravity [15] (Figure 7.1).

Figure 7.2   *The NASA STS-51A human spaceflight mission was one of the first to involve capturing malfunctioning satellites, to service them on Earth. The Westar 6 satellite is shown captured with the aid of an Apogee Kick Motor Capture Device. Image: NASA*

7.3   Applications of orbital grasping

The NASA Space Shuttle and the ISS have demonstrated the basic applications of orbital target capturing, in operations of payload deploying, capturing, and berthing. The increasing interest in orbital grasping comes from emerging applications that need greater autonomy.

7.3.1   On-Orbit Servicing

On-Orbit Servicing (OOS) has been proposed as a concept shortly after the first satellites were launched. Servicing refers to any maintenance operation required to extend the lifetime of a payload, such as refuelling, orbit raising, and repairing of malfunctioning subsystems. While the end vision is autonomous robotic servicing operations, OOS has been initially performed by astronauts. Notable examples are the NASA STS-51A mission which involved capturing two satellites with an Apogee Kick Motor Capture Device (Figure 7.2), the multiple servicing missions of the Hubble Space Telescope (HST), and the servicing activities conducted routinely on the ISS.

Such missions were designed to service large payloads, with humans performing the actual servicing. As a result, it required careful design and planning, with increased cost. However, the rise in the number of satellites being launched annually and the ongoing reduction of the launched satellite size mean that servicing will need to become less expensive, faster, and repetitive. Automated servicing using robotic arms and grippers has been proposed as the solution. A review of the most important studies and mission concepts related to OOS is given in [16].

7.3.2   In-space telescope assembly

The mirror diameter of an Earth-based optical telescope is proportional to the image resolution it can provide. The trade-off for Earth telescopes is that larger mirrors have larger supporting weights. In addition, the Earth's atmosphere also distorts optical data requiring the usage of correction techniques and digital post-processing of the image. To compensate for the atmospheric distortion, space-based optical telescopes have been designed and launched, with the most notable being the HST. HST is equipped with a 2.4 m primary mirror, small compared to some ground-based telescopes. The new James Webb Space Telescope will instead have a 6.5 m modular mirror that will be stowed during launch and unfolded afterwards. However, space telescopes are limited by the launching vehicle's volume and payload capacity, making the deployment of larger mirror difficult. To further increase the mirror size, the concept of in-space modular telescope assembly has been proposed. The modules can be stored in the spacecraft's main body, and the telescope can be assembled with the help of a robotic manipulator that can move along the spacecraft body. Various mission architectures have been proposed for this purpose [17, 18]. Figure 7.3 shows a proposed architecture example by Surrey Space Centre [19]. The robotic manipulator typically has a mechanical interface as an end-effector, and the matching part is on the module. The grasping action is realised by connecting the two interfaces and then activating a locking mechanism that secures contact.

The robot can then move to the mirror and add the module, increasing the total mirror diameter. With such an approach very large mirrors can be assembled in space, further increasing the resulting image resolution.

7.3.3   Active debris removal

It is estimated that about 34 000 pieces of debris larger than 10 cm are currently in-orbit [20]. Among others, they consist of spent rocket stages and payload adaptors, satellite collision debris, as well as other launch by-products. These larger pieces of debris can be tracked and a satellite can execute a manoeuvre to avoid them if needed. However, such manoeuvres would be costly and prohibitive for sustainable future space access. Most pieces of debris tend to deorbit naturally; however, this typically happens over the course of years, or even decades. One way to mitigate the issue of space debris is performing Active Debris Removal (ADR). In ADR solutions, a spacecraft uses a method of capturing one or more pieces of debris and then returns to the ground, removing the debris from orbit. Capturing a piece of debris is a challenging problem, due to the rotational velocity of the debris, the lack of contact points, and its inherent degradation and fragility that comes after years of exposure to space environment. For numerous types of debris, a variety of capturing

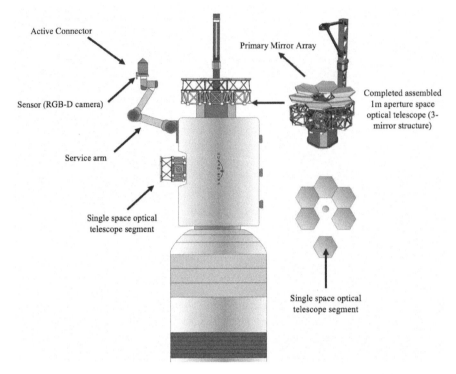

Active Connector

Primary Mirror Array

Sensor (RGB-D camera)

Completed assembled
1m aperture space
optical telescope (3-
mirror structure)

Service arm

Single space optical
telescope segment

Single space optical
telescope segment

*Figure 7.3   A proposed architecture of a spacecraft for in-space modular
telescope assembly. The robotic manipulator captures modules from
the module compartment using an active connector and connects
them radially to the main mast of the mirror. Image by [19]. Image:
STAR LAB*

mechanisms have been proposed and tested, such as nets, harpoons, magnetic effectors e.a [21]. The usage of one or more robotic manipulators that can securely grasp
larger objects has also been proposed. A notable study is the e-Deorbit mission by
ESA [22]. It is planned to use robotic arm and clamps to deorbit the 8-ton Envisat
satellite (Figure 7.4).

7.3.4   Astronaut–robot interaction

A core application of space robotics is in helping astronauts with maintenance tasks.
Currently, such robots have already been deployed on the ISS, but future plans
involve their usage in the Gateway station. Grasping actions in these environments
involve not only object manipulation but also anchoring to handles and bars on the
ISS surfaces for locomotion purposes. A small number of advanced robots capable
of object grasping have been already deployed on the ISS. The most notable examples are the NASA Robonaut2 humanoid (Figure 7.5), the Astrobee free-floating
platform [23, 24], and the Roscosmos humanoid FEDOR [25].

Figure 7.4   Illustration of the ESA e-Deorbit mission [22]. A spacecraft with a robotic arm and grapple attempts to capture and deorbit the non-operative 8-ton Envisat satellite. Image: ESA

7.4 Robotic hardware for orbital grasping

Capturing a target in orbit can be achieved with a number of methods, which include different types of hardware. The selection of the necessary hardware is typically performed according to the mission objectives and design parameters, cost of spacecraft budgets such as power and bandwidth, as well as the profile of the target to be captured. The method of capturing may not include robotic manipulators, with notable examples of such methods including harpoons and nets launched at a target. These methods are more suitable with non-cooperative targets, i.e., targets that their motion cannot be controlled. The most common application for non-robotic methods is ADR, where an object needs to be securely captured and deorbited. Robotic manipulators are more suitable for cooperative targets that require additional operations such as berthing, refuelling, or servicing.

Perhaps the most crucial component in the capturing process is the robotic end-effector, mounted at the end of the manipulator. The end-effector includes the capturing mechanism, which is used to secure a rigid contact with the target. The contact point on the target, as well as the operations to be performed, dictate the type of capturing mechanism. The most common types of end-effector mechanisms for orbital grasping are given as follows.

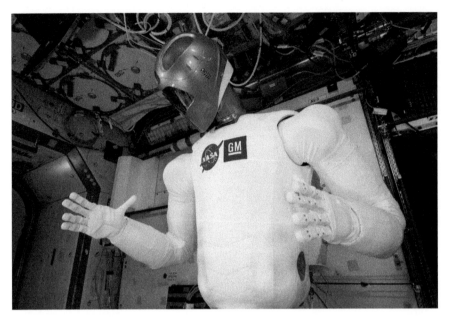

Figure 7.5   *The Robonaut2 robot was developed by NASA and General Motors.*
It has dual-arm and equipped with anthropomorphic hands. It has
performed a large number of dexterous manipulations on board the
ISS. Image: NASA

7.4.1   Coupling interfaces

The most common mechanism for capturing an orbital target is a mechanical interfacing couple, or adapter. One part of the couple is mounted on the approaching
satellite's end-effector, and it includes an actuated mechanism. The other part of the
couple is mounted on the target. When the robotic arm connects its coupling part
on the target, a mechanism is actuated and holds the target in place. Such coupling
mechanisms often include data and other transmission buses. Examples of such
interfaces are shown in Figure 7.6.

Numerous interfaces have been developed for servicing. An analytical review
and categorisation is given in [27]. For mechanical latching, the authors mention
four different types: Clamp, hook, roto-lock, and carabiner. Hook type involves
the use of hook-like appendices that are actuated and grapple the extrusions of the
matching part. Roto-lock involves one actuated male interface that is inserted into
the female part and rotates after insertion, locking the mechanism. Clamp involves
an androgynous pair that after insertion actuates latches or bars, securing the mechanism. Carabiner involves insertion of a male part in a female part, which passively
disengages an interference piece and locks. To release the male interface, actuation of the interference piece is required. Table 7.1 extracted from [27] summarises
some of the interfaces in literature and their basic properties: androgynous matching

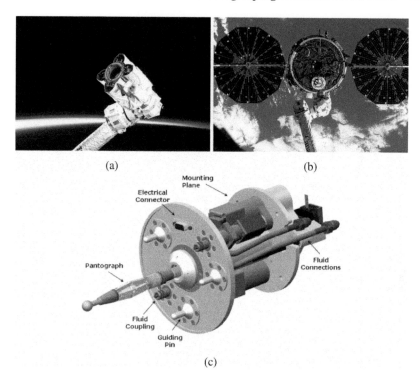

(a)   (b)

(c)

Figure 7.6   *Examples of standardised mechanical interfaces for capturing orbital payloads. (a) The end-effector of Canadarm. The wire-based capturing system enables it to capture any payload equipped with the matching grapple fixture. Image: NASA. (b) Grapple fixture compatible with the Canadarm end-effector, on a Cygnus cargo spacecraft. The end-effector wires close tightly around the central rod, securing the payload. Image: NASA. (c) A standardised connector for payload grappling [26]. The central pantograph expands when inserted in the matching part of the payload, effectively capturing it. The driving pins facilitate the insertion, and the other pins transfer data, power, and propellant to the target. Image: Elsevier*

design for mechanical latching; support for power, data, and thermal transfer; form of redundancy; and capability of orientation.

A notable interface is the end-effector of the Remote Manipulation System, also known as Canadarm, that has flown on the Space Shuttle on numerous missions [29]. The capturing mechanism on its end-effector consists of two coaxial cylinders, one stationary and one able to rotate. The matching interface consists of a grapple fixture that typically includes a rod with a sphere at its end. As the end-effector moves to the grapple fixture and the rod is inserted towards the centre of the cylinders, the movable cylinder rotates and three metallic wires start closing around the

Table 7.1   Mechanical interfaces for orbital grasping and their properties, as noted in [28]. Image: Frontiers

	Androgynous	Rigid connection	Mechanical	Latch type	Power	Data	Thermal	Redundancy	Marker	Angular orientation	No. of orientations
MTRAN				Magnetic	×	×				×	4
SINGO	×	×		Clamp				×		×	4
CAST		×	×	Hook						×	2
CEBOT			×	Hook					×		1
ATRON		×	×	Hook		×					1
Telecube	×			Magnetic				×		×	2
PolyBot	×	×	×	Rotational				×		×	4
AMAS		×	×	Hook				×		×	4
SMORES	×			Magnetic						×	2
GENFA	×	×	×	Rotational	×			×		×	4
ACOR			×	Hook	×			×			1
SWARM		×	×	Rotational	×			×			1
Pheonix Tool		×	×	Hook	×			×		×	1
Pheonix Satlet		×	×	Clamp	×	×		×		×	1
SSRMS LEE		×	×	Snare	×	×		×			1
DEXTRE (OTCM)		×	×	Clamp	×	×	×				1
iSSI	×		×	Rotational	×	×		×	×	×	4
EMI		×	×	Clamp	×	×		×	×	×	4
Berthing and Docking		×	×	Hook				×		×	∞
EM-Cube	×			Magnetic				×	×	×	2

rod, effectively trapping it. As the wires close tightly, the grapple fixture is aligned with the end-effector, completing a successful connection. The Canadarm was compatible with fixtures that supported mechanical latching as well as power and data transfer ports, enabling it to capture various payloads equipped with the appropriate fixture. The same mechanism has been used in the next version, Canadarm2, which is operational on the ISS.

One recently developed interface is presented in [26]. It includes a rod that is inserted in the target matching part, equipped with an expandable triangular pantograph that mechanically traps the target. It also includes additional conical ports driven to matching holes for further facilitation of latching, as well as connectors for power transfer, data transmission, and refuelling operations. Successful capturing of a payload with this interface has been tested on an air-bearing table.

The European Commission SIROM project has also led to the development of an interface that includes triangular shaped sides for easier mating with the coupling part [30]. As soon as the coupling is achieved, small hooks grapple the target part. It also includes ports for data, thermal, and power transfer.

7.4.2   Engine nozzle probing

A well-studied method of capturing a space target is by inserting an actuated probe at the nozzle of its main engine. The nozzle of the engine has the form of an expanding cone towards the exit, which gets narrower towards the throat. A probe is inserted into the nozzle and expands upon meeting the throat, blocking the exit and preventing slippage out of the nozzle. The target is then secured and captured in a soft way. This method can be used with most types of satellites, as well as pieces of space debris, that include an engine, such as spent rocket stages.

One of the first mechanisms to be used for this purpose was developed by NASA and used for capturing satellites that failed to reach the pre-defined orbit [31]. The mechanism included a retracting probe, which upon entering the nozzle toggled three mechanical fingers and captured the nozzle. It included a ring that was used to support the overall structure and actuators controlled by a control box. The whole mechanism was equipped with a standardised grapple fixture that enabled it to be mounted both on the Manned Manoeuvreing Unit, as well as the Space Shuttle's Remote Manipulator System.

Michigan Aerospace Corporation developed an autonomous rendezvous and docking system that is low cost and lightweight [32]. It uses the same principle of capturing a conical target interface with an expanding probe. The active component on the servicing spacecraft includes a conical boom that extends an elastic probe with a brass sphere at its end. The elasticity of the probe enabled capturing of a target at 27° angle and did not require elaborate control for sliding the probe on the bottom of the cone. The target cone is not an engine nozzle but a similarly shaped conical interface that latches on the brass sphere. After latching, the probe is pulled to perform hard capturing.

A prominent study of the dynamics of capturing nozzles by probing was conducted by Tohoku University [33]. In it, an expanding probe was used to capture

*Figure 7.7   Probe insertion on an engine nozzle [33]. The approaching
spacecraft inserts a probe on the nozzle interior, using a compliant
controller for minimum disturbance. The probe expands upon
reaching the throat, securing the payload. Image: Taylor and
Francis*

a target nozzle, shown in Figure 7.7. The interaction dynamics and the impedance
control modelling of the service arm were the main subjects of the study, and experiments were conducted using two robotic arms.

The Experimental Servicing Satellite project by the German Aerospace Agency
(DLR) also explored the idea of capturing a satellite with an expanding probe [34].
Apart from the actuation mechanisms, the capturing tool was equipped with force
sensors, so that a more elaborate compliant control could be implemented during
insertion. A visual servoing technique was implemented for the detection of the nozzle and the respective control of the manipulator. DLR also conducted experimental
results on a hardware-in-the-loop testbed of two industrial robotic arms. One arm
was equipped with the capturing tool and was controlled to emulate an approaching
spacecraft, and the second arm had a target's panel and nozzle mounted, so as to
simulate their in-orbit dynamics.

An expanding rod mechanism was also developed by the Harbin Institute of
Technology [35]. An actuated rod driven by a DC motor was used, protruding from
a conical locking surface. The mechanism has F/T sensing capabilities to detect
collision. It has a total mass of 4 kg, and it can achieve locking in about 3 s. The
capturing mechanism was tested on a robotic arm with a machined engine nozzle,
and it was able to lock to the nozzle and lift it.

7.4.3   Robotic grapples

Robotic grapples have also been proposed as simple mechanisms to capture targets
that incorporate handles. A typical design of a grapple has a joint and two fingers
that can open and close, enclosing the handle securely. Such mechanisms are used
in servicing targets that have corresponding handles preinstalled, as well as the ISS.

The JAXA ETS-VII mission [36] used a grapple tool with two fingers to capture a handle on a small satellite about 20 cm away. The handle included a marker

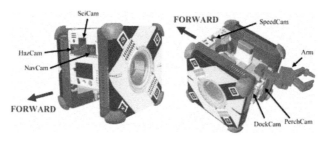

Figure 7.8   *The Astrobee robotic platform, designed for operations inside the ISS. Among other systems, it is equipped with a 3-DOF arm and gripper system that enable it to perch on the handrails of the station. Image: Sage Publishing*

for visual servoing and control of the arm. After grappling, the target was again released, achieving linear velocity drift of 1 mm/s and 0.01 deg/s in rotation.

The Astrobee robot is a platform developed by NASA to assist with operations in the ISS [23]. It consists of a cuboid free-floating platform, with a visual-based navigation and localisation capabilities, and a fan-propelled propulsion system (Figure 7.8). It is also equipped with a small 2-Degree-of-Freedom (DOF) manipulator with a 1-DOF gripper to enable it to perch on the ISS handrails [37]. All DOF are motor-controlled and the gripper system includes a torsional spring mechanism and tendons to enable the robot to hold its grasp even when the gripper motor is off. The fingertips can apply a gripping force of 2.87 N. The Astrobee has been tested in simulations and air-bearing tables in NASA Ames Research Center, as well as on board the ISS.

Apart from handles, a proposed alternative grasping point is the Payload Attachment Fixture (PAF), otherwise noted as Launch Adapter Ring (LAR) of a satellite, or Marman ring. A LAR is a ring-shaped structural extrusion on a satellite that is used as a mounting point for the payload on the launch vehicle. It is a structural part that can endure large force loads and vibrations, and its extrusion makes it ideal as a capturing and docking point. MDA Corporation has developed a LAR capture tool to deorbit the Envisat satellite [38]. The LAR tool consists of two robotic clamps in a curved configuration that can open and close, trapping the LAR section. A laser positioning system is used to determine whether the tool is in capturing range, and an integrated vision system is used for relative pose estimation. The tool has a mass of 15 kg and clamping force specifications of 1 000 N vertical and 2 000 N axial. The LAR tool was tested as part of the ESA e-Deorbit mission study [39], successfully capturing a section of the Envisat LAR mock-up. A more lightweight mechanism has been developed by the Polish Industrial Institute for Automation and Measurements [40]. This version consists of two independently actuated grapples. The capturing is performed in two phases: The first grapple applies a soft grasping force and positions the LAR in place for the second grapple. The second grapple then applies a high force that secures the target. The gripper has a mass of 3 kg and can withstand maximum loads of 200 Nm along the grasping direction. It has been tested in air-bearing tables and rendezvous testbeds with robot arms, successfully capturing LAR mock-ups. Another mechanism was presented by OHB System AG [41]. This mechanism is grapple-shaped, but with linear actuation of the jaws. One of the jaws is slightly tilted so that when it closes around the LAR section, it prevents slipping. The design was tested in simulations with the Envisat satellite.

NASA Jet Propulsion Laboratory has proposed the REMORA CubeSat design for rendezvous and attachment to large space debris [42]. The CubeSat includes a miniature stowable robotic arm. The arm has a mass of 0.75 kg and a reach of 30 cm. Two types of end-effector were tested for the arm. Both include a grapple design, with two sets of two fingers. The first design includes non-intersecting jaws and a motor with spring loading mechanism. This mechanism stores energy when the gripper is open, and swiftly exerts high grasping forces when closed. The second design has intersecting jaws, and it is motor-actuated to enable control of the grasping motion. Both designs were tested to grasp cylindrical and flat sections, which correspond to spacecraft handles and engine nozzle surfaces.

7.4.4   Dexterous hands

The tasks that an astronaut has to perform on daily basis for operations on board the ISS require dexterity in handling a multitude of tools, various handles, electrical buttons, and soft materials such as blankets and covers. Such tasks need to be performed both inside the ISS as well as on the exterior as part of Extra-Vehicular Activities (EVA). The need to automate such tasks using both autonomous and teleoperated robots has led to the development of a number of dexterous robotic hands that aim to reproduce the motions of a human hand as faithfully and smoothly as possible. The main advantages of these hands are their versatility in handling different objects and the seamless incorporation with haptic feedback gloves for teleoperation.

The Self-Adapting Robotic Auxiliary Hand robotic hand was developed by MD Robotics Limited and Laval University, to be used as end-effector for the CSA Special Purpose Dexterous Manipulator in ISS inspection and maintenance tasks [43]. It was a 3-fingered underactuated hand with 10 DOF but has only 2 motors to control them. The hand was reconfigurable and self-adaptable, through a passive mechanism consisting of linear springs that enabled the fingers to adapt to the shape of the captured object. One motor was used to control the opening and closing of the hand, and the other to control the rotation of the fingers on the palm plate. It was lightweight, weighing 10 lbs (4.53 kg), with a maximum grasping force of 50 lbs (222 N). It was capable of grasping EVA-related objects such as tools and handrails, increasing the versatility of existing orbital grippers.

A 3-fingered hand for Space Shuttle EVA operations has also been developed at Space Systems Laboratory, University of Maryland [44]. Its design was based on grasping requirements and taxonomy for 242 crew aids, tools and interfaces. Examples of such CATs include tools, restraints, handles, and storage compartments. The resulting hand was tendon driven with 12 DOF. As it was a prototype, the actuating system consisted of hand-driven pulleys. The hand was tested in grasping cylindrical objects and was found able to hold a cylinder of 20 lb (9 kg).

The NASA-DARPA Robonaut is perhaps the most notable example of dexterous robot developed for space applications. Robonaut was developed to assist ISS astronauts with EVA operations. A basic design requirement was the ability of the robot to interact with the standardised handles, interfaces, and tools on board the ISS. Its hand was 5-fingered, and it had 14 DOF, 12 in the hand and 2 in the forearm [45]. It was equipped with absolute position sensors and encoders on all its motors, and it had load cells to detect force feedback. It was also capable of tactile sensing. Even though the power and precision grasps are the basic methods of interaction with objects for the ISS tools and interfaces, the Robonaut hand had been designed to accommodate 50 per cent of the anthropomorphic grasp configurations from the Cutkosky grasp taxonomy [46], and this enabled it to perform additional dexterous manipulations. For additional safety, the hand was able to lock when power loss occurred, to prevent sudden release of objects.

The second version, the Robonaut2 hand, improves upon the original with larger joint limits, minimisation of wiring, and extended durability [47]. To increase modularity and reliability, the actuators of the hand are contained inside the forearm, making the palm-forearm element a standalone system. The finger design is also modular, enabling easier replacing and interchanging of components. Each finger is capable of applying 2.25 kg of force when extended, and a tip velocity of 20 cm/s. Its dexterity capabilities are increased as well, managing to execute 90 per cent of the Cutkosky grasp taxonomy set. The Robonaut2 hand features easier interaction with soft materials such as blanket-type coverings, and precision interaction with knobs and rotary switches.

The DLR in collaboration with ESA has developed robotic hands for orbital operations. Dexhand [48] is a 4-fingered anthropomorphic hand designed for both autonomous and teleoperated control in orbit (Figure 7.9). The main development

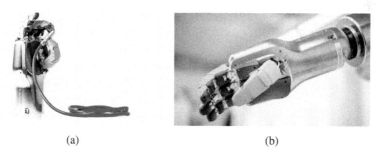

(a)   (b)

Figure 7.9   Dexterous hands are a technology for orbital grasping developed for teleoperation and applications on board the ISS. (a) The Dexhand developed by DLR. It is a dexterous 4-fingered hand designed for applications on board the ISS. Image: DLR. (b) The Spacehand is an improved version of Dexhand, designed for teleoperation in geosynchronous orbits. Its electronics are stowed in an aluminium shell for radiation hardening, and it operates under the SpaceWire data protocol. Image: DLR

requirement was the ability to grasp a large variety of tools for EVA activities, such as pliers, hammers, brush, wrenches, and pistol grip screwdrivers. An analysis of these requirements drove the technical design of the hand and resulted in the optimal technical parameters such as finger size, shape, number of DOFs, and palm geometry. The hand has 12 DOF and it is fully actuated with a combined driving system of 3 gear motors and tendons for each finger. Their fingers can produce a force of 25 N and passively resist a force of 100 N. For sensing, it uses Hall effect sensors for joint position measurement at each motor and strain gauges for torque measurement. This enables the implementation of joint position and torque controllers. Its successor, Spacehand [49], was heavily based on Dexhand but it was designed to operate for GEO missions (Figure 7.9). The main updates were the improved wire housing and routing within the hand structure, as well as the modification of the initial CAN data transformation bus to the SpaceWire communication protocol to comply with space-grade standards. It was designed to offer prolonged resistance to radiation and modified thermal behaviour.

The Beijing Institute of Spacecraft System Engineering has developed a dexterous hand for space operations [50]. The hand is 5-fingered, with 15 DOF in total. The thumb, ring, and middle fingers have three DOF and are driven by a tendon system actuated by four motors. The ring and little fingers have two DOF and require three motors for actuation. There are two additional motors on the wrist of the hand. The hand has extensive sensing capabilities, with 6-axis F/T sensors on each fingertip, a tension sensor at each tendon, and position sensors on each motor. The total system has a mass of 5 kg.

7.5   Latest R & D on orbital grasping

While some methods have been studied and tested to the point that they become standardised in orbital robotics, some recent research attempt to push the boundaries in how orbital grasping can be executed.

7.5.1   Alternative gripper designs

While most of the robotic gripping technologies for orbital targets are categorised above, some recent studies propose mechanisms with more elaborate designs that are tailored specifically for the designed target.

The mission concept presented in [51] describes the deorbiting of a 30 kg payload that is equipped with a triangular handle. For secure capturing of the handle, the authors introduced a novel gripper consisting of a triangular palm and three fingers (Figure 7.10). Each finger is designed with a distal and proximal phalange that can move independently of each other. When the finger approaches the handle, the distal phalange is actuated by a rigid tendon system that slowly closes it. The proximal phalange does not move during the first close due to a torsional spring holding it with a bar stop. As soon as the rigid tendon hits the bar stop, it releases the spring and enables the rotation of the proximal phalange, further enclosing the handle and preventing it from opening. The payload is then firmly

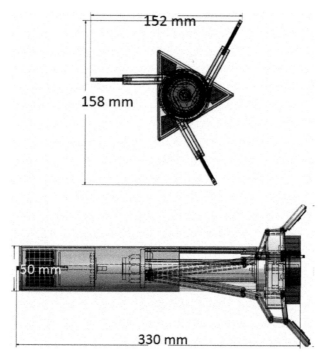

Figure 7.10   *The 3-fingered gripper presented in [51]. The fingers are designed to optimally capture a triangular handle on the target. Image: Springer Nature*

attached to the end-effector. The system kinematics and dynamics were provided, and a simulation was executed to capture a payload with impedance control of the end-effector.

A gripper for capturing the Marman ring of a spent upper rocket stage is presented in [52]. The gripper has two co-linear fingers that are able to extend to opposite directions when actuated. Each finger has a truss-like structure to facilitate retraction and extension as well as to provide extra rigidity. The end tip of each finger is V-shaped, with two metal bars protruding. The mechanism is actuated by a single motor. The fingers are stowed during the target approach phase. When the manipulator is in line with the Marman ring, the fingers are slowly extended, and they trap antipodal sections of rim. The V-shape enables capturing even in the presence of end-point placement error. The authors tested this mechanism in an air-floating system. The target was a mock-up of a payload adapter ring with a target that had 7.3 kg of mass. The chasing spacecraft was equipped with a 3-joint manipulator and F/T sensor, and its mass was 30 kg. They implemented two methods of capturing, one with an impedance controller and another with position control. The impedance control method demonstrated that the gripper captured the target with minimal interaction forces.

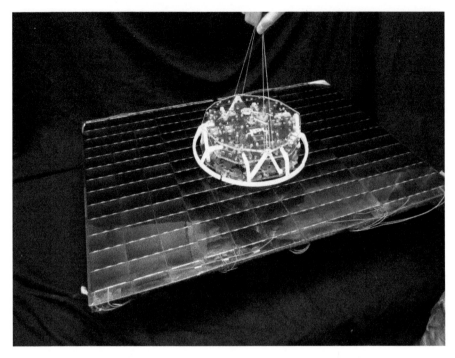

Figure 7.11   Researchers at Stanford and JPL have developed a new adhesive
grasping method inspired by gecko climbing. It has been tested as
an alternative method for capturing orbital targets. Image: NASA

7.5.2   Adhesive grasping

What is common in all capturing approaches and methods is the need to firmly
secure the payload on the gripper of the chasing arm. In most cases, the used gripping mechanism is actuated to apply mechanical locking of a specific part of the target. This is the case for the aforementioned grapple-handle and probe-nozzle pairs,
as well as the standardised interface pairs. This leads to the target having a few, or
sometimes a single, potential contact points where secure capturing can be achieved.
In addition, capturing a target this way requires elaborate impedance control algorithms that minimise collision forces, as well as accurate positioning of the end-
effector to the contact point.

To tackle these issues, researchers from Stanford University and NASA Jet
Propulsion Laboratory have developed a novel type of bio-inspired adhesive that
mimics the sticking mechanism of gecko lizard toes [53] (Figure 7.11). A typical
adhesive patch consists of microscopic flaps of about 40 nm size. When applied
on a flat or curved surface, the generation of Van Der Waals forces between the
patch and the surface securing the grasped payload. The forces are generated by
gently touching the patch on the surface and applying a tangential (shear) load on
the patch. When the same tangential load is applied again, the contact is released

with minimum disturbance on the object. The dry adhesion principle was applied with a gripper manufactured to apply loads on multiple patches simultaneously. The gripper includes a set of rectangular adhesive patches, as well as four adhesive fingers. The fingers are used to facilitate capturing of curved surfaces and the adhesive patches for capturing flat surfaces. A system of pulleys in the gripper is used to evenly distribute the load over all patches and to apply the adhesive shear force.

The gripping mechanism has been tested in JPL's Formation Control Testbed, where a 370 kg robot on an air-bearing table successfully approached and captured a second rotating robot using gecko-adhesive grippers. The chasing robot then pulled the target to a desired location and released it with minimal detachment force. Additional testing was conducted aboard NASA's parabolic flight aircraft, where the gripper was held by a human and was used for capturing a sphere, a cylinder, and a cube. The success rate was 75 per cent for the cube, 81 per cent for the cylinder, and 100 per cent for the sphere. Handheld flat grippers were deployed aboard the ISS. The results showed that behaviour of the adhesive forces in space is similar to that on the ground. The durability of the patches was also demonstrated by leaving them on top of a surface for several weeks. Continuing research on this technology shows its capability for operations outside the ISS and other large spacecraft. The LEMUR robot [54] has demonstrated a climbing robot successfully crawling across mock-up solar panels and radiator panel surfaces in a gravity off-load testbed.

Instead of relying on impedance control for soft grasping, the authors in [55] presented a capturing mechanism that includes tentacle-like fingers and adhesive patches. The end application was ADR. The grasping system consists of a gripper that is held to the spacecraft base by a tether. The gripper has eight tentacle-like 'soft' fingers that are deployed and attached to the target. They tested two designs for the fingers. The first design consists of a textile patch with long trapezoidal shape that has a spine for actuation and control. The second design consists of interconnected rigid links with pin joints to increase compliance and an internal pulley actuated to roll and unroll the tentacle. Both designs included patches of adhesive tiles that are attached to the object. They tested both gripper designs on an air-bearing testbed, capturing a mock-up target. They tested various surface types on the type to emulate different parts of the satellite to be grasped. For the adhesive tiles, they tested between 'gecko' adhesion described above, and a mixed design of gecko and electro-adhesive patches. Overall, the interconnected gripper showed better grasping capabilities by withstanding higher loads for all surface types and mixed adhesive type.

7.5.3 Affordance-based grasping

The theory of affordance has been developed in the field of psychology [56]. One way the affordance can be defined is the set of an object's properties that shows the actions users can perform with it. As a result, affordance is the association of an object with predefined actions, e.g. door handle-rotate, tennis racket-swing. The theory of affordance has been utilised extensively in robotic grasping, mostly for learning object-related actions and detecting visual information that will enable these actions, such as handles, buttons e.a [57].

The affordance theory has also been applied to orbital robotics. The ISS is an environment where most pieces of equipment have been designed to fulfil a specific purpose, with specific associated actions. Blue handles are built for grappling, stabilisation, and navigation within the ISS modules. Storage compartments have pre-designed ways of opening and stowing. A caveat is that even though the manipulations to be executed with this equipment are known in advance, a high level of dexterity is still required to execute them.

The Robonaut2 developed by NASA Johnson Space Center and General Motors has been designed for dexterous manipulations on board the ISS and can accommodate them with a high number of DOF and sensing capabilities. For executing these manipulations, Robonaut2 uses affordance-based manipulation algorithms. The *Affordance Template* Framework [28] introduces the affordance templates, which are tree descriptions of objects and corresponding end-effector trajectory waypoints. The root of the tree consists of an object, a coordinate frame that shows the pose of the object with reference to the robot, and a set of end-effector waypoints for the associated actions. By applying an action, the robot then traverses the tree graph to a new object node with associated waypoints. For example, an affordance template for Robonaut2 can consist of a ring valve. The waypoints that correspond to the valve enable the robot to plan for trajectories that will eventually place its hands on the valve, close the grasp, and rotate the valve and let it free. The templates can be modified by changing an object's location, scale, or waypoint positions. This enables the robot to execute complex tasks for a large number of objects and generalise to different types of object categories.

The Affordance Template Framework is highly incorporated into Robot Operating System (ROS) to build a shared-autonomy system for controlling the Robonaut2. Each template can be dragged and dropped from an interactive panel to the facing scene and manipulated to fit the real object's pose and scale. This way the robot knows which part of the scene is a template and where to apply the action. The waypoint planning is achieved by integration with the Moveit! planning framework. An example scene is shown in Figure 7.12.

The Affordance Template Framework has been tested in microgravity conditions using the ARGOS gravity offload testbed in NASA JSC [58].

7.5.4   Grasp synthesis

In terrestrial robotics, object grasping is a field that has received continuous attention over the years. It is an action that robots can conduct autonomously using visual raw data such as point clouds or images and for a great variety of objects. As more and more robots are deployed in space, and as the levels of their autonomy are increased, the solutions presented for all phases of orbital operations should be more versatile. Target capturing has been an activity that requires simple solutions such as grapples and clamps, or grippers exclusively tailored to work for an individual target. With the aim of increasing the levels of autonomy and versatility of the capturing process,

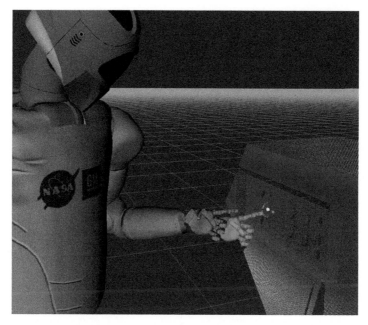

Figure 7.12   The Affordance Learning Framework is a ROS-based framework that enables shared control of the Robonaut2 humanoid [28]. A button template is placed by the user in the corresponding art of the scene and the robot plans the hand trajectories to press it. Image: TracLabs

research from STAR LAB at Surrey Space Centre has proposed one method for autonomous grasp synthesis on an orbital target [59].

The target used for capturing was an engine nozzle. The algorithm starts from a point cloud representation of the engine nozzle extracted from an RGB-D sensor. The point cloud is downsampled for faster processing, with a voxel size of 1.5 cm. The cloud is then geometrically filtered with outlier radius of 5 cm. To characterise the nozzle surface and distinguish between areas closer and farther to the rim, a moment-based surface analysis is applied. For each point in the cloud, the Zero-Moment Shift (ZMS) is calculated as a measure of the point's distance to the mean of its neighbourhood. The ZMS norm is then used to construct a 4D feature vector for each point that includes the 3D coordinates of the point with reference to the camera and the ZMS norm. The feature vectors of all points are fed to an Affinity Propagation clustering algorithm. The algorithm results in clustered nozzle surface patches that share similar surface characteristics and are close to each other. Each patch corresponds to a grasping surface. A grasping frame is generated by applying Principal Component Analysis (PCA) to each patch. The final grasp results from selecting the grasp that satisfies three criteria: being not very close to the rim or far inside the nozzle throat, having a patch area greater or equal to the gripper finger

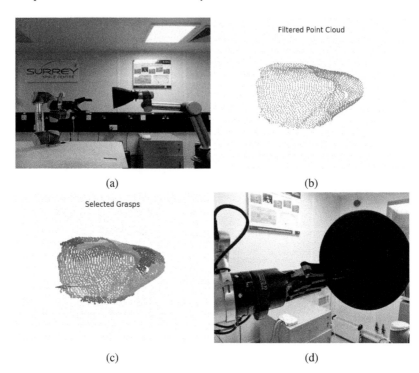

Figure 7.13   *Grasp synthesis on the surface of a rocket engine [59]. Images:*
STAR LAB. (a) The chasing arm faces the nozzle engine mock-up
mounted on the right arm. (b) The RGB-D sensor extracts a point
cloud of the nozzle that is preprocessed. (c) The result of the
grasp synthesis algorithm is a reachable pose on the nozzle. The
calculation speed can be as fast as 0.2 s. (d) The robot executes the
reaching motion and grasps the nozzle

pad, and resulting in a gripper yaw angle within specified limits so as to capture the nozzle as straight as possible.

The algorithm is very fast and can extract 30 grasps in 0.2 s, which makes it suitable for real-time grasp synthesis. It was tested in a V-REP simulation and the Orbital Testbed of STAR LAB [19]. An example execution is shown in Figure 7.13.

The stability of the nozzle grasp has also been studied, with preliminary analysis and simulations showing that the fingers yield a stable grasp on the surface of the nozzle [60].

The NASA Astrobee platform is also capable of autonomous identification of an ISS handrail, for perching with its 3-DOF manipulator [61]. The Astrobee platform is equipped with a depth sensor that enables it to extract a point cloud of the scene with the handrail. First, a plane is fitted on the point cloud using the RANSAC algorithm. The points are then classified as inliers or outliers according to their distance from the plane as well as the width and shape of their local geometry. The result of this are elongated clusters of points that correspond to the handrails. A line is fitted on each cluster using RANSAC. Since the handrails on the ISS are either parallel or vertical to the 'ground', the selected handrail for perching results from the line that has lower angle difference to the Astrobee gripper. The robot can then navigate to the handrail using an Extended Kalman Filter for pose estimation and tracking of the handrail. The algorithm was tested in simulation and on a real testbed, achieving pose estimation errors of about 1 cm for the position and 1° for the orientation.

7.6   Related missions

Spacecraft docking and payload berthing are operations that have been conducted ever since the start of the Space Race of the 1960s and consist of routine operations of all transfer vehicles that reach the ISS. At the same time, orbital debris capturing has been demonstrated with the usage of fired harpoons and nets. Instead, this section aims to describe past and future milestone missions that have demonstrated autonomous or teleoperated target capturing with the use of *robotic manipulators* and various levels of autonomy.

7.6.1   ETS-7

The ETS-7 mission launched in 1997 by NASDA (now JAXA) was the first unmanned mission to incorporate a robotic arm on the main satellite body [62]. The mission consisted of a chaser spacecraft with a mass of 2 450 kg equipped with a 6-DOF manipulator with a length of 2 m, and a smaller target spacecraft, with a mass of 410 kg. The robotic arm supported both autonomous and teleoperated control. The mission had a design lifespan of 1.5 years but it was able to operate for additional 6 months.

Throughout its mission, ETS-7 performed docking experiments through its main docking mechanisms and various robotic capturing operations using its robot arm. The robotic arm used visual-based navigation to capture handles on the target spacecraft. A visual target recognition experiment was conducted in partnership with ESA and a teleoperation experiment with DLR.

7.6.2   OSAM-1

The NASA OSAM-1 (On-orbit Servicing, Assembly, and Manufacturing 1, formerly Restore-L) mission is a planned one that will demonstrate many operations on satellite servicing and in-space assembly [63]. The mission is scheduled to launch in 2024. The main objective is to perform rendezvous and docking with a US government-owned satellite to extend its life by transferring propellant. In addition to the refuelling objective, OSAM-1 will assemble a Ka-band communications antenna that consists of seven modular parts and has a diameter of 3 m. The antenna will be used to establish successful downlink with the ground station. OSAM-1 will also assemble a composite beam.

To achieve the servicing and manufacturing objectives, OSAM-1 will utilise a range of novel robotic technologies. Two 7-DOF lightweight manipulators will be used for target capturing and refuelling. One will be equipped with a clamp-like mechanism that will detect the Marman ring of the target satellite and grapple it. After grasping, the target satellite will be pulled towards the OSAM-1 satellite. The second manipulator equipped with a propellant transfer system, flexible hose, and oxidiser nozzle tool, will transfer propellant to the target.

The 5-m, 7-DOF Space Infrastructure Dexterous Robot will be used as a third manipulator for antenna and beam assembly.

7.6.3   ELSA-d

The ELSA-d (End-of-Life Service by Astroscale-demonstration) mission was launched in 2021 by Astroscale and will demonstrate a number of technologies for end-of-life spacecraft services [64]. The mission consists of a chaser spacecraft equipped with a 1-DOF actuator with a magnetic plate as end-effector and a target satellite equipped with a magnetic docking plate with localisation markers.

The mission will demonstrate both far-range and short-range target detection and capturing with the magnetic docking system. The capturing will be performed on both non-tumbling and tumbling target scenarios. For the tumbling target, the chasing spacecraft will execute a manoeuvre to come in close proximity and match the target attitude.

7.6.4   MEV-1

The MEV-1 mission by Northrop Grummann is the first to perform docking of two commercial satellites [65]. Launched in 2019, the mission was designed to service Intelsat-901, which had been put in graveyard orbit. MEV-1 was able to locate Intelsat-901, dock with it, and place it back to a geosynchronous orbit. MEV-1 is tasked with performing the station-keeping of Intelsat-901 until it is time to place it again on the graveyard orbit.

MEV-1 used an extending probe to execute a soft capturing of Intelsat-901. The probe was inserted in the bell of Intelsat-901 liquid apogee engine and the end point expanded after the insertion, successfully locking the target. The probe was then retracted and the two spacecraft docked. For releasing the captured satellite, the operations will be executed in reverse, with the probe extending, releasing the target, and retracting again.

7.6.5   ClearSpace-1

The ClearSpace-1 mission, developed by the Swiss company ClearSpace and contracted by ESA, is poised to be the first mission to detect, capture, and deorbit a piece of space debris [66]. The mission will be launched in 2025. The target is the Vespa payload adapter from the launch 2013 ESA Vega flight VV02.

The spacecraft will detect the target and execute an approaching manoeuvre. For grasping the target, ClearSpace-1 will be equipped with four long-reach robotic manipulators that will capture the target in an enclosing manner, preventing any slippage. The spacecraft will then deorbit while holding the target.

7.7   Technical challenges of orbital grasping

Making secure contact with an orbital target is an inherently challenging and complicated problem. The processes of autonomous orbital docking and berthing have been crucial elements of space activities in the last century. As a result, their challenges have been widely mapped, and the mechanisms and operations for docking and berthing have been heavily standardised. The difference in autonomous orbital grasping is that it is applied on non-typical targets, and the related studies and experiments are rare in comparison. Some attempts have been made to map out challenges and articulate standardised requirements for robotic grasping in orbit [67], and general examples of the challenges of orbital grasping are given below.

7.7.1   *Algorithmic modelling – design challenges*
7.7.1.1   Target state estimation
Several properties of the target's state need to be estimated prior to capturing, such as its pose in relation to the chasing spacecraft, angular velocity, and inertial parameters. When the target is cooperative, estimation may not be necessary, as such properties can be communicated to the chasing robot by the ground operations team or the target satellite. The state of the target can also be controlled to facilitate the grasping process. When the target is non-cooperative there is a need for implementation of estimation algorithms. Such algorithms typically make use of visual and geometrical information of the target expressed as RGB images and point clouds, and use analytical or data-driven models to estimate the current and predict the future state.

7.7.1.2   Identification of grasping point
Grasping an orbital target needs one or more contact points for the chasing robotic arm. The selection of a contact point is not a trivial task. A contact point needs to be load-resistant to ensure that the capturing is stable. The grasping point should not risk compromising any of the target's operations, and so subsystem parts such as solar panels and antennas are not possible candidates. Most prominent targets are the PAF and the engine nozzle; however, the mentioned reasons make the standardisation of contact points for orbital grasping more difficult.

7.7.1.3   Grasp analysis and modelling
In terrestrial robotic grasping, the analytical models that describe the relationship between the wrench applied from the fingers and the total wrench resulting in the grasped object's motion have been well developed over the past 40 years. The *Grasp Map* is a matrix that encodes this information. Similarly, notions such as *grasp stability* and *force-closure* have been well defined and studied in terrestrial grasping. A

variety of *grasp quality* metrics have been proposed to evaluate the grasp in terms of resistance to disturbances and efficiency in object handling. Most of the aforementioned properties describe the performance of a grasp on external and applied forces, and object weight is the primary force to be resisted. In orbital robotics, the target weight becomes negligible and the dominant forces that risk breaking a grasp are reaction forces at the moment of contact and inertial forces while the target is handled. The models and metrics that characterise a terrestrial grasp should be redefined for the case of weightless grasping.

7.7.1.4   Machine learning

The recent application of machine learning algorithms has led to a rapid growth of terrestrial grasping research and solved problems related to grasp modelling, synthesis, and planning with leaping progress. In orbital robotics, the application of machine learning has not been widely attempted. Similarly, to terrestrial robotics, machine learning can be used to provide fast solutions to many of the challenges mentioned above.

7.7.1.5   Gripper design

The gripper selection and design needs to be on par with the mission requirements, operations, power and communications budgets, and economic cost. This is important for all phases of the gripper design, e.g. grasping principle selection, mechanism design, material selection, actuation choice, degrees of freedom e.a.

7.7.2   *Physical challenges*

7.7.2.1   Space environment

The harsh nature of space affects grasping and target capturing. The material parameters of a target surface such as friction and elasticity are affected by the low orbital temperatures and the temperature gradient that a target may be subject to. Solar radiation may degrade the target material, making it more brittle and prone to breaking. Collisions with space debris particles may further damage the structural integrity of the target. Such parameters are unique to each potential target for capturing and need to be taken into account in the design phase of the grasping algorithm or hardware.

7.7.2.2   Impact mitigation

When two bodies make contact in microgravity, their total angular momentum remains constant. This can cause the target to drift away when a momentary impulse is applied as a result of the grasping process. To mitigate these effects, most existing studies use elaborate impedance controllers on the chasing arm. This enables the arm to increase its compliance and act as a shock absorber, preventing potential drift. Another suggested method is to coat the robot fingers with compliant material so that the contact is softer.

7.7.2.3 Debris generation

The grasping operation should be executed with minimum risk of producing space debris. Damaging and breaking up of a target's component is a rare but existing danger of orbital grasping. A more apparent risk is the generation of micro-debris such as paint flecks and grind dust material that may result from non-compliant contact.

7.7.3 *Operational – verification challenges*

7.7.3.1 Post-capture operations

The grasping method should be designed in a way to accommodate post-capture operations, such as easier rigidisation and target detumbling. In addition, it should be able to facilitate the chaser manipulator motion planning and control for the completion of the mission objectives.

7.7.3.2 Standardisation and benchmarking

The sheer volume of research in terrestrial robotic grasping has led to attempts in benchmarking of the development and testing phases through the use of common object datasets where algorithms can be tested and compared, as well as commonly used open-source grasping software and experimental protocols. The results from this benchmarking have led to some platforms, datasets, and algorithms becoming the standard for the community, enabling accelerated development of novel grasping methods and the comparison with the existing ones. In orbital robotics, similar initiatives are fairly new, with attempts being made to provide industry-accepted standards for docking and manipulation tasks related to OOS and debris removal.

7.7.3.3 Verification and validation

The development of verification and validation methodologies for orbital grasping is another field that needs to be developed to accelerate the field and attract wider industrial interest. As each mission has unique objectives and design parameters, a bespoke combination of verification methods needs to be applied such as formal verification, extensive simulation, gravity offload testbed prototyping, and runtime verification through requirement monitoring. Such studies are still in preliminary stage [68].

Orbital robotic grasping is one of the fields that have been designated as strategically important by space agencies and organisations. The 2020 NASA Technology Taxonomy lists Dexterous Manipulation, Grappling Technologies, Capture Mechanisms, and Fixtures, as well as their modelling, control, and simulation as key technologies for space robotics [69]. The European Commission PERASPERA project also lists Dexterous Manipulation as a technology to be developed for its Orbital Track. With such organisations increasingly including orbital grasping activities as one of the priority technologies to be developed, the challenges will be defined and standardisation will occur.

References

[1]   Plaisier M., Smeets J. 'Mass is all that matters in the size–weight illusion'. *PloS One*. 2012:7–8.

[2]   Ross H.E., Reschke M.F. 'Mass estimation and discrimination during brief periods of zero gravity'. *Perception & Psychophysics*. 1982;**31**(5):429–36.

[3]   Flanagan J.R., Wing A.M. 'Modulation of grip force with load force during point-to-point arm movements'. *Experimental Brain Research*. 1993;**95**(1):131–43.

[4]   Hermsdörfer J., Marquardt C., Philipp J., *et al.* 'Moving weightless objects: grip force control during microgravity'. *Experimental Brain Research*. 2000;**132**(1):52–64.

[5]   Augurelle A.-S., Penta M., White O., Thonnard J.-L. 'The effects of a change in gravity on the dynamics of prehension'. *Experimental Brain Research*. 2003;**148**(4):533–40.

[6]   White O., McIntyre J., Augurelle A.-S., Thonnard J.-L. 'Do novel gravitational environments alter the grip-force/load-force coupling at the fingertips?' *Experimental Brain Research*. 2005;**163**(3):324–34.

[7]   Zatsiorsky V.M., Gao F., Latash M.L. 'Motor control goes beyond physics: differential effects of gravity and inertia on finger forces during manipulation of hand-held objects'. *Experimental Brain Research*. 2005;**162**(3):300–8.

[8]   Crevecoeur F., Thonnard J.L., Lefèvre P. 'Forward models of inertial loads in weightlessness'. *Neuroscience*. 2009;**161**(2):589–98.

[9]   Giard T., Crevecoeur F., McIntyre J., Thonnard J.-L., Lefèvre P. 'Inertial torque during reaching directly impacts grip-force adaptation to weightless objects'. *Experimental Brain Research*. 2015;**233**(11):3323–32.

[10] Opsomer L., Théate V., Lefèvre P., Thonnard J.-L. 'Dexterous manipulation during rhythmic arm movements in Mars, Moon, and micro-gravity'. *Frontiers in Physiology*. 2018;**9**:1–10.

[11] Bock O., Howard I.P., Money K.E., Arnold K.E. 'Accuracy of aimed arm movements in changed gravity'. *Aviation, Space, and Environmental Medicine*. 1992;**63**(11):994–8.

[12] Berger M., Mescheriakov S., Molokanova E., Lechner-Steinleitner S., Seguer N., Kozlovskaya I. 'Pointing arm movements in short- and long-term spaceflights'. *Aviation, Space, and Environmental Medicine*. 1997;**68**(9):781–7.

[13] Jüngling S., Bock O., Girgenrath M. 'Speed-accuracy trade-off of grasping movements during microgravity'. *Aviation, Space, and Environmental Medicine*. 2002;**73**(5):430–5.

[14] Schepers A. *Andreas The GRIP experiment* [online]. Note: European Space Agency. 2018. Available from https://blogs.esa.int/alexander-gerst/2018/06/21/deutsch-fingerfertigkeit-im-all/ [Accessed 21 Mar 2020].

[15] NASA. *Gravitational References for Sensorimotor Performance: Reaching and Grasping* (GRASP) [online]. Available from https://lsda.jsc.nasa.gov/Experiment/exper/13911 [Accessed 21 Mar 2020].

[16] Li W.J., Cheng D.Y., Liu X.G., *et al.* 'On-orbit service (OOS) of spacecraft: a review of engineering developments'. *Progress in Aerospace Sciences.* 2019;**108**:32–120.

[17] Nanjangud A., Underwood C., Bridges C., Saaj S. 'Towards robotic on-orbit assembly of large space telescopes: mission architectures, concepts, and analyses'. *Proceedings of the International Astronautical Congress.* 2019:21–5.

[18] Saunders C., Lobb D., Sweeting M., Gao Y. 'Building large telescopes in orbit using small satellites'. *Acta Astronautica.* 2017;**141**(4):183–95.

[19] Hao Z., Mavrakis N., Proenca P., *et al.* 'Ground-based high-DOF AI and robotics demonstrator for in-orbit space optical telescope assembly'. *Proceedings of the International Astronautical Congress.* 2019.

[20] ESA SDO. 'ESA's annual space environment report'. *International Journal of Advanced Robotic Systems.* 2019;**3**(2).

[21] Shan M., Guo J., Gill E. 'Review and comparison of active space debris capturing and removal methods'. *Progress in Aerospace Sciences.* 2016;**80**(A6):18–32.

[22] Biesbroek R., Innocenti L., Wolahan A., Serrano S.M. 'e.Deorbit—ESA's active debris removal mission'. *Proceedings of the 7th European Conference on Space Debris.* 2017:10.

[23] Bualat M., Barlow J., Fong T., Provencher C., Smith T. 'Astrobee: developing a free-flying robot for the International space Station'. *Proceedings of SPACE Conference and Exposition.* 2015:4643.

[24] Diftler M., Mehling J., Abdallah M., *et al.* 'Robonaut 2-the first humanoid robot in space'. *Proceedings of International Conference on Robotics and Automation.* 2011:2178–83.

[25] Ackerman E. *Russian humanoid robot to pilot Soyuz capsule to ISS this week* [online]. IEEE Spectrum. 2019. Available from https://spectrum.ieee.org/automaton/robotics/space-robots/russian-humanoid-robot-to-pilot-soyuz-capsule-to-iss-this-week [Accessed 21 Mar 2020].

[26] Medina A., Tomassini A., Suatoni M., *et al.* 'Towards a standardized grasping and refuelling on-orbit servicing for GEO spacecraft'. *Acta Astronautica.* 2017;**134**(4):1–10.

[27] Yan X.-T., Brinkmann W., Palazzetti R., *et al.* 'Integrated mechanical, thermal, data, and power transfer interfaces for future space robotics'. *Frontiers in Robotics and AI.* 2018;**5**:64.

[28] Hart S., Dinh P., Hambuchen K. 'The affordance template ROS package for robot task programming'. *Proceedings of IEEE International Conference on Robotics and Automation.* 2015:6227–34.

[29] Aikenhead B.A., Daniell R.G., Davis F.M. 'Canadarm and the space shuttle'. *Journal of Vacuum Science & Technology A: Vacuum, Surfaces, and Films.* 1983;**1**(2):126–32.

[30] Jankovic M., Brinkmann W., Bartsch S., Yan X. 'Concepts of active payload modules and end-effectors suitable for standard interface for robotic manipulation of payloads in future space missions (SIROM) interface'. *Proceedings of the IEEE Aerospace Conference.* 2018:1–15.

[31] Harwell W.D. 'AKM capture device'. *NASA Technical Reports Server*. 1987.
[32] Hays A., Tchoryk P., Pavlich J., Ritter G. 'Advancements in design of an autonomous satellite docking system'. *Proceedings of Spacecraft Platforms and Infrastructure*. 2004;**5419**:107–18.
[33] Yoshida K., Nakanishi H., Ueno H., Inaba N., Nishimaki T., Oda M. 'Dynamics, control and impedance matching for robotic capture of a non-cooperative satellite'. *Advanced Robotics*. 2004;**18**(2):175–98.
[34] Reintsema D., Landzettel K., Hirzinger G. 'DLR's advanced telerobotic concepts and experiments for on-orbit servicing'. *Advances in Telerobotics*. 2007:323–45.
[35] Zhang Y., Sun K., Zhang Y., Liu H. 'Expansion rod design method of capturing nozzle device for non-cooperative satellite'. *Proceedings of IEEE International Conference on Robotics and Biomimetics*. 2013:1947–52.
[36] Inaba N., Oda M. 'Autonomous satellite capture by a space robot'. *Proceedings of the IEEE International Conference on Robotics and Automation 2*. 2000:1169–74.
[37] Park I.W., Smith T., Sanchez H., Wong S.W., Pedro P., Ciocarlie M. 'Developing a 3-DOF compliant perching arm for a free-flying robot on the International Space Station'. *Proceedings of IEEE/ASME International Conference on Advanced Intelligent Mechatronics*. 2017:1135–41.
[38] Ratti J. 'Launch adapter ring (LAR) capture tool: enabling space robotic servicing'. Workshop on the Next Generation of Space Robotic Servicing Technologies, IEEE International Conference on Robotics and Automation; 2015.
[39] Estable S., Pruvost C., Ferreira E., *et al.* 'Capturing and deorbiting Envisat with an airbus Spacetug: results from the ESA e.Deorbit consolidation phase study'. *Journal of Space Safety Engineering*. 2020;**7**(1):52–66.
[40] Jaworski J., Dudek Ł., Wolski M., *et al.* Grippers for launch adapter rings of non-cooperative satellites capture for active debris removal, space tug and on-orbit satellite servicing applications. *Proceedings of the ESA Advanced Space Technologies in Robotics and Automation Symposium*; 2017.
[41] Jaekel S., Lampariello R., Panin G., *et al.* Robotic capture and de-orbit of a heavy, uncooperative and tumbling target in low earth orbit. *Proceedings of the ESA Advanced Space Technologies in Robotics and Automation Symposium*; 2015.
[42] McCormick R., Austin A., Wehage K., *et al.* 'REMORA CubeSat for large debris rendezvous, attachment, tracking, and collision avoidance'. *Proceedings of the IEEE Aerospace Conference*. *2018*:1–13.
[43] Rubinger B., Brousseau M., Lymer J., Gosselin C., Piedbœuf J. 'A novel robotic hand-SARAH for operations on the International Space Station'. *Proceedings of the ESA Advanced Space Technologies for Robotics and Automation Symposium*. 2002:1–8.
[44] Akin D.L., Carignan C.R., Foster A.W. 'Development of a four-fingered dexterous robot end effector for space operations'. *Proceedings of the IEEE International Conference on Robotics and Automation*. 2003:2302–8.

[45] Lovchik C., Diftler M.A. 'The robonaut hand: a dexterous robot hand for space'. *Proceedings of the IEEE International Conference on Robotics and Automation*. 1999;**2**:907–12.

[46] Cutkosky M.R. 'On GRASP choice, GRASP models, and the design of hands for manufacturing tasks'. *IEEE Transactions on Robotics and Automation*. 1989;**5**(3):269–79.

[47] Bridgwater L.B., Ihrke C.A., Diftler M.A., *et al.* 'The Robonaut 2 hand – designed to do work with tools'. *Proceedings of IEEE International Conference on Robotics and Automation*. 2012:3425–30.

[48] Chalon M., Wedler A., Baumann A., *et al.* 'Dexhand: a space qualified multi-fingered robotic hand'. *Proceedings of IEEE International Conference on Robotics and Automation*. 2011:2204–10.

[49] Chalon M., Maier M., Bertleff W., *et al.* 'Spacehand: a multi-fingered robotic hand for space'. *Proceedings of the ESA Advanced Space Technologies in Robotics and Automation Symposium*; 2005.

[50] Zhao Z., Li D., Gao S., Yuan B., Wang Y., Yang Y. 'Development of a dexterous hand for space service'. *Proceedings of IEEE International Conference on Robotics and Biomimetics*. 2016:408–12.

[51] Genta G., Dolci M. 'Robotic gripper for payload capture in low earth orbit'. Proceedings of ASME International Mechanical Engineering Congress and Exposition; 2016.

[52] Hirano D., Kato H., Tanishima N. 'Caging-based GRASP with flexible manipulation for robust capture of a free-floating target'. *Proceedings of IEEE International Conference on Robotics and Automation*. 2017:5480–6.

[53] Jiang H., Hawkes E.W., Fuller C., *et al.* 'A robotic device using gecko-inspired adhesives can GRASP and manipulate large objects in microgravity'. *Science Robotics*. 2017;**2**(7):eaan4545–12.

[54] Parness A., Abcouwer N., Fuller C., Wiltsie N., Nash J., Kennedy B. 'Lemur 3: a limbed climbing robot for extreme terrain mobility in space'. *Proceedings of IEEE International Conference on Robotics and Automation*. 2017:5467–73.

[55] Narayanan S., Barnhart D., Rogers R., *et al.* 'REACCH-reactive electro-adhesive capture cloth mechanism to enable safe grapple of cooperative/non-cooperative space debris'. *Proceedings of AIAA Scitech Forum*. 2020:2134.

[56] Gibson J.J. 'The ecological approach to visual perception: classic edition'. *Psychology Press*. 2014.

[57] Andries M., Chavez-Garcia R.O., Chatila R., Giusti A., Gambardella L.M. 'Affordance equivalences in robotics: a formalism'. *Frontiers in neurorobotics*. 2018;**12**:26.

[58] Farrell L., Strawser P., Hambuchen K., Baker W., Badger J. 'Supervisory control of a humanoid robot in microgravity for manipulation tasks'. Proceedings of IEEE International Conference on Intelligent Robots and Systems; 2017. pp. 3797–802.

[59] Mavrakis N., Gao Y. 'Visually guided robot grasping of a spacecraft's apogee kick motor'. *Proceedings of the ESA Advanced Space Technologies in Robotics and Automation Symposium*; 2019.

[60]   Mavrakis N., Gao Y. 'Stable robotic grasp synthesis on a spent rocket stage for on-orbit capturing'. *Proceedings of International Symposium on* Artificial Intelligence, Robotics and Automation in Space; 2020.

[61]   Lee D.-H., Coltin B., Morse T., Park I.-W., Flückiger L., Smith T. 'Handrail detection and pose estimation for a free-flying robot'. *International Journal of Advanced Robotic Systems*. 2018;**15**(1):172988141775369.

[62]   Kasai T., Oda M., Suzuki T. 'Results of the ETS-7 mission-rendezvous docking and space robotics experiments'. *Proceedings of International Symposium on* Artificial Intelligence, Robotics and Automation in Space 440; 1999. p. 299.

[63]   Reed B., Smith R., Naasz B., Pellegrino C. 'The restore-1 servicing mission'. *Proceedings of AIAA Space Forum*. 2016:5478.

[64]   Blackerby C., Okamoto A., Fujimoto K., Okada N., Auburn J. ELSA-d: an in-orbit end-of-life demonstration mission. Proceedings of International Astronautical Congress; 2018.

[65]   Redd N.T. 'Bringing satellites back from the dead: mission extension vehicles give defunct spacecraft a new lease on life – [News]'. *IEEE Spectrum*. 2020;**57**(8):6–7.

[66]   European Space Agency. ESA commissions world's first space debris removal [online]. 2019. Available from https://www.esa.int/Safety_Security/Clean_Space/ESA_commissions_world_s_first_space_debris_removal/ [Accessed 21 Mar 2020].

[67]   Mahalingam S., Sharifi M., Dwivedi S.N., Vranish J.M. 'Special Challenges of Robotic Gripping in Sace'. *Proceedings of Southeastern Symposium on System Theory*. 1988:581–6.

[68]   Farrell M., Mavrakis N., Dixon C., Gao Y. 'Formal verification of an autonomous grasping algorithm'. International Symposium on Artificial Intelligence, Robotics and Automation in Space; 2020.

[69]   NASA. 'Tx 04: robotic systems'. *NASA Technology Taxonomy*. 2020.

Part III

Astronaut–robot interaction

Chapter 8

BCI for mental workload assessment and performance evaluation in space teleoperations

Fani Deligianni[1], Daniel Freer[2], Yao Guo[2], and Guang-Zhong Yang[3]

Astronauts have to complete hundreds of hours of training with simulation systems that help them to improve their ability to operate robotic arms for docking operations. In docking tasks, performed for example on the Canadarm2, the operator does not have direct view of the International Space Station (ISS) but relies instead on visual feedback from multiple 2D camera views. Failure to accomplish the tasks on time costs millions of pounds and can potentially endanger the life of the crew members. Even in simulated tasks of the Soyuz-TMA(Transportation Modified Anthropometric) approach and docking, tension and anxiety build up quickly as the precision required is high and virtual fuels are limited. In this chapter, we investigate how simulation systems can be used as a platform to enhance and measure an operator's performance, as well as to design and evaluate semi-autonomous modes of operation that facilitates effective human–robot collaboration. Furthermore, we review how brain–computer interfaces (BCIs) can monitor workload, attention and fatigue. These systems can be evolved to provide an intuitive human–robot interaction experience that provides guidance and feedback as they are needed.

8.1 Human–robot interaction in space – what we learn from simulators

Astronauts spend hundreds of hours of training on simulators in order to acquire the necessary skills and abilities to follow procedures that are vital to their survival in space. Furthermore, they need to acquire expert knowledge to maintain and interact with complex electronic and robotic systems. In order for astronauts to retain the skills they learned from the simulator and to generalise or transfer to the operational

[1]School of Computing Science, Glasgow University, Glasgow, UK
[2]Hamlyn Centre, Imperial College London, London, UK
[3]Institute of Medical Robotics, Shanghai Jiao Tong University, Shanghai, China

domain, the training simulators used must be high fidelity [1]. This reflects the fact that crew needs to train to handle low likelihood events that are time-critical. Several simulators mimic the extreme conditions of outer space, such as a lack of gravity, oxygen and pressure, as well as very low temperatures. Examples of specialised simulation facilities include the thermal vacuum chamber at the NASA Goddard Space Flight Center and the Large Space Simulator at the European Space Agency. These systems are designed to physically and mentally prepare the crew as well as to examine material properties and tolerance in extreme conditions. On the other hand, several other types of simulators have been developed to train astronauts to use complex technology to perform surveillance, maintenance and assembly tasks, as well as how to dock modules, teleoperate robots and fly a spacecraft [2, 3]. Inherently, simulation systems provide a testbed in several aspects, which includes system testing and verification, ergonomics in human–machine collaboration and neuroergonomics in closed-loop interactive interfaces.

System testing and verification are of paramount importance in both fully automated and semi-automated systems. Real-time, complex distributed systems that combine both hardware/sensing components along with software components are notoriously difficult to design and test. Engineers rely on simulation systems to provide realistic test cases of rare and dangerous scenarios that aim to highlight shortcomings and failures. Among the most well-known space flight simulators are the STS-133 simulator and the vertical motion simulator at Ames Research Centre [3].

The aviation industry is a representative example of the challenges to address. Although the industry complies with the highest standards and there is accumulated experience of several decades, catastrophic failures continue to highlight the downside of complex automated systems. In March 2019, Boeing 737 Max 8 and 9 jets were grounded following two deadly accidents. Apparently erroneous sensor readings triggered an automated system response, which could not be controlled by the crew [4]. From one perspective, automation has helped to improve safety, but the difficulty to comprehend the complexity of the systems along with an inherent lack of transparency of current machine learning algorithms hinders a widespread adoption of fully automated systems [5].

To this end, human–machine interaction via intuitive designs [6] could empower humans and create augmented artificial intelligence(AI) frameworks that encompass expert knowledge along with data science. Simulators in these scenarios are important to perform the so-called human-in-the-loop tests [7]. In particular, docking modules on the ISS are of particular significance because they require a good spatial understanding of the environment, even when presented with limited information.

Although human error is estimated to cause over 60 per cent of fatal accidents in aviation, well-trained crew is acknowledged to be critical to the overall system's safety [7]. There is extensive research in ergonomics to design human-centred systems, and towards this end appropriate simulation environments are of paramount importance. Towards this end, cognitive workload data obtained in simulations provide valuable insight on how to design efficient interactive frameworks that minimise fatigue and improve performance.

In this chapter, we examine how realistic simulation environments used typically in teleoperated robots in ISS and docking tasks can facilitate the development of testbeds for augmented human and AI systems that sense human neurophysiology and allow them to react in real-time via intuitive designs, such as displays, haptic and auditory feedback.

8.1.1   Soyuz-TMA

The Soyuz-TMA is a spacecraft used by the Russian Federal Space Agency to launch missions from Earth to space. Currently, it is used to carry astronauts from and to ISS and it is considered to be one of the safest and most cost-effective spacecrafts in operation. Astronauts are trained to operate the Soyuz-TMA in a number of modes, which include prelaunch preparation, insertion to the orbit, orbital manoeuvering, approaching and docking at ISS, undocking, reentry to the atmosphere from orbit and landing. The onboard control system is based on a camera and a periscope view, whereas a KURS radio telemetry system provides information with regard to the relative velocity, attitude and distance of the spacecraft to the docking station. Some of the key data displayed along with the camera and periscope views include approach distance and velocity, rates of rotation for attitude stabilisation and line of sight angle for alignment. During docking, it is important to maintain rotational and translational velocity within safe limits and the docking target should also be closely aligned with the spacecraft centreline. The Soyuz-TMA can operate either in automatic mode or can be switched to manual mode when the automatic system fails to dock. In this case, the spacecraft has to abort its approach, move backwards and try again. Due to the lack of gravity once a thruster exerts a translational or rotational force on the spacecraft, the spacecraft will continue moving or rotating unless the force is counterbalanced. Therefore, successfully docking the module to the ISS requires calculated movements and precision to approach the docking station without excessive forces and to avoid running out of fuel.

Figure 8.1 shows a simulator for the Soyuz-TMA approach and docking. The docking has two levels of difficulty: (i) docking the spacecraft in the ISS hatch directly in front of it, which does not require rotational manoeuvering and (ii) docking the spacecraft in one of the left/right/bottom ISS hatches, which involves activating first the translational and rotational thruster. Rotating the spacecraft is more challenging since excessive forces could result in spinning around in a difficult to control way. Another challenge related to the camera and periscope views is that the motion is a mirrored translation of the camera's view; you move left whereas the docking target in the periscopic view is moving right, creating a perceptual conflict which can confuse the user if he is not adequately trained. Furthermore, there is no intelligent guidance to help the user understand what values could be optimal for navigating. For example, intuitive information regarding the 3D position and orientation of the Soyuz-TMA along with optimum navigation paths could allow users to learn faster.

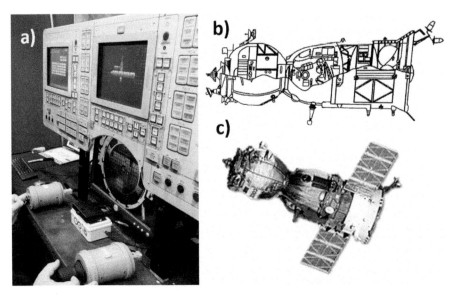

*Figure 8.1   (a) A Soyuz-TMA simulator at the UK Space Conference 2019
at Wales. (b) Schematic representation of the simulator. (c)
A photograph of Soyuz-TMA. [Photos credited to the British
Interplanetary Society.]*

8.1.2   Canadarm2 and Dextre

Perhaps the robots that have been most utilised in space have been the large robotic arms outside of the ISS, such as the Space Station Robotic Manipulator System (SSRMS), which is also known as the Canadarm2, and the Special Purpose Dexterous Manipulator (SPDM), also known as Dextre [8]. Together with the Mobile Base System, they form the Mobile Servicing System (MSS), the most used tool for On-Orbit Servicing (OOS) in space. The SSRMS, a 7 Degrees of Freedom (DoF), 17-m long robot, was largely designed to perform ISS assembly tasks and to capture visiting vehicles and other similar large-scale tasks [9]. Dextre, as a 2-armed robot, was made to perform external maintenance on the ISS. Dextre has the ability to use many different tools to robustly perform smaller-scale and more dexterous tasks than the SSRMS. Each arm has seven independently controllable DoF with an ORU/Tool Changeout Mechanism (OTCM) as their end-effector [10, 11].

One of the most effective aspects of these robots are the latching end-effectors, which were specifically designed to firmly latch onto grapple fixtures strategically placed in many locations around the ISS and on incoming cargo and spacecraft. There are several types of grapple fixtures which allow for different capabilities. For example, the Latchable Grapple Fixture (LGF) is intended for longer-term stowage on the Payload Orbital replacement unit Accommodation (POA), while the Power and Video Grapple Fixture (PVGF) additionally allows for access to data, video and power. These innovations have given versatility to the robots, allowing the SSRMS to relocate its base by 'walking' from fixture to fixture around the station. This concept has also given a mechanism for combining robots together in a macro-micro configuration, where for example the SSRMS (Canadarm2) would position the SPDM (Dextre) in the best location while the SPDM performs the smaller scale tasks [9, 11].

The SSRMS can be controlled in multiple modes: joint control, end-point control and automatic trajectory control. When using the different control modes the user must constantly consider several different frames of reference and coordinate systems [10], while looking at multiple camera angles to determine the best course of action to complete a given task. A certain level of autonomy has already been implemented into the SSRMS, such as the automatic trajectory control mode, which has allowed ground control to overcome the latency to the ISS and perform simpler tasks, including preparation and initial positioning of the robot arms [10]. The SPDM, on the other hand, is now mostly controlled from ground control on Earth despite the fact that it was initially designed to be controlled from within the ISS [12]. This is made possible through a series of tests called On-Orbit Checkout Requirements which ensure the safe operation of the SPDM even with significant amounts of latency.

Training to use the Canadarm2 and Dextre occurs at the Robotics Training Centre of the Canadian Space Agency in Saint-Hubert, Quebec. The training centre includes a replica of the Robotics Workstation that is on the ISS and sophisticated training and simulation software [13]. While it is difficult to find information about the specific software used in this training centre, research groups have developed their own similar software to investigate the best ways to train astronauts and flight controllers as how to control these robots. Belghith *et al.*, for example, developed the Robot MANinpulation Tutor (RomanTutor), which, in addition to providing a general simulation platform for practising robotic control, did automatic path planning for a particular task and considered strategies for camera and view selection [14]. These tools could then be used to show trainees what an 'optimal' solution to task performance may be, even in complex and ill-defined domains. Using a previous iteration of this simulator, Fournier-Viger developed a cognitive model of Canadarm2 task performance to break down a complex and ill-defined task into more understandable steps, evaluating spatial representations [15].

Another simulator for the Canadarm2 is available on the Canadian Space Agency (CSA) website [16]. This simulator plays more like a game, teaching the different aspects of Canadarm2 control before allowing you to attempt a 'mission' utilising the skills that were learned in the tutorials. Control is carried out with a keyboard and mouse, allowing the player to move and turn the Canadarm2 to match various given visual cues. Training tasks within this simulator include following a circular trajectory with the end-effector using the camera placed on the end-effector, rotational control for precise docking and finally a complete task in which the Canadarm2 is controlled to perform a task aboard the ISS involving the replacement of a component and carrying an astronaut to perform a final task at this component. While this simulator provides insight into many of the necessary skills for Canadarm2 operation, it also provides real-time suggestions about which keys to press to successfully achieve a given outcome. This results in minimal cognitive load and is likely to be difficult to translate to real robotic control.

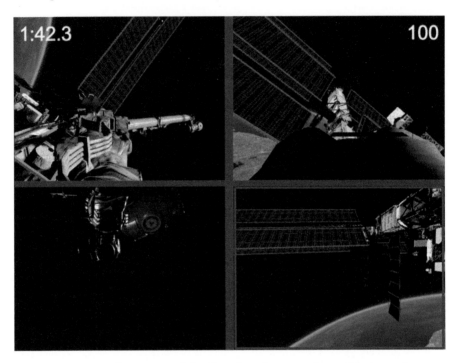

Figure 8.2   In order to control the Canadarm2, users rely on 2D views of cameras located at close proximity to the joints of the Canadarm2 and other key locations at ISS

Most recently, a Canadarm simulator was developed at Imperial College London, Figure 8.2, which aimed to provide both photorealism and increased cognitive load [2]. This simulator was built to allow for different grades of cognitive load through the addition of confounding factors such as latency, time pressure and pieces of space debris which acted as obstacles to avoid. The simulator was also developed to be compatible with physiological data collection, and thus was paired with modules to collect Electroencephalography (EEG) and eye-tracking data, as well as information about heart rate, body temperature and Galvanic Skin Response (GSR). The collected data was synchronised via Lab Streaming Layer (LSL) and analysed to determine the effect of the added workload from each of the proposed confounding factors. Measures from these sensors were additionally compared to a task performance score, which considered the user's precision at each stage of the task, time to complete the task, and any errors or collisions that occurred.

8.2 Cognitive models underlying neuroergonomics in space flight

In the field of human factors and ergonomics, there is extensive research literature on how to develop human-centred designs of technology that aim to minimise
errors, enhance performance and enable effective human–machine interaction [17].
The recent expansion in AI and the success of these systems in information retrieval,
robot vision and language processing automate low-level applications [18, 19].
These systems are gradually adapted to everyday life and have already automated
several manual tasks. However, how these systems can interact for higher-level
decision-making and whether we can trust them remains a challenge. In addition,
safety concerns and ethical considerations with relation to the underlying responsibility are profound. These factors imply that recent progress will translate several
applications from manual to semi-automatic, and thus designing supervisor control
mechanisms that take into consideration human factors and ergonomics are in high
demand.

Neuroergonomics are concerned with human brain function and performance
in a number of critical applications that range from medical interventions, aviation, driving and so on [20–22]. Experts in the field predict that within the next 20
years, neuroimaging technologies involved in human cognitive augmentation would
mature to seamlessly allow monitoring and enhancement of brain processes [20].
Brain functions, such as decision-making, cognition, attention, vigilance and situation awareness are important to complete a task successfully. In neuroergonomics,
there is a distinction between vigilance, also referred to as sustained attention, and
attention under workload. Vigilance is usually tested under low workload and it
involves the ability to detect a stimulus, which is important in applications such as
air traffic control, surveillance and inspection tasks. Usually, it is evaluated based
on the reaction time, which is the time from the stimulus presentation to a simple
motor response of the subject. On the other hand, maintaining attention and situation awareness under workload involves shifts of attention via higher cognitive
functions and executive control. In these scenarios, spatial attention, which refers
to the ability of orienting attention to a particular direction, and theories based on
working memory models have been employed to explain the information processing
that underlies brain function.

These functions are supported by interconnected circuits that involve several
brain regions and they are significantly affected under stress and workload. In particular, the prefrontal cortex (PFC) has been implicated in attention control, concentration, executive function and decision-making [22–24]. Furthermore, ventromedial
PFC and dorsomedial PFC are implicated in social processing and anxiety [25]. In
sustained attention both top-down and bottom-up networks act in parallel to facilitate selective attention [26–28]. Selective attention is referred to as the ability of the
brain to prioritise sensory information. The former originates from forebrain regions
that include the PFC, the parietal cortex, somatosensory cortices and subcortical
structures, such as thalamus and basal ganglia. It encodes brain states related to

working memory, reinforcement learning, selection of task-relevant processes and inhibition of task-irrelevant processes [29]. On the other hand, the midbrain network is thought to exert a bottom-up regulation of attention related to the saliency of the stimulus. This process is thought to have an evolutionary purpose in order to alert humans when a 'threatening' stimulus, as for example when a pop-up stimulus enters their peripheral vision, the forebrain enables humans to concentrate on a particular target.

In humans, only a small fraction of the visual field corresponding to the fovea is perceived in fine detail. Visual spatial attention refers to the ability to select relevant objects/stimulus and process information within the underlying area of the visual field [30]. Spatial attention results in increasing the gain of the mean firing rate and decreasing the noise correlations across neuronal populations related to the relevant location/objects [31]. This improves detection and discrimination of relevant stimulus and shortens reaction time. The processing of visual information is complex and involves the ventral and dorsal neuronal pathways that start from the primary visual cortex and extend to the temporal lobe and parietal lobe, respectively. The former pathway is involved in object recognition, whereas the dorsal pathway is mostly related to spatial awareness and spatial attention [30].

Damage of the spatial attention pathways could result in directional bias in orienting attention. This is a common clinical syndrome called spatial neglect. It is thought that spatial neglect is the interaction of several deficits that involve directional bias in competition for selection, spatial working memory deficits and sustained attention deficits [32, 33]. Although, the exact brain regions involved in visual selective spatial attention is under debate, the Parietal Eye Fields (PEF), Frontal Eye Fields (FEF) and the Temporal Parietal Junction (TPJ) have been highlighted as most likely to play an important role [32–35].

Eye movements and spatial attention are interrelated, since it is evident that usually eye movements follow attention (overt orienting gaze). In fact, neuronal circuits that control attention are also related to eye movements/saccades [28]. For example, animal studies have shown that after unilateral removal of the FEF in PFC, the animal could not direct gaze in the affected hemisphere. Overt orienting can be either reflexive or controlled. Reflexive movements are related to midbrain, bottom-up attention selection mechanisms rather than conscious processing of the visual field. In this context, the consensus is that there is a single mechanism that drives both selective attention and motor preparation. Nevertheless, humans are able to mentally direct their attention to spatial locations without moving their eyes (covert orienting gaze). This has been found to slower saccades and to alter the underlying perceptual processes.

8.2.1 Neuroergonomics and spatial attention

Several theories have been developed to explain and model how a brain processes the information with relation to attention and decision-making [27, 36]. These models introduce the concept of working memory, which describes the ability of the brain to hold a limited amount of information for a short period of time while it is

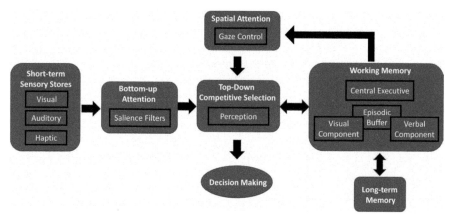

Figure 8.3   *A consensus cognitive model that shows how attention and working memory interact to process information*

processed [26]. Working memory does not only refer to the ability to memorise but also the ability to suppress irrelevant information. Stimuli compete to gain control over working memory, whereas gaze and spatial attention processes play an important role in it, Figure 8.3. Visual, auditory and haptic information enter the 'short-term sensory stores' (STSS), which retain information of up to a fraction of a second [36]. STSS is thought to have a large capacity but short duration, whereas working memory can persist for several seconds but is restricted to few items [36]. This is also referred to as an 'attentional blink' and is the reason that stimulus experimental setups allow a 300 ms gap. This restricted attentional capacity emerges within sensory modalities. In other words, concurrent attention to visual stimuli limits attention to another visual stimuli but it does not limit concurrent attention to auditory stimulus [37]. Furthermore, it was shown that processing and recognition of a scene takes 100 ms. Nevertheless, humans are able to recognise scenes better than even with rapid stimulus presentation of less than 50 ms [38, 39].

8.3   Workload and performance measures in human–robot collaborative tasks

There is an overwhelming amount of evidence to suggest that workload and performance are strongly related. In fact, when humans face increased demand their performance may deteriorate, they will perform more errors, their tasks will be less accurate, they often lose awareness of their surroundings and they become increasingly frustrated and fatigued [36]. However, the relationship between workload and performance measures is not linear. Lower workload than normal can also result in similar performance deterioration, possibly due to boredom and drifts of attention. Therefore, it seems that there is a subject-specific point of workload where performance reaches its maximum, which normally reflects that the task is challenging, yet does not overwhelm the operator. Overall, human performance depends on a number of factors that include time pressure, operator's fatigue, training level, innate abilities to adapt to the task in hand and resilience to stress and anxiety. Furthermore, workload and performance may need to be compromised to satisfy operational goals.

Performance in human–robot collaborative tasks should take into account the ability of the human and also consider the ability of the robot to adapt and optimise its actions with relation to human responses. There is considerable effort in the research community to develop human–robot collaborative strategies to ease the cognitive and physical load and thus minimise workload. However, there is a risk that with increased automation, humans become observers and they are not actively engaged in the loop. This can also cause boredom, loss of awareness and lack of the ability to comprehend the complexity of the system. All these factors can result in the inability of the human operator to control the system if the automated mode is inaccurate or fails.

Mental workload can be sub-categorised based on human sensing and cognitive processes into visual, auditory, tactile and cognitive workload [40]. Visual perception is influenced by contrast, colour, dark-adapted vision, depth perception, movement detection and glare. Hearing perception is also influenced by loudness, pitch and location. Finally, with tactile sense, we perceive differences in temperature, pressure and the frequency of vibrations of our skin. Human capacity to absorb and process information is called perception and is influenced by internal cognitive models and expectation. Human perception has the ability to fill as well as remove information based on contextual information and this mismatch between reality and perception of sensory information could lead to misinterpretations and illusions.

Characterising workload is an important yet quite complex process. Most common subcategories include mental/cognitive demand, physical demand, temporal demand, performance, frustration and effort. In fact, the NASA Task Load Index (TLX) has adopted this method to measure workload via subjective self-reports. Self-assessment reports are normally used shortly after the task in hand while user's memory is still fresh. To overcome the fact that subjects perceive types of workload differently, the TLX index requires them to rank the order of each subcategory. Subjective assessments are disruptive, they are not continuous, and they suffer from scaling problems, since most operators do not translate increases in workload, linearly.

Task-specific performance measures provide an objective way to measure workload based on the assumption that the ability to perform a task well is affected by workload. Primary task measures include task analysis, speed, accuracy and levels of activity. Task analysis includes various methods that break user's actions into sub-tasks, and they count how many are completed successfully. Measuring activity involves counting the number of steps per time required to finish the task. Actions may include control inputs, verbal responses, mental arithmetic, visual searches and decisions. Large numbers of measured activity imply a high workload. Task analysis and measuring activity requires the ability to break down the task and response, respectively, into specific modules, which might not be trivial in real-world dynamic environments. Speed and accuracy are the simpler performance measures to estimate but they cannot disassociate operator condition from system failures, such as slow response. Furthermore, workload in decision-making tasks is difficult to characterise based on speed and accuracy alone. None of these measures takes into account the skills of operators.

The workload is also modulated under single or multiple task demands. Secondary task measures estimate the workload of a task by looking into how well the operator performs a second task simultaneously. These techniques quantify how many 'spare resources' the operator has and provide more information about the condition of the operator. However, they rely on the assumption that both tasks are competing for the same resources and that the performance of the first task remains constant. Furthermore, different operators may have different strategies to complete the first or second task. Careful design of the interaction of two tasks is important to provide meaningful conclusions.

A recent study has used a dual-task design to understand the effect of engagement on workload and performance during driving [41]. The driver was instructed to maintain a specific distance from the vehicle in front, which was changing its speed at random. Primary performance measures included speed control and braking response time, while the NASA TLX index was also recorded after task completion. An auditory stimulus, which was selected to be interesting, boring or neutral was playing at random during the task. The results revealed that the time required to brake (response time) was longer while the driver was listening to an interesting stimulus. Also, the drivers perceived the interesting auditory stimulus to be less demanding, although the stimulus has been objectively chosen with similar difficulty index.

It is important to note that auditory processing channels are considered to be independent from visual processing channels. This is the reason that in ergonomics studies they have been suggested as effective communication channels when the primary task requires visual attention. Nevertheless, further processing of the information would require the allocation of more cognitive resources.

Dual-task designs are powerful as they can disentangle the influence of multiple sensing and cognitive pathways. For this reason, they have been used extensively in several studies in aviation, driving and surgery. It should be noted that the ability of astronauts to successfully perform dual tasks is affected both during the early adaptation to microgravity and towards the end of the flight. It is thought that the cause of this deficit is due to the increased fatigue and stress associated with both of these phases.

Fatigue and sleep deprivation have been associated with several serious accidents that resulted in collisions in space. In 1997, the Progress spacecraft collided with the MIS space station and caused extensive damage to the solar array modules. Although the astronauts claimed that there was a delay in the navigation system, NASA attributes the accident to workplace stress, fatigue and sleep deprivation. Space imposes unique challenges on astronauts that also result in an increased level of fatigue. For example, 60–80 per cent of astronauts will be affected by space motion sickness, and micro-gravity also affects their sleeping patterns, along with background noise, lack of adequate thermal control, lack of fresh air and so on.

Fatigue could be muscular or mental and is caused by prolonged physical or mental tasks, respectively. When it relates to emotional stress, fatigue can also be characterised as acute or chronic. There is no clear distinction between workload and fatigue. A problem of disentangling workload from fatigue is that there is no clear definition. Furthermore, most studies do not distinguish fatigue from sleepiness, since it is far more difficult to disentangle one from the other. Countermeasures of fatigue target regulation of circadian rhythm with scheduled sleep breaks along with well scheduled meals. Approximately 20–30 min sleep before night shifts helps to increase alertness along with administered caffeine and/or other pharmacological agents. Robotic exoskeletons have been suggested to tackle muscle fatigue by providing support to both the lower body and upper body [42].

Several measures have been proposed to quantify fatigue. The occupational fatigue inventory represents fatigue in five physical dimensions (lack of energy, physical exertion, physical discomfort) and two mental dimensions (lack of motivation and sleepiness). Other measures, such as the occupational fatigue exhaustion recovery scale, quantify the need for recovery. This measure describes fatigue in three scales that include acute fatigue, chronic fatigue and intershift recovery.

8.4   BCIs in workload and attention

8.4.1   EEG-based BCI

EEG is a sensing modality which measures electrophysiological signals coming from the brain. EEG sensing systems can come in many forms, from implantable systems requiring surgery [43] to dry, wearable caps that favour efficient setup as opposed to optimal signal to noise ratio. Because most healthy individuals would not require or want invasive brain surgery for the monitoring of workload, this chapter includes consideration of only wearable non-invasive EEG systems.

Much of the foundational work related to human EEG recordings was established in a series of 15 reports by Hans Berger [44]. Berger investigated many of the fundamental questions related to EEG recordings, such as types of electrodes, recording locations, artefact removal and Fourier analysis of the recorded signals. There has been significant development with regard to each of these topics in the last century, but there is still much to learn about what EEG signals can indicate and how they can be used.

One of the first considerations for EEG data recording is the removal of unwanted artefacts that are picked up by the electrode which may contaminate the signal. There are several known artefacts which originate from the eyes, muscles, heart and environmental factors which researchers have attempted to remove based on knowledge of the artefact [45]. One way of doing this is by simultaneously recording data from a separate reference and altering the recorded EEG signal based on the recorded reference. When using the same type of electrode, the reference signal could be simply subtracted from the recorded signal. Other methods include

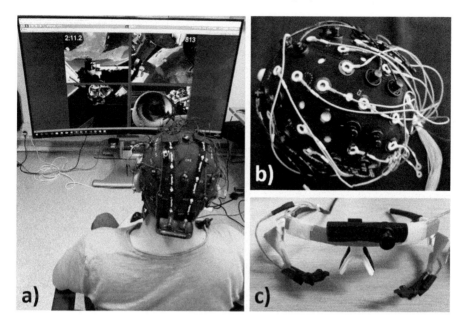

Figure 8.4   *BCIs are coupled with advanced 3D space simulator to monitor human attention, cognitive workload and spatial learning. (a) An advanced 3D simulator has developed to allow users to control Canadarm2 under realistic scenarios [2, 46]. Wearable EEG and eye-tracking technology is combined to monitor neurophysiological responses. (b) Combined Functional Near-Infrared Spectroscopy (fNIRS) and EEG technology promises to improve our understanding on learning processes and cognitive workload. (c) An example of wearable eye-tracking device*

the use of unsupervised learning algorithms such as principal component analysis (PCA) and independent component analysis and the use of filters to remove either known or learned noise from the environment [45].

Wearable EEG systems typically use multiple electrodes which can provide spatial information about a person's ongoing cognitive activity. The 10–20 systems is an international standard for electrode placement, which labels each electrode placement based on region such as Frontal (F), Central (C), Parietal (P), Occipital (O) and Temporal (T) regions. The system also labels electrodes based on the lon-gitudinal fissure, where all electrodes along the fissure have a z placed after their regional marker. An example of a labelled wearable EEG system can be seen in Figure 8.4a. This labelling standard makes it easier for studies to easily compare their results even if they are using different EEG recording platforms.

Consideration of the spatial location of EEG electrodes is often not straight-forward because the source of a particular brain response is not always known. For this reason, the spatial relationship between different electrodes can be numerically defined, which could provide a more intuitive understanding of the meaning of changes in various features. For example, researchers may want to define a spatial filter that only allows for consideration of the frontal regions of the brain. Therefore, in this instance electrodes nearest to the frontal region would have most influence over the signal, while electrodes that are far away may be entirely removed from the signal. These spatial relationships between electrodes can also be learned through algorithms such as common spatial patterns, which can reduce the feature space for a classifier. However, the feature space should only be reduced in such a way that maximises its practical utility. For example, using miniaturised EEG caps with few channels may not need any further spatial filtering, but could provide information that is unclear due to noise and cannot be verified by comparison with neighbouring electrodes. From this perspective, utilising more electrodes is more practical, despite the fact that features may not be directly extracted from each electrode.

As compared to other brain-sensing modalities, EEG is desirable because of its high temporal resolution, with recording frequency of up to 1 000 Hz. Such high rates allow for analysis of not only temporal features but also features that are calculated in the frequency domain, such as Power Spectral Density (PSD). Use of the Fourier transform in EEG processing is common, as the signal's changing frequency profile is often indicative of changes in mentality.

EEG signals are commonly separated into several broad frequency bands such as the delta (1.5–4 Hz), theta (4–7.5 Hz), alpha (7.5–12.5 Hz) and beta (12.5–30 Hz) bands. While approximate values are given here, different studies may use slightly different values for the boundaries of these frequency bands. It has been indicated, for example, that a decrease in alpha power is associated with an increase in mental arousal, resource allocation or workload, while theta power tends to increase along with task requirements [47]. These power changes are also most noticeable in predefined regions, with alpha decreases being mostly noted in parietal regions, and theta increases being noted in frontal regions. However, research has indicated that mu and alpha rhythms increase in power for humans in a microgravity environment, so previous assumptions about changes in EEG signals may need to be adapted based on new research into the effects of microgravity [48, 49].

EEG has been used to measure mental workload in studies of air traffic controllers, airline pilots, drivers and a wide range of humans performing cognitive tasks, such as memory or visuospatial tasks [50]. One of the most common tasks for workload measurement has been the n-back task, in which users have to remember whether the currently displayed stimulus was the same as the stimulus shown n trials before. The difficulty of this task can be modulated simply by changing the value of n, so different methods of measuring mental workload can be easily validated and compared [47, 51]. Other ways of validating workload include relatively simple predefined simulated tasks. Because the tasks are simulated, parameters can be easily changed to influence task difficulty. The signals for each task difficulty level can then be compared to make data-based theories about how brain function changes with regard to task workload.

One typical way of validating a method of workload measurement is via classification, where preprocessing leads to feature selection, and eventually the selected features are fed through a classifier such as Support Vector Machines (SVM) [47] or discriminant function analysis (DFA) [50]. The ability of the model to determine workload is thus evaluated based on the accuracy of the classifier and other related metrics. If the model is evaluated to work well in classification, then the features that led to higher accuracies can be analysed to draw conclusions about brain function with increased workload.

8.4.2 fNIRS-based BCI

Functional Near Infrared Spectroscopy (fNIRS) is a non-invasive, optical neuroimaging technique that allows the measurement of oxygenated haemoglobin (HbO) and deoxygenated haemoglobin (HbR). fNIRS light sources are normally arrays of LEDs or lasers that emit light in at least two wavelengths. The light penetrates the scalp and cortical regions and its relative absorption is measured by the fNIRS detectors. This allows the detection of relatively small changes in near-infrared light absorption, which relates to changes in HbO and HbR according to Beer-Lambert law [52].

fNIRS is resilient to eye-movement artefacts and this is one of the reasons it has been particularly common to measure prefrontal activation, which is related to executive function and its function has been found to be modulated with workload and training. Furthermore, miniaturised sensing technology has allowed fNIRS systems to become portable and wireless. This facilitates the continuous acquisition of brain signals in real-world settings. However, fNIRS electrodes only measure light within a few centimetres from the scalp, which limits the application of fNIRS to cortical regions only.

Several fNIRS studies aimed to replicate and confirm findings of functional magnetic resonance imaging (fMRI), which is the gold standard in functional neuroimaging and has shown average increases in oxygenation with increased workload/difficulty. Typically, in these studies, workload is modulated by n-back tasks [53]. In these scenarios, fNIRS data are recorded while the participant is serially presented with stimulus and he/she is instructed to respond when the stimulus matches the nth stimulus ago. n-back tasks have been found to activate the dorsolateral and the ventrolateral PFC brain regions. Furthermore, these studies show increased frontal-parietal connectivity [52].

In neuroergonomics, fNIRS exploits basic neuroscience principles to assess the design of new systems in terms of workload, parameter optimisation to achieve best cognitive capacity and training. The application scenarios span from air traffic control (ATC), aviation pilots and surgery [22]. In ATC, several studies based on fNIRS aim to assess the ergonomics of new human–machine interaction designs along with the number of aircraft that can safely operate in an airspace. In these cases, fNIRS activation would increase with workload up to a safety critical point, where activation is plateaued. It is evident that any further workload will not be safely handled from the operator and the probability of an accident increases dramatically. It should be highlighted that this point cannot be detected based on self-reported measures.

fNIRS activation in the prefrontal lobe with relation to workload is also modulated by the expertise level of the operator. Training results in cognitive adaptation processes that optimise attention control and problem-solving and thus it releases cognitive resources. The PFC activity in experienced operators is reduced compared to novice operators under the same tasks. This finding has confirmed both in ATC as well as in surgical tasks [22]. Therefore, neuroimaging studies offer a way to track the efficiency of training. A few days training in unmanned aerial vehicle piloting highlighted distinct phases of learning. Initially, increased activity in fNIRS reflected increased performance. In later stages of training, increased performance was associated with reduced activity in fNIRS. Similar results have been observed in surgical residence, where it was also shown that training phases are also modulated by the complexity of the underlying task.

8.4.3   Eye-tracking-based BCI

Eye-tracking data (e.g. eye movements and the pupillarity response) can elucidate visual interaction with complex user interfaces, such as where and what the operator looks at, how long the operator looks at it and which eye movement happens when looks at it. This information also reflects the cognitive workload of the operator during teleoperation. Recently, eye parameters have gained extensive popularity in the estimation of mental workload for those needing to perform complex tasks under stress, such as pilots [54], drivers [55, 56] and surgeons [57, 58]. The most significant workload metrics based on eye parameters can be categorised into pupillary response (meaning pupil diameter and pupil diameter deviation), fixation (number of fixations, fixation duration and fixation frequency), saccades (speed and amplitude of saccades) and blinks (blink frequency, number of blinks, blink duration) [59].

8.4.3.1   Point of gaze and eye movements

The 3D eye model is typically simplified as the model demonstrated in Figure 8.5a, which consists of eyeball, cornea and iris. From this model, the optical axis can be defined as the line that passes through the centres of the eyeball, cornea and iris, and the visual axis indicates the line from the Point Of Gaze (POG) to the corneal centre. There exists a constant deviation, namely *Kappa* angle, between the visual axis and the optical axis. The final POG is determined by averaging the estimated gazes of left and right eyes while both eyes are gazing at the same point. In addition, the direction of the optical axis and the rotation of the eyeball can be characterised by Listing's law, which describes the 3D orientation of the eye and its rotation axes by defining a Listing's plane [60]. Specifically, the vertical and horizontal axes of rotation formulate this Listing's plane as illustrated in Figure 8.5b, and the optical axis that is orthogonal to the plane indicates the torsional rotation. For eye-tracking research in space, it should be pointed out that gravity has a critical impact on eye movement and head-eye coordination [61]. It has also been proven that the orientation of Listing's plane significantly changes under microgravity, where the

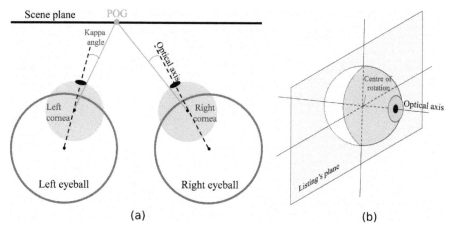

Figure 8.5   *(a) Demonstration of the top view of two 3D eye models and the point of gaze (POG) on the scene plane. (b) Illustration of the listing's plane and the 3D orientation of the eye and its axes of rotation*

elevation can be tilted backwards by approximately 10 degrees during a parabolic flight experiment [62].

With the successful tracking of the movement of eyeballs, three basic eye movements can be additionally defined as illustrated in Figure 8.6, including saccades, smooth pursuits and fixation [63]. A saccade indicates the rapid movement of the gaze point from one position to another, which can also be regarded as shifts between fixations [64]. Fixation indicates the gaze fix or pause on a small region of interest [64]. It can be typically detected when the POG is within a particular area or if the gaze velocity is smaller than a threshold. Smooth pursuit represents the eye movement that follows a moving object [63]. For the teleoperated task described in Freer *et al.* [2], for example, the saccade movement occurs when the user switches the activated camera or avoids the debris by observing different cameras. The user will achieve fixation while adjusting the robot arm in a fine manner or thinking about the control strategy. During the teleoperation, the users will tend to have smooth pursuit while performing the translation of the end-effector.

8.4.3.2   Eye-tracking systems

The existing techniques for eye movement and gaze detection include magnetic search coil, Electro-Oculography (EOG) and Video-Oculography (VOG).

Magnetic search coil For human eye-movement tracking, the subject needs to wear the contact lens that contains coils of wire, which are also known as Helmholtz coils. In the experiments, the subject sits inside a specified area with a magnetic field, then eye movement can induce a variation of voltage in the contact coils. Compared to other systems, the search coil system can achieve high detection precision of eye movement in both spatial and temporal resolution, though it typically

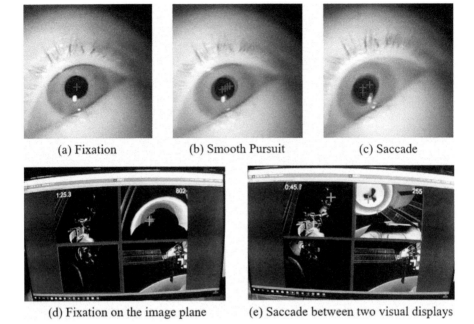

(a) Fixation   (b) Smooth Pursuit   (c) Saccade

(d) Fixation on the image plane   (e) Saccade between two visual displays

Figure 8.6   *Illustration of three basic eye movements and the gaze points of fixation and eye saccade on the image planes. Accordingly, features such as fixation duration and saccade speed can be extracted*

causes some discomfort for the human due to its semi-invasive nature and additionally involves a complicated setup procedure [65].

Electro-oculography Another popular technique for tracking eye movement is to measure the corneo-retinal standing potential differences between the front and the back of the human eye, which is known as EOG [66]. EOG is advantageous in measuring the horizontal/vertical rotation of the eyeballs by attaching two surface electrodes to the edges of the orbits along the horizontal/vertical direction. However, the EOG signal is susceptible to noise and cannot measure the pupil diameters.

Video-oculography Recent advances in computer vision technologies have led eye-tracking systems to adopt video-based techniques. Most video-based gaze-tracking systems focus on the estimation of gaze direction, which can be categorised into remote and wearable gaze trackers. The basic idea is to detect the 2D or 3D parameters of the near-circular pupil from single/stereo camera. Then the rotation of the eye with respect to the camera can be determined after the appropriate calibration procedures. VOG eye-tracking accuracy heavily depends on the quality of the calibration process before recording and the pupil detection during the recording [67]. Pupil detection is more challenging with remote eye-tracking systems as compared to wearable ones due to the corneal reflections, occlusions and eye blinks. We refer the readers to [65, 66] for more details on various eye-tracking techniques.

8.4.3.3   Eye-tracking-based mental workload detection

In the following, we give a brief review of the relationship between mental/cognitive workload and different commonly used eye-tracking parameters.

In the last century, pupil diameter has already been used as an index of cognitive workload. Extensive research has revealed that the pupil diameter increases with the task difficulty, which is highly related to the mental workload [68, 69]. One of the remaining challenges is that various confounding factors unrelated to workload, including changes of luminance condition and emotional arousal, may also affect the pupillary response [70]. The Index of Cognitive Activity (ICA) [71] can provide an estimation of cognitive workload level by disentangling pupil dilations caused by cognitive activity (small rapid dilations). The ICA is determined by the changes of pupil dilation while performing a specified complex task. The core idea of ICA is to perform wavelet analysis to identify the abrupt pupil dilation in the eye-tracking data [71]. When the rapid pupil dilation is larger than a specified threshold, it reflects the effect of cognitive activity. Higher ICA levels per second represent a higher degree of cognitive workload [57].

With the recent advances in gaze tracking technologies, studies have been conducted to find the relationship between eye fixations and cognitive processes since the 1970s. The authors of [72] have demonstrated that the locus and duration of eye fixations are all closely related to the activity of the central processor. Following this, extensive studies have proven that the duration of fixation has a negative relationship with the mental workload during various complex tasks [54, 73, 74]. In other words, higher fixation duration will be observed during lower mental workload condition and lower fixation duration will occur during higher mental workload tasks. Results in [74] also emphasised that the fixation duration is the most suitable metric among various parameters to estimate mental workload. Furthermore, in a study on visual attention of pilots, researchers have found that expert pilots will have more fixations on different instruments/displays with shorter duration time [54].

Eye saccades have become a popular metric for studying motor control, cognition and memory [75]; however, a significant relationship between the number of saccades and mental workload has not been observed [73, 74]. The authors of [74] conducted the n-back memory experiment with four difficulty levels to induce mental workload, in which 17 eye parameters were evaluated to investigate the relationship with mental workload. Results have demonstrated that the fixation duration and eye blink parameters show a significant relationship with workload, while the saccade related parameters failed to show a significant relationship. In [76], driver distraction was evaluated by detecting eye saccade movements. They have found that the older group shows worse performance with mental workload under the distracted driving conditions. The authors of [54] also suggested that expert pilots have more saccades on different instruments compared to novices.

Previous studies also found that blink frequency and blink duration showed a significant positive relationship to mental workload [59, 74]. The authors of [55] demonstrated that blink duration, compared to blink frequency, is a more sensitive and reliable indicator for workload detection. Similar to blinks, the percentage of eye closure over time can be used as a measure of fatigue, which has been extensively applied for driving fatigue detection [56].

8.4.3.4   Eye-tracking-based skill assessment

Eye-tracking data can also be adopted as an objective tool for skill assessment [58], with potential applications in training for improving performance. Recent research has demonstrated that the eye movements have significant differences between novices and experts in aviation [54] and medicine [57]. The author of [54] has shown that expert pilots tend to have more fixations on different instruments and shorter dwell time on each instrument. Meanwhile, expert pilots manage to extract more relevant information from their peripheral information. For surgeons, differences in eye metrics reflecting focused attention can also be found between junior and senior surgeons. Senior surgeons have higher fixation rates because they know what they are looking for and where to locate it, and, simultaneously, they have lower ICA values over junior surgeons as they do not experience the same degree of cognitive workload in surgical procedures [57, 58].

8.4.4   Neuroimaging in space

Neurophysiological responses are altered in space due to several factors listed in Figure 8.7 that include physiological, psychological and habitability issues. Some of the physiological changes include cognitive/neurological alterations, increased fatigue, changes in circadian rhythm, changes in stress hormone levels and immune function [77]. Most of the variations in brain neurophysiology occur as a result of microgravity. Microgravity results in a shift of body fluid towards the head and this has been implicated in neurophysiological adaptations that last several weeks after the space flight [78]. In fact, both structural and functional brain changes have been demonstrated in studies [48, 49, 78, 79].

The brain adjusts to changes such as a shortened sleep cycle and microgravity, which has been reflected in changes in heart rate variability (HRV), changes in brain rhythms and regulatory brain connectivity networks, such as the default mode network (DMN) [78]. The DMN emerges during rest and its function manifests from interactions across several brain regions. It is thought to regulate the autonomic nervous system and its interaction with other major brain networks, such as the Salience Network, reflects shifts in focused attention [78, 80]. In space, the DMN plays an important adaptation role that is also reflected in changes in the human oscillatory brain activity [48, 49]. Furthermore, the close relationship of the DMN to the autonomic system results in changes observed to the HRV.

It is important to realise that most of the scientific evidence for these changes does not come from studies conducted in space. Instead, some of the studies compare neuroimaging/physiological data obtained from astronauts before and after space flights. Other studies use simulation environments to imitate some of the common conditions encountered in space. The most notable conditions are confinement and isolation along with microgravity. Some of these simulation experiments take place in cages to resemble the extreme psychological and physical

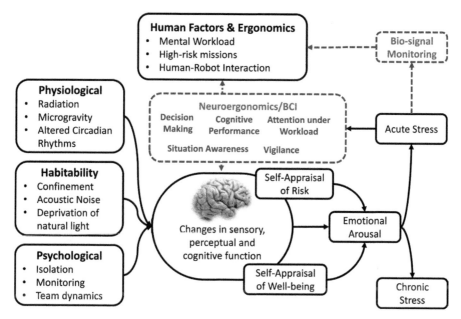

Figure 8.7   *Human factors and neuroergonomics in space flights. Current research needs to be seen in light of the homeostatic adaptations that take in space due to physiological, psychological and habitability factors in space*

space conditions, such as isolation and confinement. For example, the Space Studies Department at North Dakota has a specialised facility to resemble 'Mars' missions (Figure 8.8) and to test new spacesuit technologies along with how they affect mental workload [81]. There are also human activities, such as Antarctic expeditions, that in some cases provide a close analogue to space missions. On the other hand, microgravity effects can be simulated with bedrest approaches, in which the subject is asked to lie on a bed that is inclined downwards by roughly 6 degrees [77].

These experiments exert mental and physical pressure on individuals and they are difficult to complete. One major limitation is the small number of subjects and thus the inability to extract statistically significant results. This problem is more profound in early studies and it is exacerbated by the fact that experimental conditions between studies and exact protocols differ significantly from mission to mission. As the quality of portable neuroimaging equipment improves and becomes more practical and less cumbersome, BCI will be adapted to more realistic scenarios in space and aviation. In addition, space agents like NASA have already developed technology to reduce motion artefacts and improve the accuracy of similar systems [82].

*Figure 8.8   University of North Dakota space analogue simulations (photo
provided by Dr Travis Nelson and Prof. Pablo De Leon)*

8.5   Artificial intelligence in BCI-based workload detection

Typically BCIs rely on signal preprocessing to remove artefacts, which depend on
the nature of the signal, followed by feature extraction and classification, Figures 8.9
and 8.10. There is large literature that has accumulated knowledge derived from
well-controlled lab experiments on tasks that modulate workload, such as the n-back
task mentioned earlier. However, there is far less work to address challenges in real
environments [85]. This is partly because BCIs are sensitive to motion artefacts and
interference. Even in the most well-controlled environments subject variability in
BCIs hinders the robustness of the device and high classification rates are difficult to
achieve for more than two classes [83]. Furthermore, reliable signal detection entails
good coverage of key brain regions, which results in cumbersome equipment that
are not pleasant to wear.

Currently, there are three notable categories in workload detection and modelling that drive research forward: (i) work that extracts cognitive models derived
from neuroimaging data that allow us to explain differences in the signal between
workload conditions [86]; (ii) Multi-modal fusion, which exploits machine learning
algorithms to extract features from several modalities that include EEG, fNIRS,
eye-tracking and physiological measures to improve classification rates [85]; and
(iii) BCIs that are coupled with realistic paradigms of tasks that account for the
complexity of the task in hand [2].

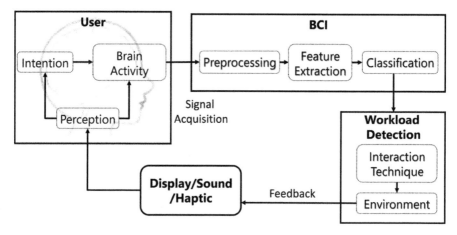

*Figure 8.9   Classical BCI framework involves signal acquisition followed
by machine learning techniques to preprocess the signal, extract
features and perform classification. Intuitive interactive approaches
can increase the robustness of the BCI [83, 84]*

Cognitive models of workload shed a light on the neurophysiological origins of the signals and can help eliminate spurious results due to motion and physiological-related artefacts, such as breathing and heart rate. Growing evidence shows that breathing patterns change under mental workload [86, 87]. This implies that occasionally BCIs classification rates may reflect motion artefacts and therefore do not generalise across subjects. Furthermore, this would hinder detection of peaks of performance with relation to optimum workload.

*Figure 8.10   A photo-realistic 3D simulator of the ISS developed at Imperial
College London [1]. The simulator allows the user to interact with
the Canadarm2 robot based on four camera views as shown at
Figure 8.2*

Information fusion of neurophysiological, physiological and behavioural data is important in real-world experiments because they have the potential to enhance reliability and sensitivity of workload detection, while they reduce uncertainty and address inter-subject variability. Neurophysiological recordings refer to modalities that directly measures brain signals. For example, combining neurophysiological modalities such as EEG and fNIRS can enhance temporal and spatial coverage [88, 89].

On the other hand, physiological modalities normally include eye-tracking, heart-rate measurements and electrodermal activity (EDA) are indices of acute stress, Figure 8.7 and they are sensitive to mental states [90]. Normally, measurements can be obtained with far less sensors than classical BCIs and they are more discreet and comfortable to wear. However, they are also affected by physical activity and their ability to detect fine changes in mental workload is inherently limited when they are used alone.

Behavioural measures include computer mouse movements and clicks and the reason they are used in workload detection is the fact that they reflect engagement and attention. Human pose tracking is among the most useful behavioural measure in human–machine interaction as well as mental state detection [85, 90, 91]. Body posture reveals attentional engagement and it also relates with the difficulty and complexity of the task in hand [91]. For example, with increased workload the person may lean forward and the distance from the monitor is smaller.

It is well known in signal-processing that information fusion from independent sensors improve signal-to-noise ratio and reduces bias of confounding factors. Multimodal fusion is normally implemented in four levels: sensor level, feature level, decision level and hybrid models. The level indicates at which point information fusion takes place along the processing pipeline [85]. At sensor level, fusion takes place after preprocessing and it is suited when the raw data reflect the same physical aspect. At feature level, fusion takes place after feature extraction and thus extracted features are estimated independently from each modality [88]. At decision level, feature extraction and classification has performed independently across modalities and fusion takes into consideration classification outcomes to reach a final decision.

8.6   Cognitive workload estimation during simulated teleoperations – a case study

To investigate neurophysiological indices of cognitive workload, eye-tracking and EEG data were simultaneously acquired from 10 healthy volunteers during teleoperations on the photo-realistic Unity simulator of the Canadarm at the ISS shown in Figure 8.10 [2, 92]. The aims of the study were to understand how exogenous factors such as time-pressure and time-latency affect the performance of operators and how these changes link to neurophysiological indices related to spatial attention and cognitive workload. Under the time-pressure condition, users need to complete the task within a certain time (4 min), whereas in time-latency condition, 0–1.5 s delays were added to the operator motion control to reflect the round-trip time that it takes for the control signal to reach the robot and the visual feedback to travel back to the operator. Time-pressure is common in safety critical operations and the ability of a user to operate under limited time can have a profound influence in the success of the operation. The relationship between time-pressure and performance has been highlighted in Section 1.3 and is supported by several studies in teleoperated robotics, such as surgical robots and other systems [86]. On the other hand, time-latency is very critical in space/teleoperation applications due to the considerable time it takes for the signal to cover large distances in space. Communication delay can also reflect hardware, design or software limitations and it is a well-known problem in master/slave robotic systems.

The simulation task has been picked to reflect a real-life scenario of the seven degrees of freedom teleoperated Canadarm at the ISS and is described in detail in previous work [2, 92]. It requires the user to locate a new module close to the ISS, navigate the robotic arm to the grapple fixture of the module so the end-effector can attach to it and then move and dock the new module next to the Columbus module. As in real-life it is important that the operation is performed safely so that the Canadarm and the module do not collide with the ISS or other objects. To ensure that the task is sufficiently difficult and it requires enhanced spatial abilities, debris objects were added to the scene so that the user would have to find a way to navigate around them.

Users were asked to complete three blocks that included: (i) a familiarisation block (data from this part was not included in the analysis), which gave them the opportunity to familiarise themselves with the simulator without any added time-pressure or time-latency. To start the main experiment, users were required to complete a simple version of the task without obstacles under 4 min. (ii) A block of nine trials with randomised order that included three trials of time-pressure, three trials with a time-delay of 0.5 s and three times with neither factors. Users in this block had the additional difficulty of navigating the robot around obstacles. (iii) A latency block that included two trials for 0.5, 1.0 and 1.5 s of latency with and without time-pressure, a total of six trials. This block also required the users to navigate around obstacles.

Several performance measures were taken into consideration that included (i) grasp time (time between the beginning of the experiment and the grasping of the module), (ii) dock time (time from the grasping of the module to the docking to another part of the ISS), (iii) grasp distance error (distance error from the optimum grasping point), (iv) grasp angular error (angle error from the grasping angle between the module and the robotic end effector), (v) dock distance error (distance error from the optimum docking point), (vi) dock angle error (angle error from the optimum docking angle between the module and the docking station), (vii) Grasp score (score after grasping the module), (viii) Dock score (score from grasping to docking the module) and (ix) Number of collisions with obstacles per trial (Figure 8.11). One-way Analysis of Variance (ANOVA) results for pair-wise comparisons showed that the grasping and docking time between low workload and time-pressure conditions is statistically significant with a p-values smaller than 0.05 and 0.005, respectively. Furthermore, the number of collisions with obstacles per trial between time-latency condition and low workload condition was also statistically significant with p-value smaller than 0.05. No significant differences were found between the precision of docking under different conditions.

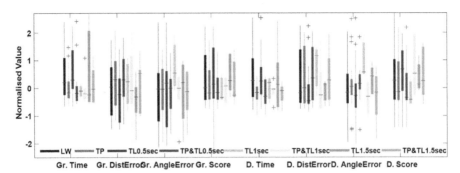

Figure 8.11   *Normalised values of performance measures during teleoperations via VR simulation of the Canadarm at the ISS that include: the grasp time (Gr. Time), grasping distance error (Gr. Distance Error), grasping angular error (Gr. Angle Error), grasping score (Gr. Score), docking time (D. Time), docking distance error (D. DistError), docking angular error (D. AngleError) and docking overall score (D. Score). These performance measures are examined in conditions that induce varying cognitive workload: The 'Low Workload' condition refers to performing the task without additional time-pressure and time latency (LW), under time-pressure only (TP), under time-latency of 0.5 s only (TL), under both time-pressure and time-latency of 0.5 s (TP & TL0.5 s), under time-latency of 1 s only (TL1 s), under both time-pressure and time-latency of 1s (TP & TL1 s), under time latency of 1.5 s (TL1.5 s) and under time-pressure and time-latency of 1.5 s (TP & TL1.5 s).*

On the other hand, eye-tracking features reveal a number of statistically significant differences. The most prominent features include saccade frequency, mean saccade speed, mean pupil diameter, mean fixation duration and the Index of Pupillary Activity (IPA) [92]. Pupillary response, such as pupil diameter has been related to cognitive workload factors. However, it is also sensitive to other factors including the brightness of the scene. The IPA, which reflect small variations in pupil diameter has been suggested as a more robust measure of cognitive workload. Mean frequency duration, mean pupillary diameter and IPA reveal statistically significant differences between the low-workload and time-pressure conditions. Significant differences between time-pressure and time-latency conditions are identified based on IPA only, whereas significant differences between low-workload and time-latency conditions were identified based on mean frequency duration and mean pupillary diameter. Finally, mean saccade speed and saccade frequency were more sensitive to time-latency conditions (Figure 8.12).

It is observed that in this scenario eye-tracking based features are more sensitive to different experimental conditions than performance measures. Furthermore, performance measures need to be redefined for each different task. For example, in this study we broke down the task into subparts of the grasping and the docking phase.

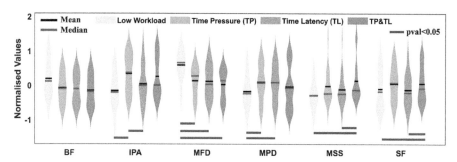

Figure 8.12   *Normalised values of eye-tracking features with relation to different cognitive workload conditions. Blink Frequency (BF), Index of Pupillary Activity (IPA), Mean Fixation Duration (MFD), Mean Pupil Diameter (MPD), Mean Saccade Speed (MSS) and Saccade Frequency (SF).*

This approach is ad-hoc, requires specific expert knowledge of the task at hand and thus performance measures are not generalisable to new situations. Another important limitation is that only a few simplistic performance measures, such as time duration, can be measured outside the simulation environment. Therefore, there are not suitable for closed-loop feedback mechanisms that could help the operator to improve his/her performance and alert the team if unexpected situations emerge. This indicates how important is to acquire neurophysiological indices during teleoperations and in this way facilitate the development of dynamic human-in-the-loop systems.

Further analysis based on two-class classification results of with/without Time-Pressure and with/without Time-Latency, respectively, is demonstrated in Figure 8.13. The figure shows the results of SVM based on radial kernel functions across different time windows of 2, 5, 10 and 20 s as well as for the whole trial data [92]. Classification accuracy is shown for a number of eye-tracking features that include blinks, fixations, saccades, pupil characteristics, all features and ANOVA features with a significant difference extracted from each trial. Blinks and eye saccades features performed well in identifying time-pressure even with very small window size of 2 s. These eye-tracking characteristics could enable real-time evaluation of cognitive workload related to time-pressure. Pupil characteristics show less stable performance, whereas concatenating all features does not increase performance. Perhaps, this reflects that there is small number of samples compared to features. On the other hand, ANOVA features perform best in terms of classification in both with/without time-pressure and with/without time-latency. They work by evaluating the variance of the predictive variables on the response. However, their performance is low with small window sizes and reaches its maximum with the whole-trial data. Classification of with/without Time-pressure performs significantly better than classification of with/without Time-Latency condition. Further validation is required to understand whether this difference in performance reflects uneven sample sizes or whether time-latency is inherently more difficult to identify based on eye-tracking features.

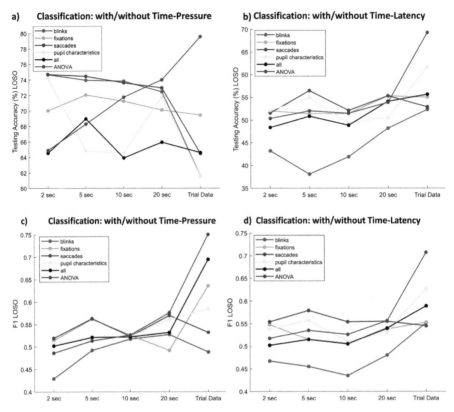

Figure 8.13   *Two-class classification results on with or without time-latency and with and without time-latency under the Leave-One-Subject-Out cross-validation protocol. Testing accuracy and F1 scores are shown for window sizes of 2, 5, 10 and 20 s as well as for the whole trial data for each classifier, respectively. The classifier is based on SVM with radial basis kernel functions. A number of eye-tracking features are examined to determine whether they can reliably identify cognitive workload.*

EEG data were also acquired simultaneously and analysed to extract features and neurophysiological indices that are related to differences in cognitive workload [2]. Previous work hypothesises that specific spectral powers, such as theta, alpha and beta as well as derived indices such as ratios of spectral power in these bands can robustly identify workload [93]. It is common for EEG data to be band-passed into five bands, which roughly categorised as delta (1–4 Hz), theta (4–8 Hz), alpha (8–13 Hz), beta (13–30 Hz) and gamma (30–70 Hz) [94]. These definitions relate to fundamental properties of human brain function that reflect distinct roles and underline communication between different brain regions. The theta band has been correlated with mental fatigue, higher demands of working memory and increased cognitive workload. It has been also

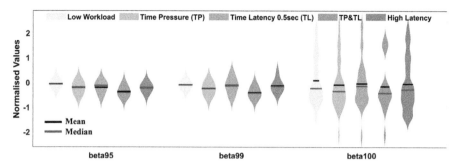

Figure 8.14   *Beta power as a neurophysiological index of cognitive workload.*
EEG data are z-normalised. A large variability is observed and for
these reasons outliers have been excluded by estimating confidence
intervals at 95 (beta95) and 99 per cent (beta99), respectively.

linked to tasks that require sustained attention and it is anti-correlated with alertness and
mental vigilance. On the other hand, occipital alpha oscillations are very well recognised
during relaxed states and eyes closed. Alpha power is also anticorrelated with vigilance
and attentional resources, whereas suppression of alpha waves in occipital and parietal
regions relates to more challenging and difficult tasks [93]. Beta power is also linked
to cognitive workload as it correlates well with increased workload intensity and high
levels of concentration and visual attention.

EEG data are prone to artefacts, which result from muscle activity, eye blinks
and movements. Independent component analysis was used to decompose the EEG
signal in statistically independent sources and reject the sources that are unlikely to
have neuronal origin. Since beta power has been demonstrated as a strong predictor
of cognitive workload, the EEG signal was band-passed in beta band and the beta
power was compared across different conditions. EEG data were z-normalised and
segmented in 2 s windows. Figure 8.14 uses violin plots to provide a sense of the
data distribution and shows a high variability and non-Gaussian distribution when
all data are included (beta100). To alleviate this problem outliers are removed by
retaining the 99 per cent (beta99) and 95 per cent (beta95) of data, respectively,
within the confidence interval of each class.

Finally, Figure 8.15 shows how changes across conditions in beta power compare with changes in Riemannian distance between covariance matrices. Covariance
matrices reveal brain connectivity and their geometric properties can provide a
better estimation of distance compared to standard Euclidean distance [80, 95].
Riemannian distance has been used also in BCIs as a robust way to classify different
brain states [46]. Here we hypothesise that cognitive workload increases will be also
associated with changes in brain connectivity. This is an argument supported also
from previous work on cognitive workload that revealed significant connectivity
changes both across condition and across expertise in surgeons [86]. Statistically
significant differences were identified based on the Kruskal-Wallis one-way analysis
of variance, which is a non-parametric method that accounts for the fact that some of

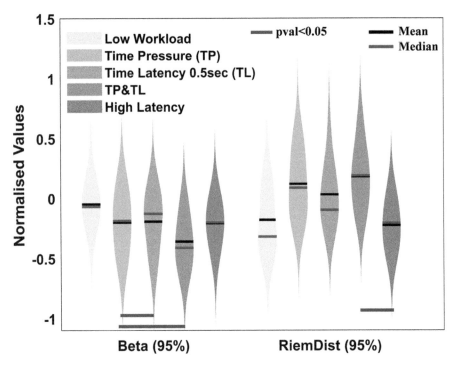

Figure 8.15   *Beta power and Riemannian distance as measures of cognitive*
workload. Statistically significant differences are identified based
on Kruskal-Wallis pairwise comparisons

the EEG data distributions are not normal even after the z-normalisation and outliers
removal.

8.7   Recommendations and future work

Realistic virtual reality environments along with augmented reality paradigms provide unique opportunities towards novel AI approaches that improve performance
in teleoperations by leveraging human factors information obtained via multi-modal
neurophysiological indices. Human in the loop systems are challenging to design
and require multi-disciplinary approaches that couple our understanding in human
brain neurophysiology with powerful computational approaches. Future applications should exploit the rich information derived from numerous neuroscientific and
psychology studies to develop real-time adaptive systems. These systems should be
able to account for the complexity in real-life scenarios via data-driven approaches
and provide subject-specific support. Therefore, future systems should focus on:

- The communication between key brain regions such as the PFC and motor cortex is relatively unexplored. The function of the PFC is key in decision-making

and as a high functional centre it is affected by anxiety and cognitive workload. Evidence shows that there are statistically significant differences in brain connectivity between conditions of time-pressure and self-paced tasks as well as between novice and expert users. The development of advanced machine learning techniques that exploit this information in real-time and provide continuous measures of workload is of paramount importance. The Riemannian distance of the covariance matrix of the EEG signal has shown some promise in characterising cognitive workload differences. Further work is required to develop intuitive and interpretable continuous measures that can detect subtle differences in brain connectivity and characterise cognitive workload conditions.

- Cognitive workload estimation exploits advances in BCIs both in terms of miniaturised, portable sensing devices and intelligent algorithms. Current systems do not generalise well across subjects due to high inter-subject and inter-session variability and require training to be adjusted to each new user. Typically, BCIs require subject-specific training and calibration prior to use. Nevertheless, there are significant advances in machine learning that pave the way towards subject-independent BCIs. These technologies will play a significant role in transferring knowledge from highly specialised neuroimaging technologies to wearable headsets thus enabling reliable systems to be developed for safety critical applications.

- It is also important to understand the differences in cognitive workload induced due to task-specific difficulty and exogenous factors such as time-pressure and time-latency. To this end several neurophysiological indices were explored as well as activity-depended measures that shed some light on the interaction level between the user and the system. Simulations environments can also provide opportunities to couple user's actions with robotic motion in a task-independent way. These causal relationships give intuition behind performance and errors, which should be predicted and avoided in space applications.

- Another important aspect of teleoperated systems with a human in the loop is semi-autonomous modes of operations. Semi-autonomy can enhance performance and reduce errors while it empowers users to be in control of the system. It is important to develop systems that are able to predict humans intentions' and learn to collaborate to achieve a task successfully. Furthermore, how a system can go safely from one level of autonomy to another is of paramount importance in safety critical applications in space. In this scenario, transparency of the system achieved by intuitive, interactive designs and continuous measures of cognitive workload will play a critical role both during the cycles of system development as well as during the operation.

- Computational complexity can be prohibitive in translating current AI success stories to space applications. Therefore, it is important to develop emulation systems that mimic the hardware capabilities in space and allow researchers to test their solutions.

References

[1]   Pieters M.A., Zaal P.M.T. 'Training for long-duration space missions: a literature review into skill retention and generalizability'. *IFAC-PapersOnLine*. 2019;**52**(19):247–52.

[2]   Freer D., Guo Y., Deligianni F., Yang G.-Z. 'On-orbit operations simulator for workload measurement during telerobotic training', submitted. *ACM Transactions on Human Robot Interaction*; Available from: https://arxiv.org/abs/2002.10594.

[3]   Clark T.K., Stimpson A.J., Young L.R., Oman C.M., Natapoff A., Duda K.R. 'Human spatial orientation perception during simulated Lunar landing motions'. *Journal of Spacecraft and Rockets*. 2014;**51**(1):267–80.

[4]   The Verge Website. Available from https://www.theverge.com/2019/3/15/18267365/boeing-737-max-8-crash-autopilot-automation [Accessed 2019].

[5]   Weller A. (eds.) 'Transparency: motivations and challenges' in Samek W., Montavon G., Vedaldi A., Hansen L.K., Müller K.-R. (eds.). *Explainable AI: Interpreting, Explaining and Visualizing Deep Learning*. Springer; 2019.

[6]   Murray-Smith R. 'Control theory, dynamics, and continuous interaction' in Oulasvirta A., Kristensson P.O., Bi X., Howes A. (eds.). *Computational Interaction*. Oxford University Press; 2018.

[7]   Kramer L.J., Etherington T.J., Bailey R.E., Kennedy K.D. 'Quantifying pilot contribution to flight safety during hydraulic systems failure'. *Advances in Human Aspects of Transportation*. 2018;**597**:15–26.

[8]   Coleshill E., Oshinowo L., Rembala R., Bina B., Rey D., Sindelar S. 'Dextre: improving maintenance operations on the International Space Station'. *Acta Astronautica*. 2009;**64**(9-10):869–74.

[9]   Visinsky M. *Robotics on the International Space Station (ISS) and Lessons in Progress*. NASA Technical Reports Server; 2017.

[10]   Fong T., Rochlis Zumbado J., Currie N., Mishkin A., Akin D.L. 'Space telerobotics: unique challenges to human–robot collaboration in space'. *Reviews of Human Factors and Ergonomics*. 2013;**9**(1):6–56.

[11]   Callen P. *Robotic Transfer and Interfaces for External ISS Payloads*. NASA Technical Reports Server; 2014.

[12]   NASA Website. Available from https://www.nasa.gov/mission_pages/station/structure/elements/special-purpose-dextrous-manipulator [Accessed 2018].

[13]   Canadian Space Agency Website. Available from https://www.asc-csa.gc.ca/eng/iss/robotics/robotics-training-centre.asp [Accessed 2018].

[14]   Belghith K., Nkambou R., Kabanza F., Hartman L. 'An intelligent simulator for telerobotics training'. *IEEE Transactions on Learning Technologies*. 2012;**5**(1):11–19.

[15]   Fournier-Viger P., Nkambou R., Mayers A. 'Evaluating spatial representations and skills in a simulator-based tutoring system'. *IEEE Transactions on Learning Technologies*. 2008;**1**(1):63–74.

[16] Canadian Space Agency Website. Available from https://www.asc-csa.gc.ca/eng/multimedia/games/canadarm2/default.asp [Accessed 2013].

[17] Stirling L., Siu H.C., Jones E., Duda K. 'Human factors considerations for enabling functional use of exosystems in operational environments'. *IEEE Systems Journal*. 2019;**13**(1):1072–83.

[18] Ravi D., Wong C., Deligianni F., *et al.* 'Deep learning for health informatics'. *IEEE Journal of Biomedical and Health Informatics*. 2017;**21**(1):4–21.

[19] Andreu-Perez J., Deligianni F., Ravi D., Yang G.Z. Artificial intelligence and robotics. 2018. Available from arxiv.org/abs/1803.10813.

[20] Cinel C., Valeriani D., Poli R. 'Neurotechnologies for human cognitive augmentation: current state of the art and future prospects'. *Frontiers in Human Neuroscience*. 2019;**13**:13.

[21] Navarro J., Reynaud E., Osiurak F. 'Neuroergonomics of car driving: a critical meta-analysis of neuroimaging data on the human brain behind the wheel'. *Neuroscience & Biobehavioral Reviews*. 2018;**95**:464–79.

[22] Modi H.N., Singh H., Yang G.-Z., Darzi A., Leff D.R. 'A decade of imaging surgeons' brain function (part II): a systematic review of applications for technical and nontechnical skills assessment'. *Surgery*. 2017;**162**(5):1130–9.

[23] Cléry-Melin M.-L., Jollant F., Gorwood P. 'Reward systems and cognitions in major depressive disorder'. *CNS Spectrums*. 2019;**24**(1):64–77.

[24] Hiser J., Koenigs M. 'The multifaceted role of the ventromedial prefrontal cortex in emotion, decision making, social cognition, and psychopathology'. *Biological Psychiatry*. 2018;**83**(8):638–47.

[25] Xu J., Van Dam N.T., Feng C., *et al.* 'Anxious brain networks: a coordinate-based activation likelihood estimation meta-analysis of resting-state functional connectivity studies in anxiety'. *Neuroscience & Biobehavioral Reviews*. 2019;**96**(2):21–30.

[26] Knudsen E.I. 'Fundamental components of attention'. *Annual Review of Neuroscience*. 2007;**30**(1):57–78.

[27] Knudsen E.I. 'Neural circuits that mediate selective attention: a comparative perspective'. *Trends in Neurosciences*. 2018;**41**(11):789–805.

[28] Moore T., Zirnsak M. 'Neural mechanisms of selective visual attention'. *Annual Review of Psychology*. 2017;**68**(68):47–72.

[29] Clayton M.S., Yeung N., Cohen Kadosh R. 'The roles of cortical oscillations in sustained attention'. *Trends in Cognitive Sciences*. 2015;**19**(4):188–95.

[30] Gilbert C.D., Li W. 'Top-Down influences on visual processing'. *Nature Reviews Neuroscience*. 2013;**14**(5):350–63.

[31] Ruff D.A., Ni A.M., Cohen M.R. 'Cognition as a window into neuronal population space'. *Annual Review of Neuroscience*. 2018;**41**:77–97.

[32] Walle K.M., Nordvik J.E., Espeseth T., Becker F., Laeng B. 'Multiple object tracking and pupillometry reveal deficits in both selective and intensive attention in unilateral spatial neglect'. *Journal of Clinical and Experimental Neuropsychology*. 2019;**41**(3):270–89.

[33] Husain M. 'Visual attention: what inattention reveals about the brain'. *Current Biology*. 2019;**29**(7):R262–4.

[34] Bartolomeo P., Thiebaut de Schotten M., Chica A.B. 'Brain networks of visuospatial attention and their disruption in visual neglect'. *Frontiers in Human Neuroscience*. 2012;**6**.

[35] Bogadhi A.R., Bollimunta A., Leopold D.A., Krauzlis R.J. 'Spatial attention deficits are causally linked to an area in macaque temporal cortex'. *Current Biology*. 2019;**29**(5):726–36.

[36] Kanki B., Clervoy J.-F., Sandal G. *Space safety and human performance*. Butterworth-Heinemann; 2017.

[37] Duncan J., Martens S., Ward R. 'Restricted attentional capacity within but not between sensory modalities'. *Nature*. 1997;**387**(6635):808–10.

[38] Boucart M., Fabre-Thorpe M., Thorpe S., Arndt C., Hache J.C. 'Covert object recognition at large visual eccentricity'. *Journal of Vision*. 2001;**1**(471).

[39] Potter M.C., Wyble B., Hagmann C.E., McCourt E.S. 'Detecting meaning in RSVP at 13 MS per picture'. *Attention, Perception, & Psychophysics*. 2014;**76**(2):270–9.

[40] Jamison H., Julie A. 'Multi-dimensional human workload assessment for supervisory human–machine teams'. *Journal of Cognitive Engineering and Decision Making*. 2019;**13**(3):146–70.

[41] Horrey W.J., Lesch M.F., Garabet A., Simmons L., Maikala R. 'Distraction and task engagement: how interesting and boring information impact driving performance and subjective and physiological responses'. *Applied Ergonomics*. 2017;**58**(2):342–8.

[42] Varghese R.J., Freer D., Deligianni F., Liu J., Yang G.Z. 'Wearable robotics for upper-limb rehabilitation and assistance: a review on the state-of-the-art, challenges and future research'. *Wearable Technology in Medicine and Healthcare*. Elsevier; 2018.

[43] Velliste M., Perel S., Spalding M.C., Whitford A.S., Schwartz A.B. 'Cortical control of a prosthetic arm for self-feeding'. *Nature*. 2008;**453**(7198):1098–101.

[44] Berger H. 'The electro-cephalogramm of humans and its analysis'. *Naturwissenschaften*. 1937;**25**:193–6.

[45] Jiang X., Bian G.-B., Tian Z. 'Removal of artifacts from EEG signals: a review'. *Sensors*. 2019;**19**(5):987.

[46] Freer D., Deligianni F., Yang G.Z. 'Adaptive Riemannian BCI for enhanced motor imagery training protocols'. *IEEE BSN*. 2019.

[47] Brouwer A.-M., Hogervorst M.A., van Erp J.B.F., Heffelaar T., Zimmerman P.H., Oostenveld R. 'Estimating workload using EEG spectral power and ERPs in the n-back task'. *Journal of Neural Engineering*. 2012;**9**(4):045008.

[48] Cheron G., Cebolla A.M., Petieau M., *et al.* 'Adaptive changes of rhythmic EEG oscillations in space implications for brain-machine interface applications'. *International Review of Neurobiology*. 2009;**86**:171–87.

[49] Leroy A., De Saedeleer C., Bengoetxea A., *et al.* 'Mu and alpha EEG rhythms during the arrest reaction in microgravity'. *Microgravity Science and Technology*. 2007;**19**(5-6):102–7.

[50]   Berka C., Levendowski D.J., Lumicao M.N., *et al.* 'EEG correlates of task engagement and mental workload in vigilance, learning, and memory tasks'. *Aviation, Space, and Environmental Medicine.* 2007;**78**(5):B231–44.

[51]   Murata A. 'An attempt to evaluate mental workload using wavelet transform of EEG'. *Human Factors: The Journal of the Human Factors and Ergonomics Society.* 2005;**47**(3):498–508.

[52]   Curtin A., Ayaz H. 'The age of neuroergonomics: towards ubiquitous and con-tinuous measurement of brain function with fNIRS'. *Japanese Psychological Research.* 2018;**60**(4):374–86.

[53]   Liu Y., Ayaz H., Shewokis P.A., Patricia A.S. 'Mental workload classification with concurrent electroencephalography and functional near-infrared spec-troscopy'. *Brain-Computer Interfaces.* 2017;**4**(3):175–85.

[54]   Ziv G. 'Gaze behavior and visual attention: a review of eye tracking studies in aviation'. *The International Journal of Aviation Psychology.* 2016;**26**(3-4):75–104.

[55]   Benedetto S., Pedrotti M., Minin L., Baccino T., Re A., Montanari R. 'Driver workload and eye blink duration'. *Transportation Research Part F: Traffic Psychology and Behaviour.* 2011;**14**(3):199–208.

[56]   Gao X.-Y., Zhang Y.-F., Zheng W.-L., Lu B.-L. 'Evaluating driving fatigue detection algorithms using eye tracking glasses'. *2015 7th International IEEE/EMBS Conference on Neural Engineering (NER).* IEEE; 2015.

[57]   Richstone L., Schwartz M.J., Seideman C., Cadeddu J., Marshall S., Kavoussi L.R. 'Eye metrics as an objective assessment of surgical skill'. *Annals of Surgery.* 2010;**252**(1):177–82.

[58]   Tien T., Pucher P.H., Sodergren M.H., Sriskandarajah K., Yang G.-Z., Darzi A. 'Eye tracking for skills assessment and training: a systematic review'. *Journal of Surgical Research.* 2014;**191**(1):169–78.

[59]   Van Orden K.F., Limbert W., Makeig S., Jung T.P. 'Eye activity correlates of workload during a visuospatial memory task'. *Human Factors: The Journal of the Human Factors and Ergonomics Society.* 2001;**43**(1):111–21.

[60]   Crawford J.D., Martinez-Trujillo J.C., Klier E.M. 'Neural control of three-dimensional eye and head movements'. *Current Opinion in Neurobiology.* 2003;**13**(6):655–62.

[61]   Clément G., Ngo-Anh J.T. 'Space physiology II: adaptation of the central nervous system to space flight--past, current, and future studies'. *European Journal of Applied Physiology.* 2013;**113**(7):1655–72.

[62]   Clarke A.H., Haslwanter T. 'The orientation of listing's plane in micrograv-ity'. *Vision Research.* 2007;**47**(25):3132–40.

[63]   Kowler E. 'Eye movements: the past 25 years'. *Vision Research.* 2011;**51**(13):1457–83.

[64]   Salvucci D.D., Goldberg J.H. 'Identifying fixations and saccades in eye-tracking protocols'. *Proceedings of the 2000 Symposium on Eye Tracking Research & Applications*; 2000.

[65]   Shelhamer M., Roberts D.C. 'Magnetic scleral search coil'. *Handbook of Clinical Neurophysiology.* **9**; 2010. pp. 80–7.

[66] Haslwanter T., Clarke A.H. 'Eye movement measurement: electro-oculography and video-oculography'. *Handbook of Clinical Neurophysiology*. **9**; 2010. pp. 61–79.

[67] Nyström M., Andersson R., Holmqvist K., van de Weijer J. 'The influence of calibration method and eye physiology on eyetracking data quality'. *Behavior Research Methods*. 2013;**45**(1):272–88.

[68] Kahneman D., Beatty J. 'Pupil diameter and load on memory'. *Science*. 1966;**154**(3756):1583–5.

[69] Hess E.H., Polt J.M. 'Pupil size in relation to mental activity during simple problem-solving'. *Science*. 1964;**143**(3611):1190–2.

[70] Wang W., Li Z., Wang Y., Chen F. 'Indexing cognitive workload based on pupillary response under luminance and emotional changes'. *Proceedings of the 2013 International Conference on Intelligent User Interfaces*; 2013.

[71] Marshall S.P. 'The index of cognitive activity: measuring cognitive workload'. *Proceedings of the IEEE 7th conference on Human Factors and Power Plants*. IEEE; 2002.

[72] Just M.A., Carpenter P.A. 'Eye fixations and cognitive processes'. *Cognitive Psychology*. 1976.

[73] Schulz C.M., Schneider E., Fritz L., *et al.* 'Eye tracking for assessment of workload: a pilot study in an anaesthesia simulator environment'. *British Journal of Anaesthesia*. 2011;**106**(1):44–50.

[74] Volden F., Edirisinghe V.D.A., Fostervold K.-I. 'Human gaze-parameters as an indicator of mental workload'. *Congress of the International Ergonomics Association*. Springer; 2018.

[75] Leigh R.J., Kennard C. 'Using saccades as a research tool in the clinical neurosciences'. *Brain*. 2004;**127**(3):460–77.

[76] Son L.A., Hamada H., Inagami M., Suzuki T., Aoki H. 'Effect of mental workload and aging on driver distraction based on the involuntary eye movement'. *Advances in Human Aspects of Transportation*. 2017;**484**:349–59.

[77] Mogilever N.B., Zuccarelli L., Burles F., *et al.* 'Expedition cognition: a review and prospective of subterranean neuroscience with spaceflight applications'. *Frontiers in Human Neuroscience*. 2018;**12**:407.

[78] Otsuka K., Cornelissen G., Kubo Y., *et al.* 'Circadian challenge of astronauts' unconscious mind adapting to microgravity in space, estimated by heart rate variability'. *Scientific Reports*. 2018;**8**(1):10381.

[79] Schmitz J. 'Structural changes of the brain during a stay in space increased intracranial pressure for the astronauts'. *Flugmedizin Tropenmedizin Reisemedizin*. 2018;**25**(2):53–4.

[80] Deligianni F., Varoquaux G., Thirion B., *et al.* 'A framework for inter-subject prediction of functional connectivity from structural networks'. *IEEE Transactions on Medical Imaging*. 2013;**32**(12):2200–14.

[81] Rabbi A.F., Zony A., de Leon P., Fazel-Rezai R. 'Mental workload and task engagement evaluation based on changes in electroencephalogram'. *Biomedical Engineering Letters*. 2012;**2**(3):139–46.

[82] Harrivel A., Hearn T. *Detection of mental state and reduction of artifacts using functional near infrared spectroscopy (fNIRS)*. NASA; 2014.

[83] Williamson J.H., Quek M., Popescu I., Ramsay A., Murray-Smith R. 'Efficient human-machine control with asymmetric marginal reliability input devices'. *Plos One*. 2020;**15**(6):e0233603.

[84] Evain A., Roussel N., Casiez G., Argelaguet-Sanz F., Lecuyer A. 'Brain–computer interfaces for human–computer interaction'. *Brain–Computer Interfaces 1: Foundations and Methods*. John Wiley & Sons, Inc; 2016.

[85] Debie E., Rojas R.F., Fidock J. Multimodal fusion for objective assessment of cognitive workload: a review. IEEE Transactions on Cybernetics; 2020.

[86] Deligianni F., Singh H., Modi H.N., *et al. Expertise and task pressure in fnirs-based brain connectomes* [online]. 2020. Available from https://arxiv.org/abs/2001.00114 [Accessed January 2021].

[87] Jaiswal D., Chowdhury A., Banerjee T., Chatterjee D. 'Effect of mental workload on breathing pattern and heart rate for a working memory task: a pilot study'. *Conference Proceedings of the Annual International Conference of the IEEE Engineering in Medicine and Biology Society*. 2019;**2019**:2202–6.

[88] Aghajani H., Garbey M., Omurtag A. 'Measuring mental workload with EEG+fNIRS'. *Frontiers in Human Neuroscience*. 2017;**11**:359.

[89] Shin J., von Lühmann A., Kim D.-W., Mehnert J., Hwang H.-J., Müller K.-R. 'Simultaneous acquisition of EEG and NIRS during cognitive tasks for an open access dataset'. *Scientific Data*. 2018;**5**(1).

[90] Deligianni F., Guo Y., Yang G.-Z. 'From emotions to mood disorders: a survey on gait analysis methodology'. *IEEE Journal of Biomedical and Health Informatics*. 2019;**23**(6):2302–16.

[91] Qiu J., Helbig R. 'Body posture as an indicator of workload in mental work'. *Human Factors: The Journal of the Human Factors and Ergonomics Society*. 2012;**54**(4):626–35.

[92] Guo Y., Freer D., Deligianni F., Yang G.-Z. 'Eye-tracking for performance evaluation and workload estimation in space telerobotic training', submitted. *IEEE Transactions on Human-Machine Systems*.

[93] Rojas R.F., Debie E., Fidock J., *et al.* 'Electroencephalographic workload indicators during teleoperation of an unmanned aerial vehicle shepherding a swarm of unmanned ground vehicles in contested environments'. *Frontiers in Neuroscience*. 2020;**14**.

[94] Deligianni F., Centeno M., Carmichael D.W., Clayden J.D. 'Relating resting-state fMRI and EEG whole-brain connectomes across frequency bands'. *Frontiers in Neuroscience*. 2014;**8**(339).

[95] Deligianni F., Varoquaux G., Thirion B., *et al.* 'A probabilistic framework to infer brain functional connectivity from anatomical connections'. *IPMI*. 2011;**6801**:296–307.

Chapter 9

Physiological adaptations in space and wearable technology for biosignal monitoring

Shamas U. E. Khan[1], Bruno G. Rosa[1], Panagiotis Kassanos[1], Claire F. Miller[1,2], Fani Deligianni[3], and Guang- Zhong-Yang[4]

Space is a hostile environment for life and can induce profound changes in the physiology of astronauts. These changes mainly result from space radiation exposure, microgravity, the lack of an atmosphere and resulting body fluid shifts. The human body proceeds with various homeostatic adaptations that influence the cardiovascular, nervous, musculoskeletal, endocrine and other physiological systems. To adopt and develop wearable technologies that facilitate human–robot interaction in space and astronaut biomonitoring, the underlying physiological changes should be considered. In this chapter, we review basic concepts of these body adaptations, as well as biomarkers and biosignals for biomonitoring and how these can change due to the space environment. Finally, we review wearable biosignal monitoring technologies that can quantify sweat, heart rate (HR) and endocrine responses, which can act as indices of acute stress and increased workload.

9.1 Introduction

Space is a hostile environment for human life. The vacuum of space, space radiation and the absence of an atmosphere and gravity lead to an environment that is incapable of sustaining life. Furthermore, space shuttles and space stations force astronauts to be in relative isolation within confined spaces, while the physical and mental stress experienced before and during missions is significant. Notwithstanding these challenges, space exploration has always been pushing the boundaries of science and technology. As of May 2020, 9 years after the end of the Space Shuttle Program,

[1]The Hamlyn Centre, Institute of Global Health Innovation, Imperial College London, London, UK
[2]School of Design, Royal College of Art, London, UK
[3]School of Computing Science, University of Glasgow, Glasgow, UK
[4]Institute of Medical Robotics, Shanghai Jiao Tong University, Minhang, Shanghai, China

two astronauts became the first arriving at the International Space Station (ISS) using a private spacecraft. Technological advancements made by private companies, such as employing partially reusable spacecrafts, are reducing overall launching costs, essentially leading towards sustainable private exploration and exploitation of interstellar space and other planets. Nevertheless, many challenges must still be overcome, including the design of effective systems that allow and support life in (deep) space over long periods of time. To achieve this, the effect of the space environment on human physiology needs to be fully understood. This constitutes the focus of this chapter.

Since the first human missions to space, continuous experimentation and monitoring of astronauts pre-, during and post-flight have allowed us to gain insights on the physiological adaptations that take place in space. These insights have contributed in establishing various technologies in an effort to mitigate the devastating effects physiological changes can have on both the short- and long-term health of crew members as it can influence the success of a space mission.

A valuable tool towards this has been the monitoring of astronaut biosignals and the collection of biosamples at frequent intervals for subsequent analysis on Earth and also in space. Real-time physiological monitoring in space can provide information regarding the health status of crew members that can be used to optimize mission parameters, e.g., it can indicate when a crew member should return to Earth, whether medication or rest should be prescribed or identify specific crew members over others for challenging tasks. Monitoring acute stress and excessive workload are critical in maintaining well-being in spaceflights.

Astronauts interact with highly complex systems, and it has become evident that the next generation of artificial intelligence (AI) and robotics systems will be human-centred. In other words, AI will be able to adapt to human neurophysiology in ways that will optimize information rate to maximize system performance and prevent accidents. The presence of robots and robotically controlled instruments that can be locally or remotely controlled is increasing. The hostile space environment is a driving force behind this, as robots can be designed to freely operate within it. Robots can be used to perform tasks, complement astronaut abilities and aid them in completing challenging and hazardous operations. Therefore, the development of intelligent wearable technology is important in allowing astronauts to perform dexterous tasks and eliminate current spacesuit restrictions. Consequently, human–robot interaction in space is becoming increasingly important. Physiological parameters can be used to generate control signals and information upon which robots can act. Figure 9.1 illustrates the need for biosignal monitoring from the heart to the blood for chemical markers for prevention, prognosis, diagnosis and treatment and the consequent need for assistive devices, with the ever-increasing role of robotics, neurotechnology, biosensing and wearable sensing. This chapter provides a brief discussion on crucial physiological adaptations taking place in space (summarized in Figure 9.2) and highlights relevant biomarkers and biosignals for monitoring such changes. Finally, recent examples of wearable technologies for biosignal monitoring are presented.

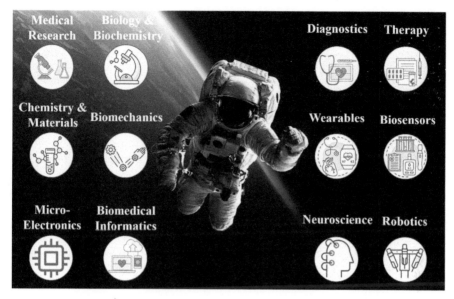

Figure 9.1   *From basic medical research to wearable diagnostic and assistive devices and therapeutics, there is a need for novel technological innovations to address current and future needs for space explorations and to sustain human life. Astronaut image by Vadim Sadovski/shutterstock.com and icons by bsd/shutterstock.com, Griboedov/shutterstock.com and elenabsl/shutterstock.com.*

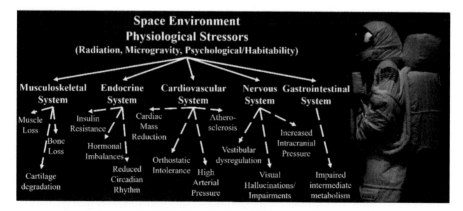

Figure 9.2   *The physiological stressors of the space environment. Astronaut image by Peter Rondel/shutterstock.com.*

9.2   Cardiovascular system

9.2.1   *Blood pressure, haemodynamic response and orthostatic intolerance*

9.2.1.1   Heart rate, blood pressure and cardiac output

The cardiovascular system integrates sensory inputs from various receptors around the body to signal the demand for oxygen and regulate body fluids [1]. Receptors include chemoreceptors, proprioceptors and mechanoreceptors. Stimulation of these receptors modulates the autonomic nervous system in the central nervous system that modifies cardiovascular parameters, including blood pressure [1, 2]. Baroreceptors sense rapid changes in blood pressure by measuring arterial wall tensions in the carotid sinus and aortic arch. For example, elevated blood pressures distend the aortic and carotid walls. Subsequently, the baroreceptors relay frequency signals to the medulla oblongata. The parasympathetic nervous system or the vagal system becomes activated and lowers the HR and blood pressure to maintain homeostasis. Decreased blood pressures invoke a vice versa response.

Body fluid redistribution and haemodynamic responses can alter the activation of such receptors depending upon their location in the human body. As astronauts launch, they position themselves supinely with their legs raised above the coronal plane, thus incurring an initial cephalic fluid shift. This initial position elevates cephalic pressures and distends the baroreceptors. Astronauts are then exposed to microgravity as they ascent into the orbit of the Earth. The absence of terrestrial gravity diminishes blood pressure gradients, which sustain the cephalic fluid shift during their entire duration in space, increasing pressure in the central circulation and causing peripheral vasodilation [3]. Astronauts complain of nasal congestion, a 'puffy face' or otherwise, flu-like symptoms [4]. This effect also influences the HR, mean arterial pressure, cardiac output and stroke volume.

Initial stages of spaceflight involve an acute phase where factors such as psychological stress, and other physiological adaptations may influence the cardiovascular system [4]. For instance, space motion sickness in astronauts occurs almost immediately during ascent. However, it resolves within 3–4 days into spaceflight, indicating acute adaptations in the vestibular system, whereas, stimulation of this system also influences HR [3]. Despite decades of research monitoring the human haemodynamic response in space, there are inconsistencies in published studies [5]. To better understand these inconsistencies, the relevant studies are considered in detail below.

Fritsch-Yelle *et al.* [6] found decreased HR, HR variability (HRV) and diastolic pressure in-flight during short-duration space missions compared to pre-flight values. They attributed this to reduced sympathetic nervous system (SNS) expression and total peripheral resistance (TPR). The reduced TPR suggests systemic vascular dilation that would, in effect, maintain the mean arterial pressure. Migeotte *et al.* [7] found that HR initially decreased when compared to both pre-flight supine and standing posture. By the end of 16 days spaceflight, both HR and HRV values resembled pre-flight supine values. Other studies also indicated decreased in-flight

HR and diastolic pressures relative to pre-flight supine values [8], during sleep [9], and also a decreased HRV [10].

Verheyden *et al.* [11] recorded an unchanged HR and blood pressure over a 6-month mission on the ISS when compared to pre-flight supine values. However, HR decreased relative to pre-flight standing posture measurements. This study stresses a significant consideration that upon gravity on Earth, different positions stimulate different, yet complex, sympatho-adrenomedullary responses [12]. For example, on Earth, HR, blood pressure and plasma catecholamines increase when changing body postures from supine, to sitting or standing. Therefore, inconsistencies in past studies can also arise from different choices of references for a comparison basis. Baevsky *et al.* [13] also reported unchanged HR and HRV in a long-duration space mission, but a decreased blood pressure was noticed relative to pre-flight supine values. The study took their first measurement after a month during spaceflight; therefore, it does not consider acute adaptations.

In contrast, Eckberg *et al.* [14] found that HR and mean arterial pressure increased significantly during a short-term space mission. They attributed this to an increased SNS activity and the derangement of vagal baroreflex function. Most short duration HRV studies, including NEUROLAB experiments, indicated increased SNS activity acutely when compared to pre-flight supine posture during spaceflight [15]. Nonetheless, short-duration studies may reveal different findings relative to long-duration ones, as the cardiovascular system would enter a transient phase before reaching a steady-state level after acute factors subside, which may take 2–3 weeks. Other factors such as astronaut's physical and genetical inter-variability, countermeasures, space radiation, social and environmental factors may also influence the response of the cardiovascular system, therefore, causing inconsistencies within studies.

By considering other haemodynamic parameters, Norsk *et al.* [16] reported an increased cardiac output by a factor of 41 per cent but unchanged HR compared to pre-flight seated values in astronauts during a 6-month space mission. Cardiac output is directly proportional to stroke volume and HR and, therefore, if the latter remains unaffected, the former must have increased. Stroke volume was also increased by 35 per cent. However, what led to increased stroke volume is unknown, yet *prima facie*, it could be attributed to cephalic fluid shift. Surprisingly, unchanged plasma catecholamine levels were found, signifying insignificant SNS activity in long-duration spaceflight. Mean arterial pressure was also reduced by 10 mmHg, though it was insignificant. Again, mean arterial pressure is directly proportional to the systemic vascular resistance and cardiac output and influenced by variations in the central venous pressure (CVP); systemic vascular resistance was reduced by 39 per cent offsetting the cardiac output increase in maintaining the mean arterial pressure. Fluctuations to the CVP in a healthy human being on Earth is insignificant; however, CVP reduces significantly in space, which, in turn, would affect venous return, and thereof stroke volume. This will be discussed ahead in more detail.

In a long-duration study by Hughson *et al.* [17], both stroke volume and HR were unchanged initially, yet found to increase later during spaceflight relative to the pre-seated position. Cardiac output also increased; however, similar to [16], the systemic vascular resistance was decreased to maintain the mean arterial pressure. Furthermore, a longer left ventricular ejection time (LVET) was likewise reported early in the flight. Theoretically, LVET is directly proportional to stroke volume but inversely to HR [18]. However, stroke volume and HR were unchanged. On Earth, the LVET is longer in the elderly compared to younger people; yet, it is found to be independent of HR and BP [19]. Changes in myocardial contractility [17] with structural changes of the heart [20] or eventual SNS decline [19] may contribute to LVET increase, which prompts further investigation.

In general, cardiac output increases, diastolic pressure decreases and HR remains either unchanged or initially declines with microgravity when compared to pre-flight supine values [21]. A probable cause for a decline in diastolic pressure could be the activation of nitric oxide-mediated pathways from the fluid shift-induced myocardial stretch, modifying diastolic properties and heart contractility [22, 23]. Whether SNS activity is increased, decreased or unchanged, it is still ambiguous; yet, it can be increased relative to supine values [21]. Surprisingly, the systemic vascular resistance in a plethora of studies is decreased, indicating complex mechanisms behind this phenomenon.

One theory is that a sustained cephalic fluid shift expands heart chambers, and this, in turn, releases vasodilatory cardiac natriuretic peptides (CNP) [5, 16]. Specifically, the atrial natriuretic peptide (ANP) hinders SNS activity in the peripheral blood vessels and suppresses the renin-angiotensin-aldosterone system [24]. Spacelab-2 flights found a sharp rise in ANP levels in astronauts; however, it decreased by 60 per cent within days due to renal blood flow [25]; therefore, sustained vasodilation may not be supported by this hypothesis.

Based on another hypothesis, Fortrat *et al.* [26] found that lower limb venous resistance decreases significantly by measuring venous function through venous plethysmography in the calves. Muscle atrophy occurs almost immediately on exposure to microgravity [4], reducing external pressures on the vessels. Additionally, less stretching of the vessels due to fluid shift may reduce contractility of the vascular smooth muscles in the legs [16] despite acute upregulation of SNS, as it has inconsequential effects on the venous vessels of the systemic system [27]. Therefore, both mechanisms in synergy may reduce peripheral resistance leading to lower systemic vascular resistance.

9.2.1.2   Central venous pressure and hypovolemia

Reduction in CVP can occur as a result of either a decrease in venous blood volume or increase in venous compliance [28]. On Earth, haemodynamic factors such as changes in systemic vascular resistance, mean arterial pressure, cardiac output or venous capacity do not directly correlate with changes in CVP, as multiple mechanisms are involved in its maintenance [27, 28]. However, the CVP strongly influences venous return and, therefore, the cardiac output. An example of water flowing in a pipe can be used to illustrate this; water can only move from one end to another if a high- to low-pressure gradient is present – the greater the difference, the higher the flow. Relatively, the low-pressure region is the CVP, and the high-pressure region is known as the mean circulatory filling pressure [29]. The latter reflects the pressure throughout the vascular system when the blood exerts stress on the vessel walls while the body is in equilibrium without the heart pumping [30]. The difference in the region is venous return. Therefore, CVP and venous return follow an inverse relationship, whereas the former is also regulated by vascular resistance [27, 31].

On Earth, venous resistance changes are insignificant [27]; however, the systemic vascular resistance reduces substantially in microgravity, indicating that venous resistance is a significant variable in the estimation of venous return. Theoretically, a reduction in CVP and systemic vascular resistance may have increased the venous return, resulting in the rise of both cardiac output and stroke volume, without changes to the HR. However, veins may compress if the CVP fall was higher than the intrathoracic pressure, which may impede venous return. Surprisingly, it was found that chest expansion, due to the fluid shift, resulted in an even larger drop of the intrathoracic pressure, thus increasing CVP transmural pressure [32]. This mechanism should increase the cardiac pre-load, stroke volume and cardiac output [16]. Nevertheless, West *et al.* [33] argue that the decrease in intrathoracic pressures was too small to attribute a significant reduction of CVP.

CVP cannot be treated as an independent variable, as though it influences the cardiac output, it also depends upon it. The relationship between CVP and cardiac output provides essential information about the heart's performance. A decreased CVP with increased cardiac output indicates a positive functioning of the right heart, improves renal function and reduces prognosis in circulatory shock [34]. However, the reduction of left ventricular end-diastolic volume due to cardiac remodelling may offset venous return to some extent.

Concerning haemodynamic variables in space, both CVP and systemic vascular resistance decrease, whereas cardiac output and stroke volume increase and HR either decreases or remains the same. A question remains as to why systemic vasodilation occurs if SNS activity increases in weightlessness. Furthermore, increasing SNS activity should raise both the mean circulatory filling pressure and CVP. The former rises because vasoconstriction and increased water reabsorption stress the vessels. However, the initial inrush of blood to the upper body induces reflexes (e.g. Gauer-Henry's reflex) that counterbalance the effects of SNS [35], leading to plasma volume reduction (hypovolemia) due to transcapillary fluid filtration. Evidence collected by Videbaek and Norsk [32] shows a rise in cardiac transmural pressure following an upward fluid shift, leading to atrial stretching (increase in atrial diameter) and cardiac distension. Consequently, the rise in the levels of circulating ANP and cyclic guanosine monophosphate(cGMP) contribute to the increase of vascular permeability, which combined with high transmural pressure favours the migration of fluid and ions (e.g. sodium) from intra- to extra-vascular compartments. The body, in turn, compensates by stimulation of the central volume carotid, aortic and cardiac receptors [36], encouraging natriuresis and diuresis, which reduce plasma volume (6–10 per cent) and body mass (2–4 per cent) until stabilization into a new steady-state level [37]. Therefore, cardiac function adapts to fluid shift by compensatory variation in the cardiac output, as thus far, no cardiac arrhythmias have been detected in astronauts that can explain the spaceflight-induced cascade of cardiac and haemodynamic events.

Early studies in space reported that hypovolemia occurs immediately upon exposure to microgravity. Surprisingly, diuresis and natriuresis were insignificant during short-term space missions [38, 39]. Thus, Diedrich *et al.* [40] hypothesized three causes of hypovolemia: lesser fluid intake and urine output, significant fluid shifts towards interstitial space or fluid shifts to interstitial space due to lower transmural pressures. As for the latter two hypotheses, post-flight data indicate an insignificant interstitial fluid increase [40], suggesting other mechanisms governing the phenomenon. ANP release can also contribute to low plasma volume by the mechanism mentioned above; however, hypovolemia remains throughout long-duration spaceflights, whereas elevated ANP reduced significantly within 10 days of spaceflight.

9.2.1.3   Orthostatic intolerance

The cardiovascular system in space undergoes complex adaptations mediated by hormonal changes, autonomic nervous system responses, reflex mechanisms and metabolic alterations induced by the space environment. An adverse effect of these upon return to Earth is orthostatic intolerance (OI). Longer spaceflight duration leads to a more significant OI [41]. This can be of great concern when astronauts need to perform emergency evacuations [5]. Mechanisms governing OI are still unknown. However, hypovolemia, inadequate cerebral perfusion, baroreflex control and cardiovascular deconditioning have all been found to play a role [5, 26]. Previously, it was hypothesized that hypovolemia induces OI; however, not all hypovolemic crew members develop OI [42]. In fact, astronauts still experienced this condition and presyncope despite the intake of fludrocortisone before landing [43], which increases plasma volume. Therefore, hypovolemia may contribute to OI, but it is not the sole factor. Women tend to exhibit greater OI than men, as well as hypovolemia, decreased baroreflex control sensitivity and hormonal response [44].

Astronauts undergo standing tests of upto 10 min upon landing to measure the intensity and onset of OI. Comparison is drawn from finishers and non-finishers of various testing methods, such as a stand or tilt testing, to understand the underlying mechanisms. Non-finishers have been found to have lower norepinephrine levels, peripheral vascular resistance and blood pressure post-flight [45], indicating impaired autonomic balance. However, the lower accumulation of catecholamine levels may be due to a shorter duration of standing tests [46]. In measuring HRV, Blaber *et al.* [46] found that non-finishers had lower sympathovagal ratio, while low-frequency spectral power remained unchanged during post-flight upon standing. Therefore, a reduced sympathetic response is considered, leading to orthostatic hypotension and, subsequently, presyncope. It was also found that astronauts with higher pre-flight vagal activity were amongst the non-finishers [46].

Blood redistribution occurs from central circulation to the lower limbs when humans stand on Earth, resulting in an instant decline in blood pressure, venous return and plasma volume due to compensatory and transient mechanisms [47]. This effect would inevitably reduce cardiac output, stroke volume and cerebral perfusion.

The human body restores venous return and blood pressure by splanchnic venocon-striction, arterial vasoconstriction through increased SNS activity and the vascu-lar recoiling effect [48]. Consequently, SNS integrity is impaired after spaceflight, also leading to OI and other related cardiovascular adaptations. Furthermore, the reported maintenance of mean arterial pressure during spaceflight [35] decreases the demand for baroreflex, which upon return to Earth remains impaired for some astronauts, contributing to OI [49].

In general, OI occurs upon inadequate brain perfusion during standing due to either delayed baroreflex response or decreased blood pressure, stroke volume and cardiac filling [5]. Cerebrovascular autoregulation impairment has also been consid-ered as a contributor to OI [50]. However, Iwasaki *et al.* found that cerebrovascular autoregulation did not impair; instead, it improved as fluctuations in blood pressure modulated the cerebrovascular blood flow velocity (CBFV) lesser than pre-flight. Blaber *et al.* [51] confirmed that astronauts with intact CBFV to pre-flight values did not experience OI. Despite this, non-finishers were unable to autoregulate the conductance upon hypotension on standing. Other factors, such as fluctuations in plasma CO_2 could also influence autoregulation. The effects of the space environ-ment on the cardiovascular system are summarized in Figure 9.2.

9.2.2 Electrocardiographic variations

Cardiovascular deconditioning can be monitored by changes in the HRV, defined as the time difference between consecutive heartbeats (RR interval). HRV is typically translated to the frequency domain by Fourier transform, and the resulting spec-trum is divided into different frequency bands. These bands are ultra-low frequency (ULF: < 0.0033 Hz), very-low-frequency (VLF: 0.0033–0.04 Hz), low frequency (LF: 0.04–0.15 Hz) and high frequency (HF: 0.15–0.4 Hz). Each band has been related to different activities and processes of the autonomic nervous system. For instance, ULF has been associated with inflammation, secretion (e.g. norepineph-rine, cytokine interleukin-6, IL-6), behavioural and intrinsic autonomic regulatory processes. VLF is reportedly associated with endocrine activities, namely involving the renin-angiotensin-aldosterone system, whereas some components of both ULF and VLF are appointed as predictors of survival rate in patients with coronary artery disease and healthy subjects.

By analysing the HRV signal of astronauts, the authors of [35] have detected a reduction of ULF power in space when compared to Earth, from which it was concluded that microgravity affected the intrinsic autonomic regulatory system of astronauts during long-duration spaceflight. A reasonable explanation for reduced ULF can be attributed to the slower rate of body movements in microgravity and the need to maintain adequate cardiac function to sustain them, leading to modifications of the intrinsic autonomic cardiovascular responses. These findings are in line with [52], where a lower HRV was found in long-duration spaceflights. Thus, more effort to the study of the lower frequency bands in HRV is needed, as opposed to HR or blood pressure studies only, which traditionally reveal minimal alterations in space due to microgravity.

HRV is extensively used to monitor the state of the heart, as well as that of the brain. High values of HRV are traditionally a sign of good health (increased workload), whereas reduced HRV is regarded as a predictor of both mortality and morbidity from cardiovascular disease (CVD). The spectral bands of HRV have also been related to the activation of several cortex areas controlling the circadian rhythm. During spaceflight, the normal earthly circadian rhythm is altered, not only due to biological factors but mostly because of environmental factors, including the day/night shift on the horizon every 90 min per spatial orbit, radiation exposure and changes to the Earth's magnetic field induced by, e.g. solar winds.

With this in mind, the authors of [53] monitored several HRV data endpoints (including VLF, LF, HF and surrogate metrics) in astronauts during long-duration missions to assess the influence of the space environment and possible connections with anti-ageing effects in humans. Moreover, the study was conducted during the days with relatively low and higher magnetic field disturbances. The increase in several endpoints, such as VLF and LF/HF ratio, on disturbed days in conjunction with more sensitive responses observed during daytime than dark led to the conclusion that a circadian stage-dependent response of HRV to magnetic activity remains functional even during spaceflight. On the other hand, the overall increase in the previous HRV endpoints may have also contributed to slowing down the ageing process during magnetic activity events due to a rise in sympathetic activation, thus predisposing the body into a 'fight-flight-or-freeze' response typical in situations of increased threat.

9.2.3 Cardiac remodelling in space

During spaceflight, the geometrical sphericity of the left ventricle (LV) of astronauts was modified to become more circular [20]. On Earth, upper body parts necessitate heart to pump blood against the direction of gravity, which demands extra force and work [48]. However, the absence of hydrostatic gradients in microgravity diminishes the demand, inducing cardiac remodelling and deconditioning.

Cardiac mass reduction in spaceflights is in line with head-down bed rest tests (HDBRs) performed on Earth, which affect the diastolic filling process due to a slower untwisting movement of the heart during early diastole [49]. Consequently, left ventricular end-diastolic volume declines with no apparent significant change detected on the injected fraction. Other consequences of abnormal diastolic properties include slower LV relaxation, loss of LV compliance and reduced active-filling phase with atrial systole [36]. Cardiac mass can be preserved by adopting counter-measures onboard ISS, such as exercise, which may help explain post-flight magnetic resonance imaging (MRI) findings, in which retained cardiac mass and diastolic/systolic volumes were observed after returning from long-duration spaceflights.

Earlier Skylab experiments reported a reduction of LV mass by approximately 12 per cent [54]. Perhonen *et al.* [55] reported LV mass reduction by almost 12 per cent after a week-long spaceflight. However, only 8 per cent LV mass reduction in 6-week horizontal bed rest (HBR) was reported in the same study. These preliminary results indicate the impact that a short-duration spaceflight can have over the cardiac structure, compared to bed rest on Earth. Left ventricular end-diastolic volume also followed a similar downward trend in the study, indicating less work exerted by the heart. Left ventricular end-diastolic volume reduction in the bedrest participants was noted as early as 2 weeks. However, LV mass reduction was not reported until 6 weeks [55]. Therefore, reduced left ventricular end-diastolic volume precedes LV mass loss and could be due to cardiac remodelling, or noticeably, hypovolemia in space.

Summers *et al.* [56] also reported an average 9.1 per cent reduction in LV mass in 38 astronauts in a short-duration spaceflight. They attributed this adaptation to dehydration-induced hypovolemia rather than cardiac atrophy since a full recovery to pre-flight values occurred within 3 days upon return to Earth. However, some HBR studies do not attribute hypovolemia as the sole contributor to reduced LV mass, since the latter takes a significantly longer duration to fully recover than the former [57].

Studies involving various methods of investigation, such as cardiovascular magnetic resonance or echocardiography (ECHO), have reported similar reductions in LV mass and LV volumes [5, 58, 59]. In a 12-week HBR experiment, cardiac atrophy was found by approximately 15 per cent [59]. A longer duration of HBR resulted in a more significant loss of LV mass [55]. Levine *et al.* [58] found a reduced distensibility of the LV; however, overall stiffness was preserved. Both microgravity-induced hypovolemia and inactivity may explain cardiac atrophy in space [58]. For example, hypovolemia precedes to reduced stroke volume and left ventricular end-diastolic volume, which shifts the pressure-volume curves to the left. Subsequently, there is reduced wall stress resulting in reduced LV mass and distensibility by cardiac remodelling. Thus, the heart is exhibiting superlative plasticity to adapt to different gravitational loadings.

9.2.4 Vascular function and cell adaptations in space

In relation to vascular function, the absence of mechanical stimuli derived from gravity has repercussions on the circumferential and shear stresses governing blood flow and associated pressures. Moreover, atrophy and hypertrophy of arteries below and above the heart, respectively, are to be expected during long-duration spaceflight based on evidence that astronaut femoral and carotid arteries upon return from the ISS were thicker and stiffer due to an increase in intima-media thickness. This leads to shorter transit time for the arterial pulse to reach the finger and ankle regions, vascular endothelium impairment and accelerated atherogenesis. Neuronal and hormonal factors are known to be altered during spaceflight, which negatively influences the internal structure of arteries. From these, hormones from the renin-angiotensin-aldosterone system promote arterial stiffness by stimulating collagen formation and matrix remodelling, as well as endothelium hypertrophy and proliferation of smooth cells [49]. In the end, the peripheral or systemic vascular resistance should increase, resembling the typical effect induced by ageing; however, due to adaptations in cardiac output, the mean arterial pressure remains constant, even if the vascular resistance has increased. Healthy vascular endothelium prevents platelet aggregation and adhesion of leukocytes. When dysfunctional, due to microgravity, radiation expo-sure or reduced cardiac workload (as in sedimentary ageing) can lead to different forms of vascular disease, from congestive heart failure to other conditions (e.g. myocardial stiffening) that share atherosclerosis as a common predecessor.

The reduction of plasma volume (up to 500 ml) results in increased concentra-tions of blood cells, metabolites and associated products. A spike in haemoglobin concentration leads to an early breakdown of newly produced red blood cells. This is also linked to the activity of the autonomic nervous system, which translates in serum erythropoietin levels being decreased by the kidneys [60]. Reticulo-endothelial cells can then phagocyte young red blood cells instead of promoting normal cell remod-elling. The increased oxidative stresses originating from exposure to microgravity and radiation further enhance the decline in red blood cell mass, increased structural fragility and anaemia [49]. As a result, haemoglobin levels decline in blood after the initial increment [61]. Decrease in workload during spaceflight, although revers-ible by countermeasures (e.g. physical exercise), contributes to a lower oxygen and nutrient supply demand for body tissues that can be sustained by the reduced number of red blood cells in circulation.

Regarding leukocytes, there is evidence of alteration in the population of this type of blood cells, including neutrophilia, eosinopenia and lymphocytopenia. The normal body immune response is also altered in microgravity due to variations in the skin microbiome, bacterial virulence and viral reactivation [37]; immunity to infections is also greatly affected. Alterations to the function of T-cells have been reported in spaceflight, whereas the stress arising from re-entering Earth's atmo-sphere seems to lead to an increase in neutrophil count, as observed in hematologic examinations after landing [37].

9.2.5   Jugular venous blood flow and thrombus formation

The normal hydrostatic pressure gradient in individuals derived from upright and supine postures on Earth creates daily variations of pressure for blood and other fluids, with volume redistribution across the human body. In upright position, blood accumulates in pools in the lower extremities [60], whereas in a supine position, cerebral venous outflow can occur due to increased blood volume of the internal jugular veins (IJVs) [62]. Whenever the external atmospheric pressure is higher than intraluminal pressure (e.g. upright position), IJVs are partially or fully collapsed, which forces the cerebral venous outflow to be redirected to the vertebral veins and plexus. This cerebral venous drainage system is important for the ordinance of intracranial pressure and fluid dynamics, with IJVs preventing excessive negative intracranial pressure on upright posture. Nonetheless, occlusion by collapse is not total in some cases, and blood interchange occurs between the cerebral and central venous systems. IJV flow stasis is one of the factors contributing to thrombus forma-tion on Earth, in conjunction with endothelial injury and hypercoagulation of blood.

In space, due to weightlessness and the consequent fluid shift and other distur-bance, it is thought that flow stasis in IJV may pose a risk for developing thrombi in healthy astronauts. The absence of daily postural changes inside ISS exacerbates the risk for venous thrombosis. To quantify both the cross-sectional area, pressure and flow velocity in the IJVs, data was collected directly from 11 astronauts undergoing space missions with average duration of 210 days in [62]. Doppler ultrasonography measurements were taken pre-, in- and post-flight to assess the evolution of the left IJV in space. A stagnant or retrograde flow was found in ~50 per cent of the astronauts during spaceflight, a partial jugular occlusion in one astronaut and a completely occlusive thrombus in another that underwent supervised treatment onboard the ISS (described later). This recently recognized spaceflight risk highlights the need for a greater understanding of the mechanisms involving venous flow alteration even when IJV's area and pressure values appear similar to supine values recorded on Earth. In the face of this discovery, lower body negative pressure was suggested to drain venous blood away from the head to the lower body parts, thus counterbalancing the upward fluid shift and improving blood circulation (including IJV flow) to reduce the risk of thrombogenesis.

The suspected thrombosis was found 2 months into spaceflight by routine examination at the ISS in a healthy astronaut, with no recorded medical condition on venous thromboembolism [63]. The decisions for the treatment undertaken onboard the ISS were then taken on Earth, making this the first-ever reported case in human history of guided telemedicine treatment in space supervised by specialists on Earth. The treatment involved the administration of enoxaparin (a blood thinner available onboard the ISS), followed by apixaban (inhibitor of blood clot formation), 42 days after diagnosis, delivered by a supply spacecraft to the ISS. Sonographic examination thereafter demonstrated a progressive thrombus volume reduction. Flow re-establishing through the affected area was observed 47 days after treatment, but no spontaneous flow was detected after 90 days. Treatment with apixaban was abandoned 4 days before the astronaut's return to Earth, where follow-up examination 24 h after landing revealed a residual thrombus volume, completely vanishing in the ensuing 10 days of recovery. This incident highlighted not only the lack of scientific knowledge about the topic but also the complexity of space medicine regarding critical decisions-making in the complete absence of direct medical observation and with limited resources. Moreover, the restricted applicability of syringes, injection needles or pumps for medical treatment due to surface-tension effects in microgravity needs to be further considered.

9.2.6 Biomarkers of cardiovascular diseases

This section provides a summary of key biomarkers, including their mechanisms, which may be relevant to monitoring astronauts during spaceflight. However, a wider number of CVD biomarkers are discussed in detail in [64, 65] and their references. Biomarkers are particularly important to detect individuals at risk of future CVD. Traditional factors include hypertension, obesity, diabetes mellitus, hypercholesterolemia and smoking [66]. These are, however, not sufficient for predicting risks in the general population [66]. Additionally, they may not be significantly relevant for highly trained and screened astronauts. Biomarkers of CVD can be detected in blood pressure measurements, electrocardiograms (ECG), imaging tests such as

ECHO and CT scans, or measured in biosamples such as blood, urine or tissues [64]. For CVD risk assessment, the traditional plasma-based markers include primarily high-density lipoprotein cholesterol (HDL-C), low-density lipoprotein cholesterol (LDL-C) and triglycerides [67].

The most common biomarkers involved in heart failure diagnosis are natriuretic peptides. These peptides are linked to sodium and water balance and regulation of the vascular tone [66]. Examples include ANP, B-type natriuretic peptide (BNP), CNP and Dendroaspis natriuretic peptide (DNP). The first two are produced by myocytes in the atria (ANP) and ventricles (BNP) [66]. Induction of BNP leads to the secretion of N-terminal pro-B-type natriuretic peptide (NT-proBNP). All these peptides (especially NT-proBNP) are overly responsive to heart failure, whereas their concentrations are related to its likelihood. BNP and NT-proBNP are clinically proven for heart failure diagnosis and heart failure exacerbation [68]. However, these markers can also become elevated due to tachycardia and myocarditis and advanced chronic kidney disease reflecting reduced renal clearance, as they are also produced by the kidneys [66]. Cardiac troponin I and T are specific and sensitive biomarkers for myocardial damage, also providing information regarding the extent of damage, being employed as biomarkers for diagnosis and risk assessment in patients with suspected acute coronary syndrome [66, 68].

Biomarkers of myocardial fibrosis that are associated with a number of collagen fibres in the myocardium include the carboxy-terminal propeptide of procollagen type I (PICP) and the amino-terminal propeptide of procollagen type III (PIIINP) [66]. Serum PICP correlates with the abundance of myocardium collagen fibres in patients suffering from hypertensive heart disease. On the other hand, PIIINP correlates with collagen deposition in heart failure patients with ischemic heart disease and idiopathic dilated cardiomyopathy [66]. Galectin-3 (Gal-3) and soluble suppression of tumorigenicity (sST2) are markers of myocardial fibrosis [66]. Gal-3 levels in plasma are increased in heart failure patients and can be used in combination with NT-proBNP, whereas the latter is produced by cardiac fibroblasts and cardiomyocytes. Elevated serum sST2 levels are also found in acute and chronic heart failure patients; however, they can also be found in other diseases such as liver disease, nephropathy, gastric and breast cancer [66].

C-reactive protein (CRP) is the most studied circulating inflammatory marker for atherosclerosis. High-sensitivity CRP (hsCRP) serum levels have been employed to assess the risk of CVD and heart attack [68], being considered as an independent predictor of peripheral arterial disease, myocardial infarction and sudden cardiac death [67]. CRP secretion from the liver is induced by interleukins (IL), such as IL-1 and IL-6. Additional atherosclerosis markers include oxidized LDL, lipoproteins, serum amyloid A, the leukocyte adhesion molecules ICAM-1, VCAM-1, P-selectin, soluble CD40 ligand, tissue-type plasminogen activator (t-PA), tumour necrosis factors (TNFs), IL-1, IL-6, IL-18, matrix metalloproteinase-9 (MMP-9) and myeloperoxidase (MPO) [66]. MPO has also been applied for the prediction of the risks for coronary heart disease [68], whereas MMP-9 is related to several stages of atherosclerosis and associated with premature coronary atherosclerosis, myocardial infarction and unstable angina [67]. Circulating levels of IL-37 are related to increased risk for chronic heart failure [68]. Additional markers of interest include cystatin C, copeptin, lipoprotein-associated phospholipase A2, fibrinogen, lactate and D-dimer [65, 67, 68]. D-dimer is a small protein fragment formed following the degradation of a blood clot by fibrinolysis. Lactate may potentially serve as a prognostic marker of late-onset heart failure and death following acute myocardial infarction [68]. Oxidative stress, defined as the excessive production of reactive oxygen species (ROS), is related to hypertension and atherosclerosis [67].

Dysfunctions in the endothelium are linked to the pathogenesis of CVD. Space radiation causes malfunctions leading to occlusive artery disease, a risk factor for CVD [69]. Around 17 proteins that can be measured in urine have been found to be associated with the function of the endothelium [69]. Several proteins measured in urine have been found to vary significantly after long-duration spaceflight in [70]. These include annexin A5, transthyretin, prostate-specific antigen (PSA), cadherin-2 (CAD-2), alpha-2-HS-glycoprotein, urokinase plasminogen activator, vitamin D-binding protein, neprilysin and gamma-glutamyl transpeptidase (GGT). Measurable protein markers in urine can also provide insights into the operation and adaptation of the cardiovascular autonomic regulatory system [71]. These include cadherin-13 (CAD-13), mucin-1, alpha-1 of collagen subunit type VI (COL6A1), hemisentin-1, semenogelin-2, SH3 domain-binding protein, transthyretin and serine proteases inhibitors [72]. The above are summarized in Table 9.1.

9.2.7 Cardiovascular disease mortality and radiation risks in astronauts

Beyond low Earth orbit (LEO), radiation is the critical factor limiting space exploration by humans. Radiation in space includes solar radiation (solar particle events consist mainly of short-duration low energy protons and electrons) and galactic cosmic radiation (primarily ionized hydrogen), comprising electromagnetic waves and charged particles (e.g. high-energy electrons, protons, α-particles, β-particles and fast heavy ions). Heavy ions have higher ionization potential, thus providing a significant contribution to the total radiation dose experienced by astronauts [73]. Protons and high-energy heavy ions forming the HZE radiation induce biological damage more effectively than X-rays or γ-rays (e.g. linear energy transfer, LET, radiation) [74]. HZE is defined as high (H) atomic number (Z) and energy (E) ions, and it is predominantly felt beyond the Van Allen belt region which provides magnetic shielding to Earth [37]. Therefore, astronauts on LEO (including ISS) are subjected to lower doses of HZE.

Currently, there are no sufficient protection methods for high-energy radiation [75]. Shielding of manned spacecraft bound for deep space missions is still the standard approach to decrease astronaut exposure to radiation. However, HZE is highly penetrating. They can produce lighter ions and neutrons in radiation showers when interacting with shielding materials. During long-duration missions onboard the ISS, overall, astronauts receive more than 70 mSv in radiation exposure, with the daily exposure being estimated to 0.48 mSv/day. On average, daily radiation levels of 1.8 mSv can be expected beyond the Van Allen belt in deep space exploration.

Table 9.1   *Summary of biomarkers useful for monitoring issues in the various systems in the body*

Manifestation	Biomarkers
Cardiovascular system	
Heart failure	ANP, BNP, CNP, DNP, NT-proBNP, Gal-3, IL-37, lactate
Myocardial infarction	Troponin-I, Troponin-C, Creatine-Kinase-MB, BNP, CRP, hsCRP, MMP-9
Sudden cardiac death	hsCRP, CRP, IL-6, cystatin-C, BNP, NT-proBNP
Myocardial fibrosis	PICP, PIIINP, sST2, Gal-3
Atherosclerosis and coronary arterial disease	HDL-C, LDL-C, VLDL-C, Triglycerides, MPO, CRP, oxidized LDL, serum amyloid A, ICAM-1, VCAM-1, P-Selectin, Soluble CD40 ligand, t-PA, TNF, IL-1, IL-6, IL-18, MMP-9, MPO, cystatin-C, blood pressure
Endothelial damage	IL-1, IL-6, TNF, hsCRP, urine samples (annexin A5, transthyretin, PSA, CAD-2, alpha-2-HS-glycoprotein, urokinase plasminogen activator, vitamin D-binding protein, neprilysin and GGT)
Cardiovascular autonomic regulatory system	CAD-13, mucin-1, COL6A1, hemisentin-1, semenogelin-2, SH3 domain-binding protein, transthyretin, serine proteases inhibitors, HR variability
Brain and peripheral nervous system	
Stress	Cortisol (salivary and serum), Cytokines IL-1, IL-2, IL-6, TNF- α, leptin, BDNF, NGF, DHEA, α-amylase, neuropeptide Y, Urinary 3-methyl-histidine HR, pulse rate
Depression	BDNF, cortisol, neuropeptide Y, serotonin, cytokines, HPAA activity
Sleep disorder and alertness	Melatonin, orexin-A
Neuroplasticity	EEG, NIRS
Endocrine and gastrointestinal system	
Insulin resistance	C-peptide, C-GMP expression, leptin
Thyroid function	TSH, T3, T4, PTH
Nutritional	Vitamins (A, B6, B12, C, D, E, K), folic acid, minerals
Urinary and renal system	
Urinary tract infection	Urinary leukocyte esterase, urinary nitrites, urine output
Renal stones	Serum and urinary calcium levels, urine output, urine citrate levels, urine pH, creatinine levels
Musculoskeletal system	
Bone resorption	CTX-1, NTX-1, tartrate-resistant acid phosphatase, osteoprotegerin, nuclear κ-B activator
Bone formation	IL-6, alkaline phosphatase, P1CP, P1NP, osteocalcin
Bone health	IL-1α, TNF-α, IFN-γ, IL-4, IL-6, IL-10, and TFG-β1
Other bone remodelling biomarkers	Cystatin C, fetuin-A, and fibronectin

In Martian orbit, this is estimated to 1.28 mSv/day [76]. In contrast, environmental radiation levels felt on Earth by humans are within 1–3 mSv/year [49]. Effects of space radiation on humans are vast and, therefore, impossible to cover entirely within this chapter. We thus highlight a limited number of issues and interested readers are referred to [75] for additional information. Space radiation can cause acute radiation syndrome, including acute skin erythema, fatigue, hair loss, loss of appetite, a decline in white blood cell levels and platelets, as well as emotional effects such as depression or anxiety [77]. Exposure to space radiation can also lead to cancer, cataracts, fibrosis, central nervous system diseases, vascular damage and immunological, endocrine or hereditary effects [75, 77, 78].

Radiation produces ROS. These can cause oxidative damage to DNA detectable by measuring the urinary levels of 8-hydroxy-20-deoxyguanosine [79], which play a key role in signal transduction [78]. Radiation-induced DNA lesions lead to genomic instability, contributing to the risk of carcinogenesis [76]. Once cancer is initiated, microgravity affects tumour progression, proliferation, migration and metastasis [76]. There is a positive relationship between spaceflight duration and chromosomal abnormalities [78]. Radiation also leads to perturbations to redox metabolism [74]. These lead to increased oxidative stress [74]. Moreover, high levels of oxygen concentration contribute to increased DNA damage when combined with γ-rays, which (together with other issues) prompted the decision to maintain only a 21 per cent oxygen level inside ISS. This contrasts with the 100 per cent oxygen level on the Apollo space capsules and extravehicular activities (EVAs) on the Moon.

Additionally, microgravity can increase radiation sensitivity by means of a synergistic effect in living organisms [76]. There is growing evidence that the interplay between microgravity and radiation is involved in numerous cellular responses, including signal transduction. Cytoplasmic (especially mitochondrial) impairment is as devastating as chromosomal damage [78]. It has been found that microgravity and radiation affect both, at varying degrees, micro RNA and long non-coding RNA that are critical for multiple cellular processes [80]. Microgravity and radiation induce additive and synergistic effects on the expression patterns of RNA, and cellular responses have also been observed in human lymphoblastoid cells [80].

Astronauts usually experience lower risks of death by chronic diseases in comparison with the general population. The high levels of health and well-being of astronauts, as a result of intense physical training, endurance adaptation and economic benefits from a hazardous occupation can explain the lower risk. However, radiation also induces endothelial dysfunctions that may lead to occlusive artery disease and pose a risk factor for CVD. CVD mortality among astronauts with missions beyond LEO (lunar flights), where the effects of space radiation are significant, was reported to be four to five times higher than that registered for astronauts who had never experienced spaceflight or those remaining within LEO [74]. Vascular endothelial damage by radiation induces inflammatory responses that accelerate atherogenesis and fibrosis formation in arteries and myocardium, thought to be associated with the higher mortality rate by CVD on Apollo astronauts. Endothelial cell dysfunction has been found to be also due to high energy radiation rather than from microgravity [74]. Animal experiments have demonstrated that radiation leads to structural degenerative injury and changes to coronary arteries [73]. Other consequences of radiation exposure include damage to smooth muscle cells and cardiomyocytes owing to increased oxidative stress, cellular senescence and death [49].

While angiogenesis is fundamental to restore the damaged areas in the microvascular system, it can also be hindered by HZE particles.

Sufficient replication of space radiation on Earth for use in controlled experiments is not possible, and there have been a limited number of astronauts that have flown beyond LEO in lunar missions. Hence, the effects of space radiation are not yet fully understood. Consequently, safe radiation exposure limits to minimize radiation effects on human physiology, such as CVD, during spaceflight have not been established, and only general guidelines are available [73].

9.3   Other physiological adaptations in microgravity

9.3.1   Gastrointestinal system and nutrition

The space environment surrounding astronauts influences the function of the human digestive system. Since this constitutes the entry point for the intake of food and processing of nutrition, the role of the digestive system is therefore essential for adequate energy balance within the body, maintenance of the musculoskeletal, immune and endocrine systems, among other factors promoting well-being. Long-duration spaceflights require a large amount of food, from which macronutrients, micronutrients and vitamins are separated along the gastrointestinal (GI) tract. For a manned mission to Mars, a 3-year supply of food may be required. The average consumption of 22 tons of supplies by a crew of six, implies a total cost of half a billion euros for space transportation alone, despite water supplies being recycled in-flight and food being dehydrated [81]. Space food is specially selected for consumption by the astronauts, following requirements for balanced nutrition. Growing food supplies inside the ISS has also been tested to lower the quantity of food carried during spaceflight while providing a source of fresh food production, though with limited success.

Some of the early symptoms arising on the first days into spaceflight reported by astronauts include anorexia, vomiting, nausea and loss of taste (and smell), which are related directly or indirectly with maladaptation of the GI tract to space environment [36]. Due to microgravity, the gaseous stomach can be experienced by astronauts, as gases within this cavity are unable to rise [82]. Alteration of the contact between the gastric content and GI mucosa can also occur, and their motility is affected by the cephalic fluid shift observed in spaceflight. Splanchnic circulation, that is, blood flow in the celiac superior and inferior mesenteric arteries (in the abdomen) is reduced, thereby affecting the circulation that irrigates the abdominal viscera, digestive tract, liver, pancreas and spleen. The venous flow of all these organs then drains to the portal vein of the liver, before returning to the heart. It is, therefore, an important energetic and nutritional exchange hub, as well as a significant autonomic and endocrine regulation centre due to the presence of liver and pancreatic secretions [83]. Consequences of inadequate splanchnic flow are ischemia, tissue damage and necrosis, with intestines becoming more permeable and allowing the passage of endotoxins and other bacterial products to lymph nodes and blood vessels [84]. Contributing to this are the alterations of intestinal flora experienced in space that impact the secretion of peptides and the normal processes of nutrient metabolism and absorption by blood vessels.

Recent studies on the dietary health of astronauts have shown a propensity for consuming 25–30 per cent fewer calories during spaceflight in comparison to the required earthly equivalent [81]. Hence, energy balance is shifted towards the negative part since energy intake does not keep up with energy expenditure in space. For instance, a mean caloric energy deficit of 5.7 MJ per day was reported during a 16-day space mission [81]. Equivalently, this value corresponds to a loss of 5 kg of body mass per month by a healthy individual. Inadequate energy intake leads to increased fatigability of muscles, susceptibility to infection, alterations to sleep quality and a reduction in general well-being on Earth. In space, these aspects are thought to be exacerbated by the harmful effects of physiological adaptation, potentially leading to psychological imbalances between crew members, with a change of mood and behaviour (including altered feeding behaviour) [36]. The required energy for spaceflight has been estimated to be around 7.5 and 8.5 MJ daily for women (55 kg) and men (75 kg), respectively. However, physical exercise and (most importantly) performance of EVA significantly increase energy demand. EVA for 6 h is estimated to require 2.2 and 2.8 MJ of energy for women and men, respectively [81].

Recommendation for macronutrient composition on Earth requires 15 per cent proteins, 30 per cent lipids and 55 per cent carbohydrates. Malnutrition is known to increase morbidity, mortality and increased hospitalization due to the associated medical conditions (e.g. diabetes and cardiovascular). Nutrition is also a precursor for the synthesis of fundamental body units (e.g. cells and components), as well as the provision of micronutrients and co-factors (vitamins and minerals) that support enzyme activity and mechanisms for tissue repair.

In space, supplement intake of vitamin D and calcium have been used to preserve bone metabolism [36, 60]. However, calcium absorption from intestines declines during spaceflight despite supplement intake of the mineral. Besides, bone is the largest reservoir of calcium inside the body, ready to be deployed during periods of inadequate dietary calcium. Spaceflight severely damages this balance, and bone loss is a critical feature of physiological impairment in space (as discussed further in Section 9.4 of this chapter). Vitamin D can also be supplemented in the astronauts' diet to compensate for the natural body production of this vitamin, occurring whenever skin is directly exposed to sunlight that is incongruent in space. Supplementation of vitamin K may help preserve bone loss in astronauts since it counteracts the reduction of bone formation by normalizing the undercarboxylated levels of osteocalcin, reported in some studies performed on Earth [81].

Space diets usually have high amounts of sodium that contribute to fluid balance adaptations in microgravity, such as liquid extravasation due to increased vascular and renal permeability [49]. It is strongly advised to avoid iron supplementation due to changes in erythrocyte mass, and spaceflight anaemia leads to increased iron storage within the body with as yet unknown consequences [85]. Unsaturated fatty acids such as omega-3 have increased benefits for human health on Earth and can be used during spaceflight to protect against the development of insulin resistance and inflammation responses. Additionally, an intake of antioxidants to counterbalance the effects of oxidative stress have been proposed for astronauts, in the form of vitamin A (including precursors beta-carotene and carotenoids), vitamin C (or ascorbate, raised from 60 to 100 mg daily), vitamin E, vitamin B12 (to support ophthalmic injuries), as well as minerals (e.g. copper, zinc, manganese and selenium) [82]. Supplementation with antioxidants may reduce oxidant damage to cells and tissues due to radiation exposure while playing a protective role against other medical conditions. In fact, vitamin A, beta-carotene, folic acid, vitamin B12, vitamin C, riboflavin, selenium and zinc present immunomodulatory actions. Deficiencies of the latter are typically linked to changes in T-lymphocytes. Moreover, glutamine, nucleotides and omega-3 can also influence immune function. Protein and aminoacidic intake have been suggested to fight muscle loss during spaceflight, but some harmful effects include a decrease of blood pH via the oxidation of sulphur amino acids, thus exacerbating bone loss [60]. High protein consumption by astronauts is also related to greater satiety but lower energy intake can induce hypocaloric absorption in space [81].

9.3.2   Respiratory system

Microgravity seems to produce no alterations of lung structure and functioning as reported by some studies on the topic [35, 37]. Nonetheless, a reduction in intrathoracic pressure is observed, with the thorax able to expand freely without gravity [5]. Abdominal organs also exhibit displacement from original positions, including the diaphragm, resulting in a 15 per cent decrease in residual capacity of the lungs and slight decay of ventilation [37]. However, this has minimal effect on the gas exchange due to alveolar ventilation-perfusion matching occurring in microgravity. Pulmonary and membrane diffusion capacities increase in microgravity due to posture changes, such as from upright to supine. This renders more surface area available for gas exchange, in combination with a more uniform distribution of blood volume through capillaries in the lungs [60]. Yet, the gradient of pleural pressures remains between the upper and lower lung areas, suggesting that this mechanism is not gravitational in its origin.

Reduction in workload also implies decrease in the respiratory frequency of astronauts, which is consonant with the decrease in metabolic rate. ISS is also maintained at a 21 per cent oxygen level, which may contribute to the decline in respiration frequency [49]. The elevated partial pressure of CO_2 derived from the accumulation of exhaled CO_2 in confined spaces can potentially affect the arterial partial pressure of CO_2 and acid–base equilibrium within blood.

Planetary exploration, such as the Moon or Mars with gravities of 0.16 and 0.36 g relative to Earth, respectively, can also damage respiratory function due to the toxicity of dust particles. Reduced gravitational sedimentation may allow such particles to penetrate deeper inside the lungs as opposed to earthly dust and remain there for extended periods [5]. Irritation of the internal walls and mucosa of the structures composing the respiratory system (e.g. trachea, bronchi, bronchioles and alveoli) can result in extreme episodes of coughing, tissue tear, bleeding, oedema or infection, in combination with the deleterious effects of particle radioactivity on the living tissues.

9.3.3 Brain and peripheral nervous system

9.3.3.1 Adaptations to neuro-vestibular, visual and somatosensory systems

The challenges posed by spaceflight in the central nervous system have been shown to take their toll on neurophysiological problems and related physical symptoms reported by astronauts. These include neuro-vestibular problems (e.g. sense of disorientation due to the absence of constant gravity input for the inner ear), optic-disc oedema and increased intracranial pressure (commonly referred to as visual impairment and intracranial pressure syndrome, VIIP), alterations to the cognitive function and sensory perception, as well as psychological disturbances [86]. The neuro-vestibular conflict arises from the different angular and linear acceleration detections by specialized organs that have evolved within Earth's gravity. The otolithic organs (utricle and saccule) situated in the vestibule (inner ear) are responsible for sensing the gravity and linear acceleration during the initiation of movement. The utricle and saccule sense acceleration in the horizontal and vertical plane of its hair, respectively, in accordance to shear forces induced by the stones on the hairs of otoliths [83]. Type II otolith hair cell synapses have been reported to increase by a factor of 100 per cent in spaceflight when compared to Earth, then decreasing significantly in the immediate hours after landing [60].

The primary somatosensory and associated cortical networks are affected by zero or near-zero gravity, producing deficits in sensory perception in astronauts. In fact, gravity experienced in space (10^{-6} to 10^{-3} g) is below the threshold for human sensing mechanisms, such as gravity receptors, directional proprioceptive and visual cues [60]. Centrally, this has repercussions on the vestibular nuclei located in the brainstem and cortical projections involved in the integration of sensory information, including the insular cortex, thalamus and temporoparietal junction. Conflicting vestibular information, altered proprioception and modified vision in space contribute to altered illusions of body positioning and orientation. Astronauts then mistakenly judge the position of the limbs in the head-trunk axis and the eye–hand distance, which are essential for coordinating movements. The postural control system as a whole is intimately tied together with the vestibular, visual and somatosensory systems, as well as the cerebral areas for motor activity, learning and memory capacities involved in human locomotion [85]. In the elderly, increased cortico-spinal drive from the motor cortex to the lower-limb efferent neurons was detected after perturbations in body posture, along with damage to the afferent sensory signals that induce muscle response delays observed in perturbed balance. Similar effects are also evidenced in astronauts during spaceflight. Therefore, the combined effects of vestibular, proprioceptive, kinaesthetic and motor changes contribute to the impairment of vision–motor coordination. Cerebellum function and connectivity can also be impaired, possibly explaining some motor discoordination during and post-flight, which can be reversibly recovered after spending some time back on Earth [87].

Regarding vision, alarmingly, 25–30 per cent of astronauts returning from the ISS exhibit near sight vision reduction and other visual impairments, such as optic-disc oedema and the perception of light flashes during dark and rest periods onboard the ISS. While the latter effect has been related to the action of heavy ions striking the retina of astronauts (radiation), the other effects are commonly associated with the adaptation of human physiology to space environment. Ocular damage includes hyperopic shift, posterior eye flattening, choroidal folding, cotton-wool spots, varying refractive index changes (often unilateral), different degrees of disc oedema and bent optic nerve sheath [85, 88]. Some of these damages are also observed in patients with idiopathic intracranial hypertension on Earth. Within the context of spaceflight, they are commonly referred to as spaceflight-associated neuro-ocular syndrome [85].

The cephalic fluid shift associated with microgravity is typically accompanied by increased intracranial pressure, venous outflow obstruction and disruption of the cerebrospinal fluid (CSF), with local effects felt on the vascularization of the structures composing the eye. Increased intracranial pressure leads to papilledema (swelling of the optic disc), where the optic nerve enters the eyeball, increasing the risk of vision loss, whereas reduced blood supply derived from tissue compression results in infracts to the nerve-fibre layer and the slower venous drain from the eye contributes to venous engorgement, responsible for choroidal thickening or folding.

The study of the functional and morphological effects of the space environment on the structure of the central nervous system has been limited, with spaceflight literature predominately reporting the impact on extra-cerebral or peripheral systems connecting to the central nervous system. Reasons for brain data absence are related to the lack of equipment to perform brain neuroimaging in situ during spaceflight (e.g. MRI scans); it is only possible before or after space missions. MRI studies on 27 astronauts after landing have shown a decline in the volume of frontotemporal grey matter and an increase in the volume of the medial primary sensorimotor cortexes [89]. More recently, another MRI study was conducted to investigate the narrowing of CSF spaces and upward displacement of brain structures between long (average of 165 days) and short (14 days) duration spaceflights [86].

Comparison between pre-flight and post-flight ventricular volumes (lateral, third and fourth ventricles) was also performed between different groups of astronauts, as well as a change in the volume of the central brain sulcus (Rolandic fissure in the central cortex). The central sulcus of the astronauts was narrowed by 59 and 19 per cent in the long- and short-duration flights, respectively. On the other hand, a mean increase of 0.74 and 0.11 mm was observed for the width of the third ventricle in the long- and short-duration spaceflight groups. Ventricular volume increased 11 per cent in the long-duration group while remaining negligible in the other astronaut group. By combining cine-clip findings from these two groups, an upward shift of the brain and brainstem was detected in 67 per cent of all astronauts, whereas, narrowing of CSF spaces at vertex (upper surface of the head) occurred in 72 per cent. At the same time, rotation of the cerebral aqueduct (within the midbrain region connecting the third with the fourth ventricles) was detected in 67 per cent, stretching of the pituitary stalk (connecting the hypothalamus to the posterior pituitary) in 61 per cent and uplifting of the optic chiasm (where the optic nerves cross) in only 33 per cent of the astronauts. Possible explanations for these findings can be attributed to the upward displacement of cerebral hemispheres and ventricular extension in space.

Rotation of the cerebral aqueduct may have caused increased resistance to CSF out-flow from the third ventricle, thereby enlarging the ventricular system, whereas the narrowing of the central sulcus likely resulted from neuroplasticity, as previously reported by the volume increase of the sensorimotor cortex [89]. Contrary to space-flight findings, loss of brain tissue by the normal ageing process is manifested by enlargement of the CSF spaces [86].

The lack of neuroimaging equipment available in spaceflight hinders investiga-tion of neuroplasticity during flights, and thus studies rely on comparing pre-flight and post-flight imaging data only. Neuroplasticity is defined as the capacity of the brain to change its structure and function according to new environmental stimuli, essential in processes such as skill learning and injury recovery. Neuroplasticity can be divided into several layers, starting by synaptic plasticity at the (sub)cel-lular level and advancing into plasticity at system and network levels. Techniques employed in neuroplasticity studies include evoked potentials analysis based on electroencephalography (EEG), structural and functional MRI and magnetic res-onance spectroscopy [87]. In spaceflight, EEG is the commonly used technique, which provides excellent temporal resolution of the electrical signals originating within the brain and captured outside the scalp. The significant advantage of EEG systems relies upon its portability, ease of use, connectivity to different recording platforms and wireless signal transmission. On the opposite side, MRI has a bet-ter spatial resolution (as opposed to temporal resolution), necessary to discrimi-nate different layers of brain tissue, which allows detailed information on brain structure and function. However, this equipment is expensive, large and heavy for deployment onboard the ISS.

Several ground alternatives to spaceflight for neuroplasticity studies have been proposed throughout the years: dry immersion, which simulates confinement, body immobilization, 'supportlessness' and body fluid shift; HDBR test, which involves a subject laying on a bed for extended periods, with up to −6 degrees head down to induce upward fluid shift; parabolic flights to simulate altered gravity; and human deployment to missions underwater, high mountain ranges, tropical forest and even the Antarctica to produce enough sensory deprivation, high-stress levels, isola-tion and temperature variations [87]. From these, HDBR is the most widely used approach as it allows the efficient control of the experimental conditions while enabling volunteers to be asked to perform mental tasks at the same time.

However, it must be noted that there are reservations surrounding the ability of HDBR tests to induce equivalent spaceflight effects on the brain. Nevertheless, we briefly summarize some of the latest neuroplasticity findings obtained by this method. In a 90-day experiment, Roberts *et al.* [90] found reduced brain activity in the motor areas related to leg representation and a decline in corticospinal excit-ability. Liao *et al.* [91] on a 72-h test reported lower thalamic connectivity observed during the resting state. Zhou *et al.* [92] found alterations in the anterior insula and middle cingulate cortex network related to the resting state in a 45-day test. Li *et al.* [93] reported declines in grey matter volume in the temporal lobes, bilateral fron-tal lobes, insula, hippocampus and parahippocampal gyrus. In contrast, the volume increased in the paracentral lobule, vermis and precentral gyrus in a 30-day test.

Finally, Yuan *et al.* [94] investigated the long-term effect of HDBR on dual-task performance and brain activation, reporting an increase in the frontal, temporal, parietal and cingulate cortexes during task execution, going back to baseline after cessation of the test. In all of these studies, the underlying mechanism had a cephalic fluid shift in its origin, altering CBFV and thereby changing the haemodynamic response of the brain: increased intracranial pressure and oxygenated haemoglobin [87]. Again, limitations of the HDBR tests such as inter-subject variability, demographics (age and gender), genetic variation and dissimilar experimental protocols compromise the validity of these neuroplasticity studies performed on Earth and their extrapolation to spaceflight. Nonetheless, these studies may provide initial insights about cortical areas to look for potential connectivity pathways, as well as identifying neurotransmitters involved in the processes of attention, long-term memory formation, cognition, learning and motor activation, when better alternatives to wearable brain monitoring become available onboard the ISS, such as EEG systems and near-infrared spectroscopy (NIRS).

9.3.4   Thermoregulation in space

The human thermoregulation system and its associated processes work to pursue and maintain optimal thermal comfort levels within the body. The wide variety of extreme conditions in terrestrial environments means that the success of the body's ability to both continuously and automatically keep temperatures within a comfortable range is essential to life. Optimal skin and core body temperature are regulated through the combination of complex and interrelated regulatory control processes, including, cardiovascular, metabolic, respiratory, osmoregulatory and thermal control systems [95]. The optimal skin temperature is 33.4 °C, differences of 1.5–3.0 °C are imperceptible, and changes of ± 4.5 °C can cause discomfort [96]. The core body temperature is maintained within a narrow window of ~37 °C when the body is at rest, rising proportionally with the intensity of physical activity [97]. However, fluctuations beyond 1.5 °C in either direction can be life-threatening [96].

To keep changes to skin and core body temperature to a minimum and to maintain a comfortable core body temperature, the body utilizes a range of thermal bodily responses that include shivering or sweating. If these are not enough in managing regulation within a safe temperature range, hypo- and hyperthermia can result. Hypothermia is caused by the body constricting blood capillaries within the skin to minimize heat loss in response to a cold environment or if clothing does not have a suitable level of thermal resistance. Contrarily, if the body cannot dissipate excessive heat, hyperthermia or heat stress will result. If the environment is too hot and the thermal resistance of clothing is too high, the body will insensibly perspire, thus dilating blood capillaries to allow evaporation of water diffused from the body. In a process called sensible perspiration, the sweat glands activate and evaporate sweat from the body in an attempt to lower the core body temperature [96]. Alongside this, there are behavioural mechanisms that an individual can do to support physiological thermoregulation, such as taking off or putting on a layer of clothing or by actively changing ones' environment. Through the interaction of these mechanisms, thermal comfort can be maintained [98].

Space exploration presents us with unique challenges in sustaining life in an environment hostile to human survival. In outer space, there is a lack of atmosphere, and temperature fluctuations are extreme. Outside ISS temperatures can be as high as 250 °F (121 °C) on the sunny side and as low as −250 °F (−157 °C) on the shady side [99]. Therefore, maintaining and optimizing both thermal control and comfort are critical issues of ongoing concern.

Being in space has implications on the body's ability to maintain thermal comfort. When engaged in EVA, outside of the controlled atmosphere of a spacecraft, the transfer of waste metabolic heat into the environment is prohibited. This is because of a combination of high thermal insulation and low vapour permeability in the extravehicular mobility unit (EMU), which is a type of EVA spacesuit [100]. Subsequently, an astronaut is unable to use behavioural mechanisms possible in terrestrial environments and must rely on their equipment and the complex garment system that makes up the EMU. As a result, there are not only physiological challenges in maintaining optimal thermoregulation but also unique psychological ones. The unfamiliar, emotionally challenging environment can be stressful for the astronaut. On Earth, core body temperature can rise by up to 2 °C as a result of psychological stress (stress-induced hyperthermia) arising from emotional or mental stressors [101]. Core body temperature elevates during spaceflight [102], and even small changes can affect physical [103] and cognitive [104, 105] performance. Recent research [106] suggests that psychologically induced stress might be one of a range of contributing factors (alongside increased physical exercise) that causes core body temperature to rise gradually at rest and higher and faster during exercise on long-duration spaceflight. Thus, there is still much to be learnt about the effects of spaceflight on thermoregulation, and, in turn, the landscape of thermoregulation garments will evolve.

9.3.5   The stress response in astronauts

Astronauts are subject to significant stress, before, during and after space missions. In space, stressors include noise, microgravity, radiation, isolation, confinement and sleep deprivation. Monitoring astronauts' stress is essential as it has severe consequences in their performance and overall health. Human body response to stress involves the release of specific chemical messengers (hormones from the endocrine system and particular neurotransmitters from the nervous system) to mediate its effect. Epinephrine (related to the autonomic nervous system emergency response and increased HR and blood pressure) and norepinephrine (a stress hormone linked to anxiety, high glucose, high blood pressure and elevated HR) and proinflammatory cytokines, including leptin, are the primary mediators of chronic psychological stress [107].

Catecholamines (monoamine neurotransmitters such as epinephrine, norepinephrine and dopamine) and leptin interact with cortisol. Cortisol is a glucocorticoid anti-inflammatory and immunosuppressive hormone secreted by the hypothalamo-pituitary-adrenal axis (HPAA), particularly in response to stress [107, 108]. Cortisol is often measured in blood serum, saliva or urine [108]. Salivary cortisol is an established approximation of the free unbound plasma cortisol [108]. Cortisol is found in large concentrations in all bodily fluids, but it is highest in sweat and then in blood [109]. According to [108], sweat contains cortisol concentrations similar to saliva, while [109] highlights that salivary cortisol has 100× less concentration (~1 ng/ml) in contrast to other biofluids. It was found in [108] that there is a significant correlation between cortisol in sweat and saliva, suggesting that it can also reflect acute HPAA activity. Cortisol is related to acute stress, depression, low blood glucose, obesity and diabetes [109]. Nevertheless, attention should be paid to the complexity and wide range of factors that can lead to HPAA activation, which can thus influence measured cortisol concentrations [110].

Stress also affects the concentration of several proteins, such as nerve growth factor (NGF) and brain-derived neurotrophic factor (BDNF) [107]. BDNF is found in both saliva and blood at lower levels than cortisol, and it is related to depression and schizophrenia [109]. Another stress marker is dehydroepiandrosterone (DHEA), an antagonist to cortisol. DHEA is another glucocorticoid steroid hormone released by the adrenal glands. Additional stress-related markers are α-amylase (a digestive enzyme and indicator for chronic stress, found in high concentrations in saliva, ~1 mg/mL) and neuropeptide Y (related to depression, anxiety, food intake, stress response and cardiovascular function) [109]. Other relevant markers include cardiac troponin T (associated with heart disorder), orexin-A (controls consciousness and is related with sleep disorders), oxytocin (related with social bonding and anxiolytic effects), serotonin (associated with mood regulation and appetite; it is an antidepressant) and TNF-α (related to immune cell regulation and immune response) [109]. Stress results also in increased cytokine activity [111, 112]. HR and cortisol are the primary measurands used to assess stress [113]. More information on markers and their concentrations in different biofluids can be found in [109].

Cytokines such as IL-1, IL-2, IL-6 and TNF aid in limiting the extent of injury or stress [112]. They mobilize host immune responses and initiate a restorative process to bring the organism back into homeostatic balance [112]. IL-6 has a critical role in these responses, being the most stable of cytokines [112]. Exercise-related IL-6 secretion is proportional to body weight, and thus, it is gravito-sensitive [114]. For example, in [112], it was found that urinary IL-6 and cortisol excretion were elevated on the first day of spaceflight. IL-6 remained within standard levels before and during spaceflight. After a 9.5-day space mission, two of the astronauts had markedly elevated levels of IL-6. While cortisol levels followed the response of IL-6 before and during flight, this was not the case post-flight. Stress is also a contributing factor to space motion sickness, which is associated with increased glucocorticoid activity, the release of insulin (detected via increased urinary C-peptide excretion), muscle proteolysis, changes in the pulse rate, anorexia, cold sweating and subjective warmth symptoms [112].

It has been found that (simulated) EVA increases the stress levels (HR measurements and salivary cortisol) in astronauts [113]. In [107], it was found that environmental stress affects metabolic and stress responses. This 105-day confinement-simulated environment study monitored circulating biomarkers of psychophysical stress (cortisol, NGF and BDNF) and glucometabolic impairment (fasting plasma glucose, FPG, insulin, C-peptide and leptin). Inflammatory phenomena lead to metabolic changes that eventually cause insulin resistance [107]. FPG, plasma insulin and insulin resistance changed only during the first 5 weeks of the study and then slowly returned to baseline levels. C-peptide levels remained relatively unchanged, and BDNF showed a late (5–10 week) increase, which normalized at the 15th week. These might be due to the exercise routine followed during the study by subjects. Leptin rapidly reduced and remained at that low level throughout the study, while cortisol steadily increased from the earlier weeks. The authors concluded that the above suggests that stress affects the neuroendocrine system through metabolic adaptations [107].

Stress is a mediating factor for the impaired immune function of astronauts, and both physical and psychological stresses lead to decreased virus-specific T-cell immunity and reactivation of Epstein-Barr virus (EBV), which can cause tumours [115, 116]. Urinary concentrations of cortisol and catecholamines increase post-flight, while cortisol also increases during missions [115]. The effect of 6-month ISS missions on stress markers (the salivary cortisol and DHEA) and salivary anti-microbial proteins (AMPs) was discussed in [111]. It was found that several AMPs were elevated with more pronounced responses in less experienced crew members (reduced levels of salivary IgA, increased levels of α-amylase, lysozyme and LL-37 during and after the mission), in comparison to control subjects that remained on Earth. Moreover, latent viral reactivation (cytomegalovirus, CMV; EBV; herpes simplex virus-1, HSV-1; and Varicella zoster virus, VZV, that can cause chickenpox during primary illness and shingles upon reactivation [117]) is related to increased cortisol levels and distinctive among crew members that performed EVA. In 43 per cent of the subjects examined in [116] all three EBV, VZV and CMV were reactivated. It was also reported that the ratio of cortisol and DHEA was significantly higher during spaceflight relative to the values before and after the mission. However, no significant relationship between this ratio and viral reactivation was found. CMV is most often asymptomatically acquired during childhood, but with compromised immune systems, it can lead to encephalitis, gastroenteritis, pneumonia and chorioretinitis, while contributing to immune-suppression [116, 118]. Elevated shedding of CMV in urine was also found in [118], while glucocorticoids have been shown to enhance CMV replication.

9.3.6 Lymphatic and urinary systems

The effects of spaceflight on the urinary system include increased calcium excretion, which, combined with faecal excretion, contribute to a negative calcium balance within the body [60]. Unloading of the skeletal system and increased bone turnover in microgravity leads to a persistent loss of calcium. Hypercalciuria due to variations in bone metabolism and loss protects the organism from calcium overloading. It leads to an increase of luminal calcium levels in the collecting duct, which, in turn, impairs liquid handling [119]. Besides, according to [119], impaired water handling could also be derived from reduced aquaporin 2 (AQP2) response to vasopressin, resulting from hypercalciuria. Reports from the late Skylab missions refer to a calcium loss of 200 mg/day [36]. This excessive calcium excretion, in turn, increases the risk for renal stone formation in long-duration spaceflights, as urine contains more crystal-forming substances than the fluid can dilute [5]. Calcifying nanoparticles or nanobacteria, which may pose a greater hazard in microgravity, as they reproduce more rigorously in microgravity, could also contribute to the pathogenesis of human calcific diseases since they create external shells of calcium phosphate [120, 121]. Thus, these can also be a contributing factor in addition to hypercalciuria and urolithiasis. Screening for antigen and antibody levels before and after spaceflight is required to obtain more evidence and assess their effect [121]. The salty diet given to astronauts to counterbalance the activity of the renin-angiotensin-aldosterone system also favours renal stone formation. Salts and minerals (calcium) further supersaturate stone-forming deposits inside the kidneys that can affect any part of the urinary system that they pass through – from kidneys to the bladder – causing significant pain if not detected early. The options available for handling severe renal stone issues onboard the ISS are limited due to the lack of on-site medical assistance, proper equipment and pharmaceuticals. Finally, apart from the physiological issues, bone loss has led to the failure of the urine processor assembly on ISS in 2009. The primary reason was calcium, which had increased by 50 per cent [122].

Microgravity leads to fluid shifts to the upper body, significant extravasation (interstitial and intracellular spaces), resulting in a rapid decrease in plasma volume [119]. The latter is further enhanced due to space motion sickness and the resulting temporary reduction of fluid intake [123]. Spaceflight disturbs the genitourinary (GU) system and renal function and, as mentioned, the water-sodium balance is affected [119]. Sodium retention takes place in an attempt to increase plasma volume; however, this is not sufficient in space. Renal diabetes insipidus that corresponds to a vasopressin-regulated water reabsorption deficiency can explain the phenomenon [119]. In addition, sodium-retaining hormones (norepinephrine, renin, aldosterone) are activated to compensate for the sufficient blood deficit [119]. Vasopressin is also activated. Nevertheless, blood volume remains reduced, and these hormones remain at abnormal levels.

Urinary tract infections (UTIs) have also occurred while in space. Dehydration, physical and mental stress, hygiene (especially regarding urine collecting devices) and delayed access to voiding and urinary stasis have been attributed as causes for UTIs [120]. These are verified by the symptoms and positive findings of leukocyte esterase or nitrites in urine. Another potential issue is hydronephrosis, characterized by kidney swelling due to urine draining failure or blockage. Some reports of urinary retention occurring in astronauts during the first days of spaceflight have been self-resolved by natural body processes after the course of time in space. However, there were also reports of urinary retention requiring catheterization [120]. These could have been the result of delayed voiding, pharmacologic side effects, microgravity and impaired sensorium [120]. Self-catheterization in space is always risky due to the surface-tension impacts in microgravity, microtrauma and associated risks for development of infections [37].

Other changes to the normal functioning of the urinary system include decreased urinary output, pH level, citrate and magnesium concentrations, in opposition to phosphate. Countermeasures adopted by astronauts to prevent this situation include reasonable ingestion of fluids (also preventing space dehydration) and potassium citrate [5]. Reproductive problems are also probable [120]; nevertheless, there have been no significant issues reported on the topic so far. However, the high occupational stress and delay of conception may be an issue in female crewmembers and their fertility [120]. Moreover, space radiation, oxidative stress [124], microgravity, psychological and physical stresses, as well as disruptions in the normal circadian rhythm are potential hazards for the reproductive system of astronauts [125].

Proteins are present in minimal concentrations in urine. These concentrations can change due to various pathological conditions [126]. When compared with the urine collected from control subjects on Earth, the urine of astronauts who stayed over extended periods in space also contained afamin (related with hypercalcemia), aminopeptidase A (associated with tubular dysfunctions, caused by intermittent renal hypoxia) and AQP2 (related to damaged intracellular recirculation of aquaporin and increased renal water reabsorption) [126]. In addition, albumin (a prominent indicator of renal function) is significantly reduced in urine [119, 127]. Finally, cellular and molecular alterations in the immune system may lead to impaired bacteria clearance [121], and more detailed discussions on GU issues can be found in [123].

9.3.7   Endocrine system

The endocrine system is sensitive to the space environment, responding mainly through hormone variations in circulation as part of the retroaction against weightlessness and radiation exposure. These variations may be reflected in the promotion of muscular and bone loss, assistance to the immune system or regulation of body fluids in the cardiovascular and urinary systems [88]. One of the most reported harmful effects of spaceflight is the development of decompensated insulin resistance in astronauts [49]. It was initially thought to be the consequence of cephalic or upward fluid shift and sarcopenia. Recent investigations highlighted the activity or expression of insulin β-subunit, Akt, GLUT4, AMPK and other intracellular signals as being responsible [85]. Insulin resistance occurs when cells in the liver, muscle and fat tissue cease to respond to insulin stimulation, making the process of glucose intake more difficult for use as energy or storage as fat. The pancreas responds by producing more insulin to facilitate glucose entry in the cells and avoid the build-up of glucose in the bloodstream, a situation that can lead to hyperinsulinemia. Insulin resistance can lead to prediabetes or type 2 diabetes when increased glucose levels start to produce characteristic body symptoms such as lethargy (tiredness), hunger, high blood pressure and high cholesterol levels. The effects of insulin resistance in bed-rest studies on Earth have shown reduced activity of proteins involved in glucose transport and storage as glycogen, as well as damage to nonoxidative glucose metabolism, reduced lipolysis and impaired microvascular function [81]. During the Apollo missions, astronauts developed insulin resistance together with microvascular injuries and endothelial dysfunctions. At the time, it was assumed upon examination that excess C-peptide excretion (involved in the binding between insulin's A and B chains in the proinsulin molecule) and reduced cGMP (a second messenger of NO associated with heart disease and diabetes) were indicators of insulin resistance. In contrast, vascular impairment was associated with a reduction of endothelial growth factor and platelets (radiation effect) and increase in IL-6 excretions (an indicator of inflammation).

Other hormonal imbalances found in spaceflight after analysis of the collected data from missions included an increase in catabolic hormones (e.g. cortisol and glucagon) on the bloodstream and excretion of 3-methyl-histidine in urine, associated with metabolic stress [82]. Elevated metabolic rate (hypermetabolism) leads to catabolism in combination with higher protein usage as supplementary energy sources. Negative energy imbalance due to deregulation of eating habits and improper physical workload in space may be appointed as the causes of metabolic stress, contributing to the progression of muscle injury and atrophy. However, the biological pathways established during spaceflight between metabolic stress, insulin resistance, inflammation and oxidative stress still need to be fully determined.

Adrenal gland hormones such as epinephrine and norepinephrine are typically increased as part of the body's response to situations involving stress, wakefulness and concentration during the first instants of spaceflight and before landing due to microgravity and re-entry to Earth's atmosphere, respectively [85]. They are also involved in glucose plasma and fatty acid elevation and lipolytic activity in adipose tissue, concordant with the promotion of alertness and vigilance, and focus attention when required during spaceflight. The elevated activity of renin-aldosterone and catecholamines (epinephrine, norepinephrine and dopamine) are likewise related to fluid balance in space, being part of the sodium-retaining endocrine system [88].

Thyroid hormones regulate important biochemical reactions, enzymatic activity and metabolism. Thyroid function is regulated by the thyroid-stimulating hormone (TSH) secreted by the pituitary gland in the brain. In early flights, increased levels of TSH led to the assumption of a microgravity-induced hypothyroid state [85] that was contradicted afterwards, when iodine was removed from the drinking water of astronauts following the SpaceLab missions. Being essential to the synthesis of thyroid hormones (particularly T3 and T4), iodine insufficiency leads to even higher levels of TSH secretion to trap more iodine from the circulation into the production of these hormones, thus explaining the increased TSH levels observed during spaceflight. By its turn, parathyroid glands produce parathyroid hormone (PTH) involved in the regulation of the calcium level in the blood, increasing it either by stimulation of calcium release from bones, reduction of calcium loss by urine or increment in absorption in the intestines. It also promotes vitamin D activation via the kidneys. Enhanced PTH secretion has been detected in early flights and, apart from the harmful effect in bone destruction, all the other actions are justifiable as functional adaptations to the space environment.

Finally, the decrease in melatonin secretion during night time is observed during the ageing process and spaceflight [85]. Melatonin is a natural hormone produced by the pineal gland in the brain involved in the control of the circadian rhythm: its levels increase after sunset (low light) and drop in the morning during sunrise, therefore helping the body to identify the sleep and awake states. The reduced melatonin level recorded on astronauts leads to improper sleep and rest periods, which contribute to a feeling of tiredness during the awake state. Reasons for this impairment in melatonin and other sleep markers are presumably due to the day/night cycle disruption arising from continuous exposure to dim lights, noise produced by electronic equipment onboard the ISS and operation of the spaceship itself, as well as the 90-min orbit with constant variations of light/dark scenes observed through the window of the ISS. Muscle and bone unloading are also appointed as contributors to chronic circadian misalignment by changing the workload schedule not only in space but also on Earth [85].

9.3.7.1   Sweat as a biosignalling fluid

Sweat is the most easily accessible biofluid. It is slightly acidic (pH range of 4.0–6.8) and is mainly composed of water (99 per cent) [128]. There are more than 100 glands/cm^2 distributed over the body, providing a large area for non-invasive monitoring of a wide range of physiological parameters. It was found in [129] that spaceflight has no negative impact on the skin parameters. Nevertheless, sweat needs to be secreted to be measured. This can be achieved during physical exercise, under stress, due to thermal heating [130] or artificially [131–135]. Sweat contains many physiologically relevant analytes. It contains metabolites such as lactate, glucose, urea, ethanol and cortisol and electrolytes such as potassium, sodium, chloride, magnesium, calcium, phosphate and ammonium, pH, as well as trace elements including copper, zinc and large molecules (e.g. proteins, nucleic acids and cytokines) [132].

Another relevant marker is orexin-A, a neuropeptide biomarker related to alertness [136]. The concentration of several cytokines, including IL-6, in sweat, is the same as in blood [136]. A recent study has highlighted the potential use of sweat for the measurement of a wide range of immune biomarkers, including a broad range of IgA, IgM and IgG antibodies, IL-1α, IL-1β, IL-2, IL-6, IL-8, IL-13, IL-31, TNF-α, and transforming growth factor-β, insulin-like growth factor-1, interferon (IFN)-γ, monocyte chemoattractant protein-1, macrophage inflammatory protein-1 δ, stromal cell-derived factor-1, epidermal growth factor (related to interstitial cystitis and psoriasis vulgaris), angiogenin (linked to heart failure, bladder and prostate cancer), apolipoprotein D, clusterin, prolactin-inducible protein and serum albumin [137]. Consequently, sweat analysis can provide valuable information regarding physiological processes, homeostasis, physical and psychological stresses, body hydration, ischemia and anaerobic threshold.

The monitoring of these can be critical as, e.g., the imbalance of electrolytes can cause muscle cramps [136] that can be problematic during long and challenging tasks or EVA. The possibility of extracting all these without the need for frequent blood sampling (and related issues), while leaving the skin intact, is of high importance, particularly for long-duration spaceflights. However, more research is needed to investigate the correlations between sweat and blood concentrations of many of these markers. Sweat arises from eccrine glands composed of a secretory coil, where it first emerges. The most abundant ion species in sweat are sodium and chloride.

The active transportation of these two ions between blood and the secretary coil creates a difference in osmolality, forcing water into the sweat gland [138]. A dermal duct then carries sweat to the skin surface [138]. The concentration of these two ions in the secreted sweat typically increase with the rate of sweating [138]. Passive (i.e. diffusion) and active mechanisms allow analytes to reach the sweat, while some are also generated within sweat ducts [132]. Understanding these mechanisms is vital to shine more light on the sweat/blood concentration relationship [132, 138, 139]. Nevertheless, the usefulness of sweat-based sampling in space is yet to be proven, although sweat-based sensors for space are emerging [140].

9.4   Musculoskeletal system modifications in space

9.4.1   Muscle atrophy in space

Early spaceflight data from Skylab missions revealed musculoskeletal adaptations involving muscle atrophy and bone resorption [141]. The skeletal muscle of the lower limbs and the lumbar spine was significantly affected, whereas changes to the upper limbs were unremarkable. The lower limbs and the lumbar spine consist of the anti-gravity muscles that provide postural support against gravity on Earth. Cephalic fluid shift together with the absence of gravity aggravates muscle volume and mass loss, both accounting for muscle atrophy observed in lower limbs [142].

In the lower limbs, muscles involved in human locomotion on Earth were identified as undergoing muscle atrophy faster and more severe when compared to other muscle groups [143]. These include quadriceps femoris, gastrocnemius and soleus. As much as 6 per cent of muscle loss has been reported in these muscles only 8 days into spaceflight [144]. In line with [144], 2 weeks of spaceflight reported up to 16 per cent of muscle atrophy in the same muscle groups [145]. In comparison to pre-flight values, post-flight data indicated muscle volume loss of up to 20 per cent in the gastrocnemius and soleus in long-duration spaceflight [146]. In another study, soleus demonstrated the most significant muscle loss, followed by gastrocnemius of up to 20 and 12 per cent [147]. However, there is high degree of inter-variability between astronauts regarding the intensity, timing and location of muscle atrophy.

Human ground models replicating spaceflight such as HBR have reported both muscle atrophy and loss of muscular strength. However, similar duration spaceflight to HBR equivalents reported higher intensity of muscle atrophy and loss without countermeasures [145]. Despite early interventions, lower-limb muscle strength loss of up to 26 per cent has been reported in astronauts during long-duration spaceflight [141, 148]. In fact, reports from Mir missions and ISS indicated muscle strength and performance loss of up to 30–40 per cent in long-duration spaceflight [147, 149]. Though it is evident that muscle atrophy accounts for muscle strength loss, there are various underlying mechanisms, such as muscle fibre shift, influencing the latter.

Muscles are composed of different muscle fibre phenotypes, known as myosin heavy chains (MHCs) that function according to energy consumption [150]. MHC I fibres are known as slow oxidative fibres that can contract for more extended periods, however with lesser contraction force. By its turn, MHC IIA fibres are fast oxidative fibres that provide greater force in shorter periods. Both fibres function aerobically, requiring oxygen supply and, therefore, have significantly more mitochondria. Aerobic exercises improve endurance and increase MHC I and IIA population.

The third category is known as MHC IIX fibres, which operate through anaerobic glycolytic mechanisms. These entail an abundant amount of glycogen and are large in diameter, thus producing the most significant forces and contractions; however, they fatigue very quickly. These fibres engage according to human locomotion activity. For instance, maintaining posture, walking or sprinting involves MHC I, MHC IIA and MHC IIX fibres, respectively [150]. They also vary in quantity and composition in different muscles. Soleus contains up-to an average of 80 per cent of MHC I fibres, which engages during maintaining posture against gravity or during slow walking [151]. In contrast, gastrocnemius contains up to an average 57 per cent of MHC I fibres with a larger quantity of glycolytic fibres. Thus, this muscle engages during sprinting to provide the muscle with large amounts of energy. These fibres also exist in hybrid forms of three phenotypes (I/IIA, IIA/IIX and I/IIA/IIX).

In astronauts, muscle biopsies indicated a transition from MHC I fibres to MHC II fibres in various muscle groups. In a long-duration spaceflight study [149], gastrocnemius and soleus had an average of 12 and 17 per cent lower type I fibres relative to pre-flight values, respectively. On the other hand, there was an increase of type IIA fibres by up to 9 and 5 per cent, respectively, in each muscle. However, growth in the latter was insignificant. Though, hybrid fibres increased in both muscles, with non-significant changes to the type IIX fibres. This was in line with previous studies indicating that hybrid fibres increase significantly, with a decrease in MHC I fibres for all muscles [152]. Several studies found an increase in hybrid MHC IIA/IIX in the vastus lateralis and soleus. Therefore, microgravity induces fibre phenotypic remodelling through altered protein synthesis mechanisms. As previously discussed, the soleus acts as a postural stabiliser on Earth. In a more detailed study, Fitts *et al.* [153] discovered that the most considerable fibre atrophy and power loss was in soleus followed by gastrocnemius, in the order of MHC I followed by MHC II. Hence, disuse of this muscle in weightlessness may exhibit a more significant effect than other muscles and contribute to OI after re-entry to Earth. In general, peak power decreased in all fibres post-flight, as well as peak force and fibre diameter in MHC I, resulting in lower muscle strength.

Electromyographic (EMG) studies during spaceflight and HBR reported that muscle strength and muscle explosive power loss are more significant than muscle mass loss, signifying alterations in neural-muscular recruitment [154, 155]. Lamberts *et al.* [156] found a decrease up to 40 per cent of EMG activity in the plantar flexors after long-duration spaceflight despite countermeasures. These results are in line with bed rest studies. During unilateral lower limb suspension (ULLS) experiments [154], it was found that EMG activity decreased in subjects without any countermeasure, compared to the ULLS-countermeasure groups in the knee extensors and plantar flexor muscles. The ambulatory-countermeasure group revealed higher EMG activity suggesting further recruitment of motor units.

Clark *et al.* [157] revealed a 30 per cent slower activation of compound motor neurons (M_{max}). In this study, the atrophy of the soleus decrease was relatively lower than the M_{max} duration; thus, fibre atrophy per se is not responsible. The authors attributed this change to a reduction in overall muscle fibre conduction velocity, which impacts the temporal dispersions of neural firing [157]. There may be a non-linear relationship between muscle atrophy and motor neuron innervation resulting in a more significant muscle strength loss and alteration to the force-frequency relationship.

9.4.2 Bone demineralization

Bone and muscle function in synergy to provide support and movement to the body on Earth. The maintenance of bone strength and integrity depends upon the exertion of mechanical stimuli on the bones. In addition to mechanical forces provided by muscle activation on bone, gravity exerts inherent and constant mechanical loading to the musculoskeletal system. Forces generated in response to gravity are known as ground reaction forces (GRFs), whereas muscles provide joint reaction forces [158]. In microgravity, there is no static and dynamic loading of the musculoskeletal system through the former, integrally reducing strain to bones. Bodyweight-induced mechanical loading is an important stimulus for the human musculoskeletal system [159]. Therefore, mechanical unloading due to microgravity causes adverse adaptations to bone remodelling balance that results in bone mineral loss.

Bone tissue remodelling is a continuous process involving a set of biochemical processes, which ensure the maintenance of the mechanical properties of the skeletal system under changing environmental conditions. The regulation of this process is also poorly understood. Calcium homeostasis plays an important role. Upon entering weightlessness, calcium and other minerals are almost immediately released into the bloodstream and urine [122]. Calcium release from bone suppresses the PTH, which in turn is associated with the reduced circulation of vitamin D and calcium absorption from the GI tract [160]. Spaceflight leads to alterations in bone and calcium metabolism leading to health risks, which include renal stone formation due to the increased urine calcium levels and dramatic bone loss due to the decreased mechanical loading of the skeletal system. This undoubtedly increases the risk of fragility fractures, osteoporosis and potential life-long disabilities in astronauts [161]. However, to date, no data indicate an increased fracture rate in astronauts [162]. Negative calcium balance (and increased calcium excretion) during space-flights were initially reported in the Gemini and Apollo missions in the 1960s and 1970s [160]. Urinary and faecal calcium excretion was initially shown to increase in short-duration spaceflights (1–2 weeks), further validated through measurements of bone loss in the more extended space missions from the Skylab program in the 1970s.

Studies on ISS crew members have indicated a reduction in bone mineral density (BDM) between 1.7 and 10.6 per cent in different subjects and bones (lumbar vertebrae, femur and femoral neck) [161]. The rate of bone loss differs for different bones, which depends upon their location [163]. However, on average, 1–2 per cent loss of BDM occurs per month of spaceflight [160, 162], while astronauts in longer space missions (>6 months) are subjected to a 10 per cent loss in BDM [161]. This is ten times greater than that observed in postmenopausal women [161]. Bone loss is primarily observed in weight-bearing bones [122], including the feet, legs and lower spine, whereas it does not take place in the arms [162]. Past research indicated subjects losing up to 25 per cent of the bone mass in their distal tibia of the lower limbs within a 6-month mission [162]. In another study, a hip strength loss of more than 50 per cent in a 6-month spaceflight was reported, a similar magnitude of loss to ageing across a lifetime [164]. Loss of bone tissue is a significant problem for long-duration space missions; e.g. a round trip to Mars with current technologies would take a minimum of 1 year [165]. There are also reports of more considerable trabecular bone loss than cortical bone in femur. Thus, bone resorption is non-uniform, and bone biomarkers associated with bone remodelling should also take careful considerations [166].

Bone loss within the human body resembles its distribution to the cephalic fluid shift. Thus, early studies hypothesized that cephalic fluid shift might be responsible for such observation [167]. The fluid shift may reduce the intramedullary pressure (ImP); hence, inhibiting bone-formation mediated mechanotransduction that results in a negative balance for bone remodelling [168, 169]. Mechanical loading of bone increases ImP through improved vascularization. As a result, this augments the interstitial fluid flow [170]. Subsequently, bone-formation-mediated channels are activated to increase mineral content through osteocytes and osteoblasts, commonly known as bone-forming cells. Furthermore, pharmaceutical and physical countermeasures in space have shown to reduce the rate of bone loss despite the sustained cephalic fluid shift, which weakens this hypothesis and indicates other possible underlying mechanisms [148]. Interestingly, the only part of the skeleton that demonstrates BDM increase is the skull [159].

In weightlessness, human locomotion and navigation are driven by the upper limbs instead of the lower limbs. While on Earth, the weight-bearing bones of the lower limb assist human movement and support the human body against gravity. GRFs in space becomes insufficient. Though lower gravity produces lower GRF, it does not influence bone remodelling balance of upper limbs. Therefore, the idea behind exercise countermeasures is to increase GRF exertion on the weight-bearing bones around 1 g that may limit BMD loss.

There is significant variability between subjects as to how they are affected by microgravity, as well as the variability of loss in different bones, which do not seem to correlate with overall calcium loss. Determining predictive factors for bone loss has thus become relatively challenging [171]. Studies on whether there are sex-specific effects have also been inconclusive [148]. The challenges and constraints of spaceflight, as well as the small number of subjects, have not been sufficiently helpful. The rate of bone loss during spaceflight is also significant, and a more considerable concern is that the rate of recovery post-flight can take years to reach pre-flight levels [148, 172]. There are examples of bone loss that was still not recovered 2–4 years after long-duration ISS missions, with BMD measurements approaching asymptotically baseline levels. A more concerning aspect is that the recovery of some types of bone (proximal femoral trabecular) reduced or reversed after few months post-flight, while cortical bone increased [148]. In fact, some animal studies indicate that osteocytes are irreparably affected by microgravity [165].

The equilibrium between bone formation by osteoblasts and resorption by osteoclasts is significantly altered. Overall the space environment promotes aged phenotypes and phenotypes with diminished regenerative potential, leading to reduced bone-formation and elevated resorption [165]. The predominant factors affecting osteogenesis are microgravity, radiation, vitamin D deficiency, hypodynamia, changes in hormone receptor sensitivity and biological rhythm of hormone secretion [173]. Other physiological alterations occurring in space, such as acidosis and oxidative stress, can also be an influence. Additional aspects contributing to bone density loss are fluid shifts in the thoracoacromial direction and hydrostatic pressure changes, which have a considerable part in bone remodelling and mechanotransduction [173]. Animal and cell studies have demonstrated that radiation induces rapid transient osteoclastic bone resorption and harmful effects on osteoblast proliferation and function [148]. Ionizing radiation causes the formation of free radicals, DNA strand breaks and oxidative damage to bone tissue, as well as cancellous bone loss, reduced formation and increased bone resorption [159].

9.4.3   Markers of bone health

Bone resorption markers include urinary calcium excretion and hydroxyproline, released into urine following collagen breakdown [171]. Collagen crosslinks measured in urine (pyridinoline, deoxypyridinoline, N-telopeptide and C-telopeptide) are specific to the breakdown of mature bone collagen. They are not affected by dietary intakes, while they can be preserved in frozen urine samples, and are thus useful markers for bone health assessment. However, the use of such markers is also linked to other challenges, namely day-to-day and subject-to-subject variabilities. These are discussed in more detail in [171].

While bone formation remains relatively unchanged in spaceflight, studies have revealed that bone resorption increases up to 50 per cent [171]. As discussed in [173], the gene expression of mesenchymal stromal bone marrow cells is altered, while the metabolic activity of osteoblasts and osteoclasts also changes [173]. The latter is evident from changes in blood serum levels of osteocalcin, tartrate-resistant acid phosphatase, osteoprotegerin and receptor-ligand and nuclear κ-B activator. Elevated concentrations of pyridinium crosslinks in urine, in combination with the analysis of serum levels of C–terminal or N–terminal telopeptides for type I collagen (CTX–I and NTX–I, respectively) have also confirmed the rise in bone resorption activity [159]. Bone formation markers such as propeptide of type I N–terminal procollagen (P1NP), propeptide of type I C–terminal procollagen (P1CP), alkaline phosphatase and osteocalcin decline in space [159]. Finally, the gene expression of peripheral blood leukocytes is likewise altered.

Changes in the proportions of peripheral blood mononuclear cells producing osteoclastogenic cytokines IL-1α, TNF-α, and IFN-γ and those producing anti-osteoclastogenic cytokines (IL-4, IL-6, IL-10 and TFG-β1) can provide insight into bone health [114]. According to [114], IL-6 in particular is a powerful marker. IL-6 is a cytokine with an active role in both bone and muscle homeostasis, promoting bone formation by enhancing the differentiation of osteoblasts, while protecting against apoptosis and bone resorption. It also plays an essential role in bone remodelling, among other functions [114].

Osteoblasts, osteoclasts and leukocytes are involved in the formation, growth, development, functioning and metabolism of bone tissue [173]. Thus, most regulating factors associated with bone remodelling are proteins (enzymes, cytokines, growth factors, regulatory proteins, protein receptors and transponder proteins) [173]. Pastushkova *et al.* [173] reported a significant decrease in regulatory proteins, such as cystatin C, fetuin-A and fibronectin, in the blood proteome after a 6-month stay in space. Cystatin-C is directly linked to bone resorption; fetuin-A to inhibition of bone mineralization, development of bone marrow, bone formation and resorption; and finally, fibronectin to bone resorption, mineralization and formation. A rise in the expression of 36 essential genes was reported that encode markers of osteoblast differentiation regulators (bone morphogenetic proteins and IGF-1), osteoblast (tissue-specific alkaline phosphatase, ALPL, and osteocalcin, BGLAP) and osteoclast (macrophage colony-stimulating factor) differentiation markers, ossification markers, collagens, markers of adhesion and intercellular interactions, growth factors and key transcription factors [173]. The above are summarized in Table 9.1.

9.4.4   Bone health monitoring

Densitometry techniques, such as quantitative computerized tomography and dual-energy X-ray absorptiometry, are valuable tools for the assessment of specific bones pre- and post-flight; however, they can detect only relatively high bone loss (levels that take months to occur) [160]. The measurement and monitoring of biomarkers that can provide information about bone formation or resorption (N-telopeptide, bone-specific alkaline phosphatase) is a more rapid approach for assessing these processes [160]. However, the relative association between bone resorption and formation has not been established yet, therefore assessment of net bone calcium content is challenging. Monitoring the ratios of stable calcium isotopes in urine (six naturally occurring calcium isotopes: ^{40}Ca, ^{42}Ca, ^{43}Ca, ^{44}Ca, ^{46}Ca and ^{48}Ca) is a promising alternative. Bone formation exhausts soft tissue from lighter calcium isotopes, while resorption releases them back into soft tissue (more ^{42}Ca and less ^{44}Ca) [160]. The relationship between calcium isotopes and bone mineral balance is well established, thus allowing assessment of bone health. It should be noted that a limitation of all these biochemical markers is that as they reflect changes in the entire skeletal system, they may mask small localized bone changes [171].

9.4.5   Cartilage

Articulating synovial joints are of particular importance as they are responsible for accommodating large tensile and compressive loads, while facilitating smooth mechanical motion [174]. Synovial joints are comprised of subchondral bone, tendon, several ligaments, articular cartilage and meniscal fibro-cartilage [174]. These are covered by synovial fluid and enclosed in a fibrous capsule. Changes in any of these lead to joint instability and degradation, which results in pain and mobility loss, often leading to osteoarthritis [174]. Since cartilage has a weak regenerative capacity, it is crucial to understand the effect of the space environment (microgravity and radiation) on cartilage, as it can have long-term consequences. Cartilage is comprised of a dense matrix of collagen and elastic fibres. There is a limited number of studies devoted to the examination of the effects posed by spaceflight to cartilage; however, these suggest a degeneration of cartilage in the knee joint; tibial cartilage thickness decreases [159]. Microgravity leads to articular cartilage proteoglycan loss in mice, while evidence suggests that this is worsened by radiation [174]. Cartilage breakdown products released by damaged joints are detectable in synovial fluid, urine and blood. These include carboxy-terminal telopeptides of Type II collagen. Overall, cartilaginous tissue experiences physical and organizational changes, as well as alterations in cell cycle pathways and associated macromolecules [175]. More research is required to understand the effects of spaceflight on intervertebral discs and cartilage in general, with the CARTILAGE study launched by the National Aeronautics and Space Administration (NASA) aiming to address this [159, 175].

9.5   Wearable technology for space biosignal monitoring

As evident from previous discussions, space poses a challenging environment for life. Its effects on human physiology are complex and have the potential to compromise astronaut's health (both in the short and long term) as well as mission success. There is still insufficient knowledge regarding many physiological adaptations, the associated mechanisms and risks to human health. Wearable biomedical devices can potentially illuminate these processes by monitoring an astronaut's physiology, detect acute stress, increased workload or compromised health responses in real-time while facilitating human–robot interaction in space.

However, constraints posed by NASA on wearable technology makes physiological monitoring during spaceflight quite challenging. NASA's specifications for wearable devices are summarized as follows: (1) minimum physical dimension, volume and mass to meet spaceflight restrictions (e.g. transport, deployment and storage); (2) ease of operation, with minimal training required for astronauts and wireless data telemetry to onboard equipment and systems; (3) battery-powered, water/sweat-proof and rechargeable hardware, with long-running spells between charging cycles, as well as capabilities for troubleshooting and repair; (4) Food and Drug Administration (FDA) approval for clinical decision support [176].

Nevertheless, highly integrated and ultra-miniaturized electronic devices that meet criterion 1 are becoming increasingly common in academic literature, possessing multi-modal sensing capabilities on a single system (with or without battery supplies) and connected to some sort of wireless data transceiver, easily attached to external devices such as personal mobile phones, tablets or computers, with minimum user input (criteria 2 and 3) [177]. Some of these devices even employ substrates to mimic body-like tissues (e.g. skin) and able to geometrically conform to them, while being lightweight and protected against sweat or water immersion with functional chemical-sensing units [130, 135, 178, 179]. Multi-site placement of wearables allows collecting data simultaneously and in synchronization between different locations of the human body, such as chest, forehead, wrist and legs.

Recent advances in AI and machine learning algorithms allow the fusing and analysis of physiological information from multi-sensing data to provide a way to monitor mental and physical well-being in astronauts. However, their computational complexity limits their application to space flights. In the near future, it is likely that these systems will provide unsupervised decision-making assisting medical crew members in responding to any medical emergency where Earth-based resources (communication and equipment) are too far away to be of support. It is likely these systems will support multiple scenarios found in the field of space medicine [37, 180]. This constitutes the work of NASA's task centre called the Exploration Medical Capability element that aims to identify monitoring capabilities with potential added value for future exploratory missions and associated with minimally invasive medical treatments [176]. It is important to note here that standard commercial silicon chips (amplifiers, microcontrollers, etc.) cannot be used in space. Specialized devices, design methodologies for chip layout and processes based on insulating substrates (such as silicon-on-insulator and silicon-on-sapphire) are needed, in combination with specialized packaging methods and other radiation-hardening techniques. We briefly describe next some recent studies with wearable devices and recent wearable solutions.

Wearable devices have been used in several studies both in simulated and actual space missions to understand the pressure human teams face living in confined spaces for extended periods of time, characterized by microgravity, restricted water and food supplies, delayed communications while wearing spacesuits and using solar energy for powering [181, 182]. In particular, in an 8-month simulated Mars mission, wearables were used to monitor sleep activity and understand behavioural changes that reflect lifestyle choices and well-being [181]. Quality of sleep has been related to physical activity, mental well-being, respiratory problems and neurological conditions. High-quality sleep consists mainly of deep sleep without interruption, and it does not necessarily correlate with sleep duration. However, in most wearable devices, the quality score reported is highly correlated with the latter parameter, which indicates the fact that devices cannot capture accurately sleep stages. This information comes mostly from accelerometer readings, which suffer not only from limited capability in distinguishing intervals of awakeness but also from the absence of gravity. The gold-standard in sleep monitoring is the use of polysomnography (PSG), which includes EEG, ECG, EMG, electrooculography, as well as airflow measurements and monitoring of thoracic movements [183]. Such approaches have been applied to study changes in cardiac rhythms during space explorations [184]. However, they are generally uncomfortable to wear, and they have only been used in a limited number of nights and by a few astronauts.

Ambulatory ECG has also been used in space to investigate changes in cardiovascular activity [7, 182]. In medical settings, ambulatory ECG is used to monitor heart function during activity to detect rare abnormalities, such as arrhythmia [185].

Holter monitors are used in these settings, and they allow recording of two to three channels based on wet gel skin electrodes. Although they have been used in space missions, they are relatively obtrusive, can result in skin irritation, are not water-proof and accidental disconnection of electrodes is typical [182]. Several device types exist that include patch electrode monitors (PEMs) that are small and record ECG for 7–14 days continuously.

Photoplethysmography (PPG) has likewise been used in spaceflights to detect blood volume changes affecting the microvascular tissue located below the skin sur-face [13, 186]. PPG is a simple optical technique that illuminates the skin to detect changes in light absorption induced by the cardiac cycle and breathing. Usually, it encompasses a device fitted to the index finger, providing continuous measure-ments. Traditional arterial blood pressure measurements, which involve placing an inflatable cuff around the upper arm, have also been used in spaceflights, requir-ing astronauts to take sparse measurements by themselves [11, 17]. More recently, ambulatory blood pressure monitoring has been used in parabolic flights [187].

Respiration has been measured by plethysmograph belts that are placed both on the abdomen and on the rib cage [7, 184]. This method allows continuous mea-surements of the movement of the abdomen and chest during breathing to evaluate ventilation of the lungs. In other words, the volume of air is related to the changes in the summation of the abdomen and chest circumferences.

One big challenge with wearable device operation in space is the uncertainty associated with the level of accuracy of wearables as compared with gold-standard laboratory methods that generally involve more acquisition channels, calibration, reference measurements and specialized personnel experimenting with control envi-ronments. Nevertheless, one significant advantage is the big volume of data created, in combination with the long-term monitoring capability they provide. Recently, the Canadian Space Agency developed a t-shirt for monitoring vital signals in space missions. The Astroskin Bio-Monitor system records activity level (3-axis accel-erometer), breathing rate and volume, oxygen saturation in the blood (SpO2), skin temperature, ECG (3-lead, with QRS detection, HR and HRV estimation) and sys-tolic BP [188]. The device has the capability to store, analyse as well as stream data via Bluetooth during physical exercise. Quality of each signal depends on the fit of the sensors to skin. Additionally, the quality of ECG measurements is affected by skin conductance. Experiments showed that 24 h continuous monitoring is possible, but real-time streaming significantly impacts battery life and does not allow full-day recordings [189]. On the other hand, the inability to visualize the sensing data increases the probability of data loss and poor quality. The fit of this garment to the female body has proved to be insufficient, resulting in displaced sensors, motion restriction and reduced measurement quality. Furthermore, it has been shown that in more than 45 per cent of the cases it caused some skin irritation, in more than 80 per cent it was tight, in 40 per cent it was slightly or more restrictive, in 20 per cent was uncomfortable, and in more than 30 per cent it had some unpleasant odour. The authors of [190] proposed a comfortable fabric sensing sock with integrated sen-sors and interfacing modules for data processing and communication via Bluetooth. Acquired data could also be stored and displayed on a helmet-mounted display, including PPG/pulse oximetry, skin temperature and galvanic skin response, thus contributing to the study of tissue blood perfusion, blood volume-induced changes correlated to respiration, stress, oxygenation and HR during continuous EVA operations.

Onboard biomedical analysis of blood, saliva and urine for the detection and measurement of essential biomarkers was until recently not possible. This changed with the introduction at the ISS of a lab-on-a-chip device to analyse blood and provide information about specific cell counts and proteins [191]. The bio-analyser can deliver results in 3 h, eliminating the need to freeze samples for an extended period of time, which, in turn, degrades their quality [191]. Another similar example was deployed onboard of the ISS and employed electrochemical sensors for the measurement of glucose, partial pressure of CO_2, electrolytes and haematocrit, using 100 µL samples; however, the cartridges employed had a limited lifetime of ~4–6 months. Later a similar device using dry chemistry was tested in the ISS, overcoming these lifetime limitations [178].

Kinnamon *et al.* demonstrated an impedimetric non-faradaic affinity-based label-free flexible cortisol biosensor for space applications [140]. MoS_2 nanosheets were integrated into a nanoporous polyamide membrane between two palladium electrodes, whereas cortisol specific antibodies were used to measure cortisol in perspired human sweat within the physiologically relevant range of 8.16–141.7 ng/ml in 1–5 µl samples. A dynamic range of 1–500 ng/ml (detection limit of 1 ng/ml) was achieved and the sensor was interfaced with a custom portable potentiostat. A lateral flow immunoassay-based biosensor incorporating chemiluminescence as the transduction mechanism (measured with an ultrasensitive cooled charge-coupled device camera) for measurement of salivary levels of cortisol in astronauts was presented in [192]. The device was optimized for operation in weightlessness, while withstanding mechanical stress (e.g. vibrations occurring during take-off) and onboard cabinet depressurization. The microfluidics was developed considering the processes of bubble formation, surface wettability and liquid evaporation observed in microgravity [192]. Paper-based microfluidics were used, as well as manually activated reagent reservoirs. The device could detect the cortisol concentration as low as 0.4 ng/ml in astronaut samples tested onboard the ISS. The stability of reagents (including enzymes and antibodies often used in biosensors) is of significant concern for biosensing in space. Nevertheless, it has been shown that these are sufficiently stable in the radiation environment of space [178, 192]. A wearable organic electrochemical transistor-based device for non-invasive cortisol monitoring with a molecularly selective nanoporous membrane was presented in [193].

The analysis of sweat requires the use of chemical sensors or biosensors, which are most often based on electrochemical processes taking place on microelectrodes functionalized with various biomolecular ligands or other membranes to achieve high specificity [194]. These membranes are sensitive and can be damaged when in contact with skin, and thus they are often integrated within microfluidic systems for sweat handling, generating a new class of devices termed as lab-on-skin devices [195]. Microfluidics are also important to address issues with sampling, skin surface contamination and refinement of each analyte. Skin sensors are also used to obtain the ECG and other electrophysiological signals as well as pulse oximetry to monitor blood oxygen concentration. Thus, epidermal chemical sensors can easily be combined with more traditional skin-interfacing sensors. To sustain the mechanical motions of the skin and to be seamlessly worn, these devices need to be stretchable and ultrathin [179]. Such epidermal devices thus have a significant potential for biomonitoring. Of course, sweat sampling and analysis in microgravity can be an issue; nevertheless, such technologies can potentially address the needs of physiological monitoring in space. In contrast to the previous examples, artificial receptors realized with a molecularly imprinted polymer were used instead (between sweat and the PEDOT:PSS transistor channel layer) in [193]. The device also incorporated a laser-patterned microcapillary channel to create a wearable sweat diagnostic tool for accurate sweat acquisition and precise sample delivery to the sensor interface. Recently, flexible thin-film InGaZnO (also known as IGZO) transistors for lightweight space wearables were demonstrated [196]. IGZO transistors are a viable option for space application as opposed to organic semiconductors, such as pentacene, which are unstable, or hydrogenated amorphous silicon, which has low mobility. The devices were tested in LEO-mimicking conditions using high-energetic electron irradiation with fluences up to 10^{12} e^-/cm^2, low operating temperatures of 78 K and magnetic fields up to 11 mTA [196]. These conditions simulated 278 h of spacewalk (sufficiently longer than typical spacewalks) in LEO above the South Atlantic Anomaly, where there is evidence of increased radiation activity due to the weak local geomagnetic field [196]. Some small shifts with radiation and greater ones with temperature were observed for the threshold voltage and mobility, whereas magnetic fields did not affect the devices significantly. The amorphous phase of IGZO improves its radiation hardness due to the absence of a crystalline structure. The use of an Al_2O_3 gate insulator did not demonstrate the magnetoresistive effect that is evident with other gate dielectrics (e.g. SiOx, SiNx) [196]. A label-free carbon-nanofibre-based biosensor for electrochemical impedance spectroscopy and cyclic voltammetry for cardiac troponin-I has also been presented [178]. Portable flow cytometer using optical techniques have likewise been demonstrated onboard the ISS [178]. Other systems include multiplexed biosensors for blood analysis and electrokinetic approaches for sample and analyte handling, as well as real-time quantitative polymerase chain reaction approaches [178]. Other NASA-sponsored research developed wearable devices for monitoring of ECG, respiratory frequency, blood oximetry and blood pressure, skin temperature and body acceleration [178]. Textile integration of all the above-mentioned features, together with direct epidermal interfacing and implantable approaches are promising routes for these technologies for space. We briefly describe next some wearable devices developed in our centre that can presumably be applied to physiological monitoring during spaceflight, given that the effects of altered gravity and radiation exposure are perfectly compensated or used inside radiation protective units such as some compartments inside the ISS.

A small wearable device for measurement of cardiac, electrodermal and motion parameters has been proposed by [197] in the form of a flexible chest patch and primarily employed as a mental health assessment tool for classification of multi-modal signals during the execution of daily activities. This device is shown in Figure 9.3

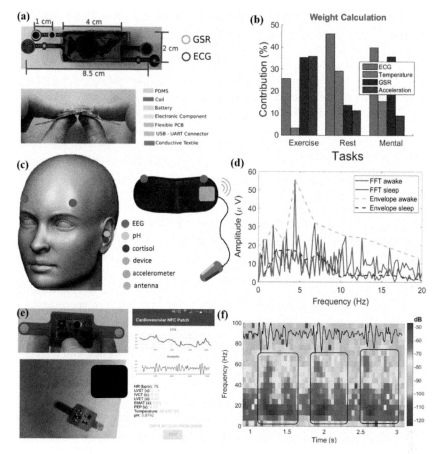

Figure 9.3   *(a) Compact flexible wearable device for measurement of ECG, GSR, body temperature and motion signals in the chest region. The maximum thickness of the device is 9.65 cm. (b) Contribution of each signal modality in the estimation of the training matrix used for the classification of different tasks (exercise, rest and mental activity). (a–b): © 2019 IEEE. Reprinted, with permission, from [197]. (c) Bluetooth Low Energy device integrated into an eyeshade for measurement of EEG, head motion and chemical analytes (pH and cortisol) in the ear. The PCB device was 2 × 2.1 cm. (d) Fast Fourier Transform of the EEG signals acquired from the same subject in the awake (relaxation) and sleep states. (c–d): © 2019 IEEE. Reprinted, with permission, from [198]. (e) NFC-powered chest patch for monitoring of cardiac, haemodynamic and endocrine (pH, temperature) parameters. (f) Spectrogram of the heart sound signal (superimposed in black) involved in the calculation of the haemodynamic systolic intervals. (e–f): © 2019 IEEE. Reprinted, with permission, from [199].*

(a) and it measures body temperature, motion, ECG and GSR signals directly over the chest of a subject, storing the data inside a flash memory that can be downloaded at the end of the recording. Memory can last up to 7 h during continuous measurements or extended to 6 days through the intermittent operation of the device, in which data acquisition occurs only once in every 10 min, therefore saving battery power and reducing the charging cycles. The design and composition of the device also limits the discomfort for the wearer as it naturally conforms to the shape of the chest, while its operation requires minimal user input. Technologies such as this can be used onboard the ISS to replace the traditional Holter recording systems, which are obtrusive and bulky in nature, besides requiring additional wiring cables over the chest to connect to the ECG electrodes, thus affecting the body movements in space. Classification in quasi-real-time of the recorded multi-modal signals can then be employed to estimate the type of activity being executed by the astronaut, as shown by the graphic in Figure 9.3 (b). Regarding the individual contribution of each signal modality, different types of tasks from the user can be identified after correct training of the classifier, from which health inferences are made relative to the level of physical and mental stresses involved, energy expenditure, cardiovascular or muscular deconditioning, etc. The device can also be attached to the fabric composing the inner spacesuit and textile-based electrodes, sensors and interconnects can be used to capture the different bio-signals more closely to their body source during space walks or EVA that require wearing an outer spacesuit.

Another wearable device for measurement of the EEG signal, head motion and chemical analytes (pH and cortisol) in the inner ear has been recently proposed by [198] and integrated into an eyeshade with wiring connections to an earplug, as demonstrated in Figure 9.3 (c). This device uses Bluetooth Low Energy for signal transmission and was originally envisioned for monitoring the circadian changes produced in the sleep cycle, which mediate brain functions such as neuroplasticity, cortical maturation, memory function and cognition, with repercussions on the EEG patterns detected externally, in particular the theta waves. Post-processing of the EEG signal by spectral analysis helps revealing the low-frequency content for these brain waves (<10 Hz), their bandwidth and the power ratio between different brain states, as exemplified by the graphic in Figure 9.3 (d). By its turn, cortisol mediates cortical maturation with peaks of the hormone normally detected in the time instants preceding normal sleep on Earth that decline during the awake state, while pH has been correlated with stress-induced mental conditions that have repercussions on bad sleep habits. The circadian alterations produced by either the spaceflight or Earth's magnetosphere disturbances can be potentially monitored in astronauts by this technology since, currently, no viable method for sleep assessment is available onboard the ISS. In a different direction, and for the cases of wireless communication loss and access to rechargeable power supplies, batteryless wearables operating from external magnetic inductive sources (e.g. NFC interface) are interesting alternatives. With this in mind, the device presented in [199] was developed to generate snippets of cardiovascular parameters directly from the human chest by an NFC-enabled mobile phone, as depicted in Figure 9.3 (e). Estimated parameters include HR, skin temperature and pH, as well as some haemodynamic parameters from the heart (systolic intervals), such as the left-ventricular systolic time, ejection time, isovolumic contraction time, electromechanical activation time and pre-ejection period. These parameters are obtained from the measurement of the so-called heart sounds by an acoustic transducer, followed by digital signal processing that reveals relevant audio features. From the spectrogram in Figure 9.3 (f), two frequency bands located predominantly between 10 to 50 Hz and 10 to 20 Hz are observed in every cardiac cycle, which correspond to the S1 and S2 sounds, respectively. Signal processing further promotes morphological changes in the audio signal around S1 and S2, thereby helping in the identification of the temporal onset and decline for each sound peak, as required by the mathematical formulas involved in the estimation of the systolic intervals. Potential applications of the device for the spaceflight include fast assessment of astronaut's vitals in emergency scenarios by other crew members (e.g. space accident or natural catastrophe) without specialized medical training, to check if body compensation response has not been completely exhausted, haemo-dynamic stability still persists and decompensation has not occurred yet, which may give enough time for evacuation to a medical centre or something similar in future space missions.

A soft stretchable device for multiparametric sensing was proposed in [135] and is shown in Figure 9.4 (a). The device incorporated electrodes for the development of amperometric (for metabolites, e.g. glucose [200] and lactate), impedimetric (for protein biomarkers [201], e.g. for heart and stress hormones, e.g. cortisol) and potentiometric (for electrolytes, e.g. sodium, potassium and pH, as in [202], chloride and magnesium) sensors for sweat analysis. Electrochemical deposition of Au and IrOx was demonstrated for enabling pH sensing. A bioimpedance sensor for tissue hydration, galvanic skin response and tissue ischemia monitoring, optimized using electromagnetic optimization techniques (discussed in [203–206]) was also incorporated. Electrochemical deposition of Pt black was used to reduce the interface impedance of the electrodes from the MΩ range to a few kΩs, as shown in Figure 9.4 (b). Figure 9.4 (c) shows scanning electron microscopy (SEM) images from the untreated bare electrodes as well as the electrodes following Pt deposition when subject to simultaneous ultrasonication. Ultrasonication during electrochemical deposition is used to remove material that is weakly attached on the electrode surface and to ensure that only well-deposited material remains on the electrode surface, thus leading to a greater lifetime and improved performance. The bioimpedance sensor was characterized using saline solutions of varying conductivity at different frequencies to obtain the cell constant of the design, which was found to be equal to 2 400.2 per m, as shown in Figure 9.4 (d). A temperature sensor was also realized with a copper meandering track used as a resistance temperature detector. The relationship between device resistance and temperature is shown in Figure 9.4 (e), demonstrating a linear relationship with a sensitivity of ~5 mΩ/°C. A second meandering track was used to implement a Joule heating element for thermotherapy. The quadratic relationship between injected current and achieved temperature is shown in Figure 9.4 (f), with the device achieving a temperature above 120 °C with a current of 600 mA. Temperature sensing and heating can be used together for localized temperature control; the temperature sensor is also important for chemical

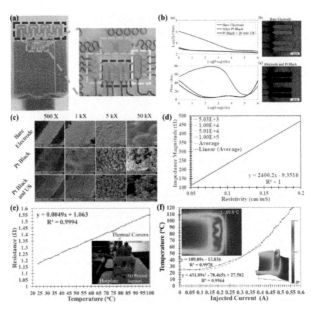

Figure 9.4   *(a) A flexible and stretchable multiparametric soft device with stretchable horseshoe interconnects and electrodes for potentiometric (orange box), amperometric and impedimetric (yellow box) electrochemical sensors, with a temperature sensor and a Joule heating element (purple box), a bioimpedance sensor (green box) and electrodes for ECG and iontophoresis (blue box). (b) Reduction of interface impedance of the bioimpedance sensor electrodes via electrochemical deposition of Pt black. (c) SEM imaging of the electrode surface before and after Pt black deposition, with and without ultrasonication. (d) Characterization of the bioimpedance sensor with saline solutions of varying conductivity at different frequencies. (e) Characterization of the temperature sensor. (f) Characterization of the Joule heating element. © 2019 IEEE. Reprinted, with permission, from [135].*

sensor calibration. The Joule heater, as well as the large electrodes incorporated on the device for current stimulation for iontophoresis (also used for ECG), were also realized to induce localized sweating and further facilitate sweat analysis. The device incorporated a thermal polyurethane substrate, and encapsulation layer and horseshoe (spring-like) interconnects to allow stretchability in the conductor layer.

Nevertheless, many of these devices, even those tested onboard the ISS, are only able to produce interesting publications, prototypes and proof-of-concept studies, with limited clinical and practical usefulness; the level of technology readiness is low. The challenges and design constraints are significant; however, with the rise of space applications, combined with an ever-growing involvement of commercial corporations and potential benefits of space exploration to humankind itself, there is significant scope for innovation.

9.5.1 Wearable systems for thermoregulation

The liquid cooling and ventilation garment (LCVG) is integral to EMU. Ventilation is required to remove metabolic heat, CO_2 and expel water vapour, but is still insufficient. A liquid cooling distribution system is also necessary and has been shown to be the most effective cooling system [207]. An LCVG example obtained from the NASA website is shown in Figure 9.5 (a). The LCVG conducts heat away from the body via a distributed network of flexible plastic tubes stitched onto a fabric substrate and held against the body. The tubing material is generally polyvinyl chloride, ethylene-vinyl acetate or silicone. The fabric is a conforming nylon-spandex mesh undergarment. This is secondary to an optional delicate nylon-tricot undergarment worn to support thermal resistance and easy donning and doffing [208]. An example of the layout of an LCVG used in the Apollo missions is shown in Figure 9.5 (b). A comprehensive overview of LCVGs can be found in [209]. A list of requirements and goals was recently developed by NASA and their collaborators [210] to support their improved design and development. These included increasing physical and thermal comfort. While there have not been significant advances in LCVG technology since the Apollo missions, incremental improvements have been made, with many still in the prototype phase. Developments have primarily focused on increasing the efficiency of existing systems: from reducing the size and weight of components [210, 211] to changing tubing coolant inlet temperature [212], reorienting tubing distribution layout [208, 211, 213] and altering tubing geometry [210, 214]. Comparatively, the fabric base layer has received little attention [215].

To address current LCVG limitations, NASA conducted a trade study to develop technologically feasible design concepts for an exploration cooling garment [210]. These concepts were based on a shortened liquid cooling/warming garment (SLCWG) [211] and the MACS-Delphi [216] prototypes. By placing tubing only in high heat transfer areas, these prototypes showed increased mobility, reduced total tubing length and overall garment weight. These areas contain high-density tissues and complex blood vessels (i.e. head and neck). The SLCWG was compared with the heavier full-body NASA LCVG, and results indicated how effective it has been in maintaining physiological comfort and subjective perception of comfort by the wearer [217]. LCVGs might be significantly improved if personalized thermoregulation garments were developed based on calculating and mapping individual thermal data profiles [211]. For example, including moisture-wicking fibres in high perspiration areas or compression that might support muscle function and thermal insulation to areas where cooling is negligible, such as the thigh [210].

The poor fit of the LCVG can be directly related to poor thermoregulatory performance. Issues pertaining to garment-body contact [208], such as gaping (see Figure 9.5 (c)), can be used as an example. A combination of relatively stiff tubing and fine base layer fabrics can result in tubing buckling away from the body during motion. Without good contact between the fabric and body, conduction is severely

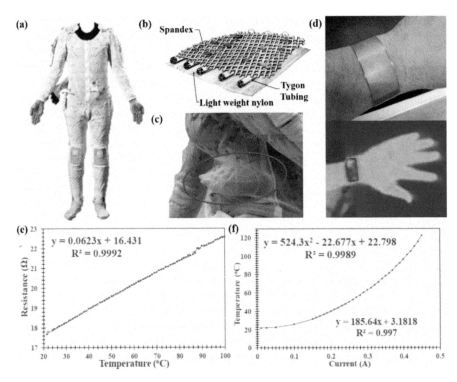

Figure 9.5   (a) LCVG worn on the body, as the base layer of the EMU spacesuit.
Image source NASA. Image obtained from the NASA website (https://
www.nasa.gov/audience/foreducators/spacesuits/home/clickable_
suit_nf.html). (b) LCVG fabric and tubing layout construction from
the Apollo missions. Image obtained from the NASA website (https://
history.nasa.gov/SP-368/s6ch6.htm). (c) LCVG gaping away from
the body during motion. From [208]. © Crystal Marie Compton
2016, reprinted with permission from the author. (d) A wrist-worn
thermotherapy and thermoregulation device for future fluidic
integration and astronaut inner suit use. The image at the bottom
was obtained from a thermal camera, and it demonstrates the device
achieving a temperature of ~42 ° C. (e) The temperature dependence
of the resistance of the copper FPC meandering track. (f) The
relationship between injected current and achieved temperature of
the FPC copper meandering track used for Joule heating. (d)–(f): ©
2019 IEEE. Reprinted, with permission, from [130].

reduced or not possible at all, thereby resulting in inefficient thermoregulation in
those areas.

As interests in spaceflight and planetary exploration increase, so do the physical
and mental demands on astronauts. While existing garments have been suitable, they

are not for more frequent and more extended duration missions. For instance, on a 6-month mission, astronauts would be required to perform around 24 more lunar EVAs than in the past using the same equipment [207]. Continued use of existing thermal garments will be detrimental to the overall efficiency and performance of crew members. Astronauts on the Apollo missions indicated that increased mobility and dexterity were required in future spacesuits [207]. By maintaining a broad understanding of comfort, which encompasses thermo-physiological, sensorial and psychological comfort with garment fit [97], a more holistic approach to the next generation of thermoregulation garments in space will be possible.

Currently, a nylon-spandex mesh is used within the LCVG for its breathability, stretch, lightweight nature and suitability as a substrate for tubing to be stitched on or woven through. Yet, both nylon and spandex are thermal insulators that do not support heat transfer through the fabric [218]. After research on 18 different fabrics, Cao *et al.* [215] found that the fabric with 80 per cent polyester and 20 per cent spandex would be an apt combination for LCVG's inner layer. This combination exhibited the lowest thermal and water vapour resistance. Smart textiles [219], such as fibres composed of lightweight and conductive nanomaterials are receiving much attention for their potential in textile-based thermo-physiological comfort [220], offering even better thermal resistance and conductivity. Trevino *et al.* [218] undertook preliminary research towards development and demonstration of the feasibility of flexible and high thermal conductivity fabrics for their application in the next generation of spacesuits. The logistical burden of auxiliary equipment [210], required as part of the LCVG assembly, could be minimized through engineered textile design by integrating components into a single textile system. This includes shoulder padding, an optional thermal undergarment [210] and the tubing itself. While functional, the current design is too bulky and overly-complex [218]. An example of an LCVG where tubing was integrated into the fabric for a better ergonomic fit can be seen in [221], where a knitted fabric construction was developed for a garment with terrestrial applications. Combining these fibre-based developments with engineered textile and functional garment design will support the specific goals and requirements laid out by NASA [16] in reducing weight while improving the thermal performance of LCVGs.

A wrist-worn device for wearable thermotherapy and thermoregulation for space applications was demonstrated in [130] and is shown in Figure 9.5 (d). The device was realized using standard flexible printed circuit (FPC) technology on a polyimide substrate and was comprised of two meandering copper tracks to be used as Joule heating elements or temperature sensors. These resistive-based devices can be used independently or together in a closed loop to induce heating for temperature regulation. Apart from local thermoregulation, the device can be used in conjunction with co-integrated textile-fluidic networks as those shown in Figure 9.5 (a–c) for thermoregulation over larger areas that can also incorporate liquid cooling. Figure 9.5 (e) demonstrates the linear relationship between device resistance and temperature with a sensitivity of ~62 mΩ/°C. An approximately quadratic relationship between injected current and achieved temperature was obtained from the characterization of the Joule heating element, as evident from Figure 9.5 (f). With a 450-mA current, a temperature above 120 °C was achieved. Two tracks were used with each device terminal to facilitate four-point resistance measurements to obtain more accurate measurements. Other approaches for thermoregulation are discussed in more detail in [130, 222].

9.6   Recommendations and future trends

Human physiology adapts rapidly to the extreme conditions in space that are mostly driven by microgravity, radiation exposure, confinement and isolation. Weightlessness results in bodily fluid shifts that immediately affect human physiology. The cardiovascular system, which circulates nutrients and oxygen to the whole body, undergoes dramatic changes to adjust to the absence of hydrostatic gradients. The reduced cardiac load results in changes in myocardial contractility and vascular function. Subsequently, the autonomic nervous system, which is involved in the regulation of HR, is triggered to maintain homeostasis. Overall, spaceflights result in a significant cardiac mass reduction and increase of peripheral vascular resistance in a way that resembles the typical ageing process.

On the other hand, head fluid shifts with the altered central venous flow can result in thrombus formation, which is a condition often associated with stroke. The neuro-vestibular system and associated cortical pathways related to the maintenance of equilibrium on Earth are also affected by upward fluid shifts, resulting in feelings of spatial disorientation and nausea at the early stages of spaceflight. Vision can be impaired with several astronauts reporting ocular problems upon returning to Earth. In fact, radiation is believed to be responsible for the perception of light flashes during dark/rest periods. Whereas these are some of the direct consequences of microgravity and radiation, further complex cellular pathways are triggered affecting the endocrine, respiratory, digestive, urinary or musculoskeletal systems. Tremendous bone loss is observed in spaceflight that can only be partly explained by muscular inactivity and reduction of calcium absorption.

Studies of physiological adaptations for space suggest the need for similar research protocols to be implemented during the assessment of cardiovascular, neurological, musculoskeletal and endocrine parameters or specific biomarkers found in blood, urine, saliva and sweat samples. Considerations also include standardization of pre- and post-flight monitoring devices, as well as in-flight, with proper documentation for all the activities engaged during space missions. Technologies should be developed in such a manner that allows real-time monitoring in both ground and space environments, with significant and selective data collection influencing decision-making on the newly emerging field of space medicine. Although largely overlooked in the past decades, researchers must also consider studying the psychological events affecting the health of astronauts in more detail, which can ultimately contribute to physiological maladaptation in long-duration spaceflights. Table 9.1 summarizes all the potentially useful markers that can possibly be monitored by wearable, implantable or point-of-care systems to assess astronaut's physiological state.

Wearable technology advances rapidly, and new devices emerge to allow the monitoring of biosignals such as ECG, EEG, EMG, PPG, body motion, metabolites and biomarkers of pathology, temperature and impedance within a Body Sensor Network [223]. Only recently have some of these devices been made available for use by astronauts in space. Nevertheless, significant technical challenges must be overcome. These include improved comfort, user acceptance and durability, as well as the incorporation of novel techniques to compensate for effects of microgravity and radiation on the various methods used in wearable and epidermal sensing. Such technologies have the potential to play a significant role in manned space exploration and the further advancement of humanity.

Implantable devices are potential substitutes for various wearable technologies, as they can have the capacity to monitor and stimulate tissues from within the body [224]. Many challenges, such as device unobtrusiveness, user comfort and a need for proximity to internal target tissues can be addressed with implantable devices. However, testing of new implantables in human subjects must still fulfil exhaustive testing and regulatory approvals on Earth, while their application in remote human explorations, where immediate access to proper medical facilities is impossible for device removal after failure or rejection by the body, currently hinders further their deployment in space applications. On the other hand, the use of pacemakers or other implantable stimulators is exclusion criteria for astronaut selection. In fact, any medical condition affecting prospective travellers denies their presence in manned space missions. Only when the proven benefits of implantable devices overly exceed the drawbacks of implantation for spaceflight can they be accepted as valid preventive, diagnostic or therapeutic tools in future space medicine. The challenges faced by humans in space, their exploration and exploitation are vast, with the support of human life being a dominant parameter. Advances in physiological monitoring devices are needed to obtain greater insight into human physiology in space and to monitor health and performance of astronauts during missions.

References

[1] Longhurst J.C. 'Cardiac receptors: their function in health and disease'. *Progress in Cardiovascular Diseases*. 1984;**27**(3):201–22.

[2] Carvil P., Baptisma R., Russomano T. 'The human body in a microgravity environment: long term adaptations and countermeasures'. *Aviat Focus*. 2013;**4**(1):10–22.

[3] Shen M., Frishman W.H. 'Effects of spaceflight on cardiovascular physiology and health'. *Cardiology in Review*. 2018.

[4] Williams D., Kuipers A., Mukai C., Thirsk R. 'Acclimation during space flight: effects on human physiology'. *Canadian Medical Association Journal*. 2009;**180**(13):1317–23.

[5] Aubert A.E., Larina I., Momken I., *et al.* 'Towards human exploration of space: the THESEUS review series on cardiovascular, respiratory, and renal research priorities'. *Npj Microgravity*. 2016;**2**(1):1–9.

[6]   Fritsch-Yelle J.M., Charles J.B., Jones M.M., Wood M.L. 'Microgravity decreases heart rate and arterial pressure in humans'. *Journal of Applied Physiology*. 1996;**80**(3):910–4.

[7]   Migeotte P.-F., Prisk G.K., Paiva M. 'Microgravity alters respiratory sinus arrhythmia and short-term heart rate variability in humans'. *American Journal of Physiology-Heart and Circulatory Physiology*. 2003;**284**(6):H1995–2006.

[8]   Karemaker J.M., Berecki-Gisolf J. '24-H blood pressure in space: the dark side of being an astronaut'. *Respiratory Physiology & Neurobiology*. 2009;**169**(SUPPL.):S55–8.

[9]   Gundel A., Drescher J., Spatenko Y.A., Polyakov V.V. 'Heart period and heart period variability during sleep on the miR space station'. *Journal of Sleep Research*. 1999;**8**(1):37–43.

[10] Hughson R.L., Shoemaker J.K., Arbeille P. 'CCISS, vascular and BP Reg: Canadian space life science research on ISS'. *Acta Astronautica*. 2014;**104**(1):444–8.

[11] Verheyden B., Liu J., Beckers F., Aubert A.E. 'Adaptation of heart rate and blood pressure to short and long duration space missions'. *Respiratory Physiology & Neurobiology*. 2009;**169**:S13–16.

[12] Tulen J.H., Boomsma F., Man in 't Veld A.J. 'Cardiovascular control and plasma catecholamines during rest and mental stress: effects of posture'. *Clinical Science*. 1999;**96**(6):567–76.

[13] Baevsky R.M., Baranov V.M., Funtova I.I., *et al.* 'Autonomic cardiovascular and respiratory control during prolonged spaceflights aboard the International Space Station'. *Journal of Applied Physiology*. 2007;**103**(1):156–61.

[14] Eckberg D.L., Halliwill J.R., Beightol L.A., Brown T.E., Taylor J.A., Goble R. 'Human vagal baroreflex mechanisms in space'. *The Journal of Physiology*. 2010;**588**(7):1129–38.

[15] Mandsager K.T., Robertson D., Diedrich A. 'The function of the autonomic nervous system during spaceflight'. *Clinical Autonomic Research*. 2015;**25**(3):141–51.

[16] Norsk P., Asmar A., Damgaard M., Christensen N.J. 'Fluid shifts, vasodilatation and ambulatory blood pressure reduction during long duration spaceflight'. *The Journal of Physiology*. 2015;**593**(3):573–84.

[17] Hughson R.L., Shoemaker J.K., Blaber A.P., *et al.* 'Cardiovascular regulation during long-duration spaceflights to the International Space Station'. *Journal of Applied Physiology*. 2012;**112**(5):719–27.

[18] Weissler A.M., Peeler R.G., Roehll W.H. 'Relationships between left ventricular ejection time, stroke volume, and heart rate in normal individuals and patients with cardiovascular disease'. *American Heart Journal*. 1961;**62**(3):367–78.

[19] Willems J.L., Roelandt J., De Geest H., Kesteloot H., Joossens J.V. 'The left ventricular ejection time in elderly subjects'. *Circulation*. 1970;**42**(1):37–42.

[20] Summers R.L., Martin D.S., Platts S.H., Mercado-Young R., Coleman T.G., Kassemi M. 'Ventricular chamber sphericity during spaceflight

and parabolic flight intervals of less than 1 G'. *Aviation, Space, and Environmental Medicine.* 2010;**81**(5):506–10.

[21]   Norsk P. 'Adaptation of the cardiovascular system to weightlessness: surprises, paradoxes and implications for deep space missions'. *Acta Physiologica.* 2020;**228**(3):e13434.

[22]   Neves J.S., Leite-Moreira A.M., Neiva-Sousa M., Almeida-Coelho J., Castro-Ferreira R., Leite-Moreira A.F. 'Acute myocardial response to stretch: what we (don't) know'. *Frontiers in Physiology.* 2015;**6**:408.

[23]   Paulus W., Shah A.M. 'No and cardiac diastolic function'. *Cardiovascular Research.* 1999;**43**(3):595–606.

[24]   Fu S., Ping P., Wang F., Luo L. 'Synthesis, secretion, function, metabolism and application of natriuretic peptides in heart failure'. *Journal of Biological Engineering.* 2018;**12**(1):2.

[25]   Leach C.S., Johnson P.C., Cintrón N.M. 'The endocrine system in space flight'. *Acta Astronautica.* 1988;**17**(2):161–6.

[26]   Fortrat J.-O., de Holanda A., Zuj K., Gauquelin-Koch G., Gharib C. 'Altered venous function during long-duration spaceflights'. *Frontiers in Physiology.* 2017;**8**:694.

[27]   Gelman S. 'Venous function and central venous pressure: a physiologic story'. *Anesthesiology.* 2008;**108**(4):735–48.

[28]   Shah P., Louis M.A. Physiology, central venous pressure [online]. Treasure Island: StatPearls Publishing. 2018. Available from https://www.ncbi.nlm.nih.gov/books/NBK519493 [Accessed 10 May 2021].

[29]   Berlin D.A., Bakker J. 'Starling curves and central venous pressure'. *Critical Care.* 2015;**19**(1):55.

[30]   Rothe C.F. 'Mean circulatory filling pressure: its meaning and measurement'. *Journal of Applied Physiology.* 1993;**74**(2):499–509.

[31]   Guyton A.C., Polizo D., Armstrong G.G. 'Mean circulatory filling pressure measured immediately after cessation of heart pumping'. *American Journal of Physiology-Legacy Content.* 1954;**179**(2):261–7.

[32]   Videbaek R., Norsk P. 'Atrial distension in humans during microgravity induced by parabolic flights'. *Journal of Applied Physiology.* 1997;**83**(6):1862–6.

[33]   West J.B., Prisk G.K. 'Chest volume and shape and intrapleural pressure in microgravity'. *Journal of Applied Physiology.* 1999;**87**(3):1240–1.

[34]   Su L., Pan P., Li D., *et al.* 'Central venous pressure (CVP) reduction associated with higher cardiac output (CO) favors good prognosis of circulatory shock: a single-center, retrospective cohort study'. *Frontiers in Medicine.* 2019;**6**:216.

[35]   Otsuka K., Cornelissen G., Kubo Y., *et al.* 'Intrinsic cardiovascular autonomic regulatory system of astronauts exposed long-term to microgravity in space: observational study'. *npj Microgravity.* 2015;**1**(1):1–9.

[36]   Vernikos J., Schneider V.S. 'Space, gravity and the physiology of aging: parallel or convergent disciplines? A mini-review'. *Gerontology.* 2010;**56**(2):157–66.

[37]   Stepanek J., Blue R.S., Parazynski S. 'Space medicine in the era of civilian spaceflight'. *New England Journal of Medicine*. 2019;**380**(11):1053–60.

[38]   Watenpaugh D.E. 'Fluid volume control during short-term space flight and implications for human performance'. *The Journal of Experimental Biology*. 2001;**204**(18).

[39]   Leach C.S., Inners L.D., Charles J.B. 'Changes in total body water during spaceflight'. *The Journal of Clinical Pharmacology*. 1991;**31**(10):1001–6.

[40]   Diedrich A., Paranjape S.Y., Robertson D. 'Plasma and blood volume in space'. *The American Journal of the Medical Sciences*. 2007;**334**(7):80–6.

[41]   SMC L., Feiveson A.H., Stein S., Stenger M.B., Platts S.H. 'Orthostatic intolerance after ISS and space shuttle missions'. *Aerospace Medicine and Human Performance*. 2015;**86**(12 Suppl):A54–67.

[42]   Blomqvist C.G., Nixon J.V., Johnson R.L., Mitchell J.H. 'Early cardiovascular adaptation to zero gravity simulated by head-down tilt'. *Acta Astronautica*. 1980;**7**(4–5):543–53.

[43]   Shi S.-J., South D.A., Meck J V. 'Fludrocortisone does not prevent orthostatic hypotension in astronauts after spaceflight'. *Aviation, Space, and Environmental Medicine*. 2004;**75**(3):235–9.

[44]   Evans J.M., Knapp C.F., Goswami N. 'Artificial gravity as a countermeasure to the cardiovascular deconditioning of spaceflight: gender perspectives'. *Frontiers in Physiology*. 2018;**9**:716.

[45]   Fritsch-Yelle J.M., Whitson P.A., Bondar R.L., Brown T.E. 'Subnormal norepinephrine release relates to presyncope in astronauts after spaceflight'. *Journal of Applied Physiology*. 1996;**81**(5):2134–41.

[46]   Blaber A.P., Bondar R.L., Kassam M.S. 'Heart rate variability and short duration spaceflight: relationship to post-flight orthostatic intolerance'. *BMC Physiology*. 2004;**4**(1):6.

[47]   Stewart J.M. 'Mechanisms of sympathetic regulation in orthostatic intolerance'. *Journal of Applied Physiology*. 2012;**113**(10):1659–68.

[48]   Hargens A.R., Bhattacharya R., Schneider S.M. 'Space physiology VI: exercise, artificial gravity, and countermeasure development for prolonged space flight'. *European Journal of Applied Physiology*. 2013;**113**(9):2183–92.

[49]   Hughson R.L., Helm A., Durante M. 'Heart in space: effect of the extraterrestrial environment on the cardiovascular system'. *Nature Reviews Cardiology*. 2018;**15**(3):167–80.

[50]   Iwasaki K.-Ichi., Levine B.D., Zhang R., *et al.* 'Human cerebral autoregulation before, during and after spaceflight'. *The Journal of Physiology*. 2007;**579**(3):799–810.

[51]   Blaber A.P., Goswami N., Bondar R.L., Kassam M.S. 'Impairment of cerebral blood flow regulation in astronauts with orthostatic intolerance after flight'. *Stroke*. 2011;**42**(7):1844–50.

[52]   Xu D., Shoemaker J.K., Blaber A.P., Arbeille P., Fraser K., Hughson R.L. 'Reduced heart rate variability during sleep in long-duration spaceflight'. *American Journal of Physiology-Regulatory, Integrative and Comparative Physiology*. 2013;**305**(2):R164–70.

[53] Otsuka K., Cornelissen G., Kubo Y., *et al.* 'Anti-aging effects of long-term space missions, estimated by heart rate variability'. *Scientific Reports.* 2019;**9**(1):1–12.

[54] Zhu H., Wang H., Liu Z. 'Effects of real and simulated weightlessness on the cardiac and peripheral vascular functions of humans: a review'. *International Journal of Occupational Medicine and Environmental Health.* 2015;**28**(5):793–802.

[55] Perhonen M.A., Franco F., Lane L.D., *et al.* 'Cardiac atrophy after bed rest and spaceflight'. *Journal of Applied Physiology.* 2001;**91**(2):645–53.

[56] Summers R.L., Martin D.S., Meck J.V., Coleman T.G. 'Mechanism of spaceflight-induced changes in left ventricular mass'. *The American Journal of Cardiology.* 2005;**95**(9):1128–30.

[57] Westby C.M., Martin D.S., Lee S.M.C., Stenger M.B., Platts S.H. 'Left ventricular remodeling during and after 60 days of sedentary head-down bed rest'. *Journal of Applied Physiology.* 2016;**120**(8):956–64.

[58] Levine B.D., Zuckerman J.H., Pawelczyk J.A. 'Cardiac atrophy after bed-rest deconditioning'. *Circulation.* 1997;**96**(2):517–25.

[59] Dorfman T.A., Levine B.D., Tillery T., *et al.* 'Cardiac atrophy in women following bed rest'. *Journal of Applied Physiology.* 2007;**103**(1):8–16.

[60] Vernikos J. 'Human physiology in space'. *BioEssays.* 1996;**18**(12):1029–37.

[61] Hodkinson P.D., Anderton R.A., Posselt B.N., Fong K.J. 'An overview of space medicine'. *British Journal of Anaesthesia.* 2017;**119**(Suppl 1):i143–53.

[62] Marshall-Goebel K., Laurie S.S., Alferova I.V., *et al.* 'Assessment of jugular venous blood flow stasis and thrombosis during spaceflight'. *JAMA Network Open.* 2019;**2**(11):e1915011.

[63] Auñón-Chancellor S.M., Pattarini J.M., Moll S., Sargsyan A. 'Venous thrombosis during spaceflight'. *New England Journal of Medicine.* 2020;**382**(1):89–90.

[64] Vasan R.S. 'Biomarkers of cardiovascular disease'. *Circulation.* 2006;**113**(19):2335–62.

[65] Dhingra R., Vasan R.S. 'Biomarkers in cardiovascular disease: statistical assessment and section on key novel heart failure biomarkers'. *Trends in Cardiovascular Medicine.* 2017;**27**(2):123–33.

[66] Ravassa S., Delles C., Currie G., Díez J. 'Biomarkers of cardiovascular disease' in Touyz R.M., Delles C. (eds.). *Textbook of Vascular Medicine [Internet].* Cham: Springer International Publishing; 2019. pp. 319–30.

[67] Ridker P.M., Brown N.J., Vaughan D.E., Harrison D.G., Mehta J.L. 'Established and emerging plasma biomarkers in the prediction of first atherothrombotic events'. *Circulation.* 2004;**109**(25_suppl_1):IV-6–IV-19.

[68] Huang Y., Gulshan K., Nguyen T., Wu Y. *Biomarkers of cardiovascular disease [online]. Disease Markers.* **2017**. Hindawi; 2017. p. e8208609. Available from https://www.hindawi.com/journals/dm/2017/8208609/.

[69] LKh P., Kashirina D.N., Kononikhin A.S., Brzhozovsky A.G., Ivanisenko V.A., Tiys E.S. 'The effect of long-term space flights on human urine proteins functionally related to endothelium'. *Human Physiology.* 2018;**44**(1):60–7.

[70]   LKh P., Kashirina D.N., Brzhozovskiy A.G., Kononikhin A.S., Tiys E.S., Ivanisenko V.A. 'Evaluation of cardiovascular system state by urine proteome after manned space flight'. *Acta Astronautan*. 2019;**160**:594–600.

[71]   Pastushkova L.H., Rusanov V.B., Orlov O.I., *et al.* 'The variability of urine proteome and coupled biochemical blood indicators in cosmonauts with different preflight autonomic status'. *Acta Astronautica*. 2020;**168**(5):204–10.

[72]   Pastushkova L.H., Rusanov V.B., Goncharova A.G., *et al.* 'Urine proteome changes associated with autonomic regulation of heart rate in cosmonauts'. *BMC Systems Biology*. 2019;**13**(1):23–31.

[73]   Baker J.E., Moulder J.E., Hopewell J.W. 'Radiation as a risk factor for cardiovascular disease'. *Antioxidants & Redox Signaling*. 2011;**15**(7):1945–56.

[74]   Delp M.D., Charvat J.M., Limoli C.L., Globus R.K., Ghosh P. 'Apollo Lunar astronauts show higher cardiovascular disease mortality: possible deep space radiation effects on the vascular endothelium'. *Scientific Reports*. 2016;**6**(1):1–11.

[75]   Jones J.A., Karouia F., Pinsky L., Cristea O. 'Radiation and radiation disorders' in Barratt M.R., Baker E.S., Pool S.L. (eds.). *Principles of Clinical Medicine for Space Flight [Internet]*. New York, NY: Springer; 2019. pp. 39–108.

[76]   Moreno-Villanueva M., Wong M., Lu T., Zhang Y., Wu H. 'Interplay of space radiation and microgravity in DNA damage and DNA damage response'. *npj Microgravity*. 2017;**3**(1):1–8.

[77]   Sihver L., Mortazavi S. 'Radiation risks and countermeasures for humans on deep space missions'. 2019 IEEE Aerospace Conference; 2019. pp. 1–10.

[78]   Yatagai F., Honma M., Dohmae N., Ishioka N. 'Biological effects of space environmental factors: a possible interaction between space radiation and microgravity'. *Life Sciences in Space Research*. 2019;**20**(8):113–23.

[79]   Hammond T.G., Birdsall H.H. 'Endocrine effects of space flight' in Pathak Y., Santos M.Ados., Zea L. (eds.). *Handbook of Space Pharmaceuticals*. Cham: Springer; 2019. pp. 1–9. Available from https://link.springer.com/referencework/10.1007%2F978-3-319-50909-9.

[80]   Fu H., Su F., Zhu J., Zheng X., Ge C. 'Effect of simulated microgravity and ionizing radiation on expression profiles of miRNA, lncRNA, and mRNA in human lymphoblastoid cells'. *Life Sciences in Space Research*. 2020;**24**(1):1–8.

[81]   Bergouignan A., Stein T.P., Habold C., Coxam V., O' Gorman D., Blanc S. 'Towards human exploration of space: the THESEUS review series on nutrition and metabolism research priorities'. *npj Microgravity*. 2016;**2**(1):16029.

[82]   Enrico C. 'Space nutrition: the key role of nutrition in human space flight'. *ArXiv161000703 Q-Bio [Internet]*. 2016.

[83]   Barrett K.E., Barman S.M., Boitano S., Brooks H. *Ganong's review of medical physiology*. 24th Edition. New York: McGraw-Hill Education/Medical; 2012. p. 768.

[84]   Jakob S.M. 'Clinical review: splanchnic ischaemia'. *Critical Care*. 2002;**6**(4):306.

[85]   Strollo F., Gentile S., Strollo G., Mambro A., Vernikos J. 'Recent progress in space physiology and aging'. *Frontiers in Physiology*. 2018;**9**.

[86]   Roberts D.R., Albrecht M.H., Collins H.R., *et al*. 'Effects of spaceflight on astronaut brain structure as indicated on MRI'. *New England Journal of Medicine*. 2017;**377**(18):1746–53.

[87]   Van Ombergen A., Laureys S., Sunaert S., Tomilovskaya E., Parizel P.M., Wuyts F.L. 'Spaceflight-induced neuroplasticity in humans as measured by MRI: what do we know so far?' *npj Microgravity*. 2017;**3**(1):2.

[88]   Demontis G.C., Germani M.M., Caiani E.G., Barravecchia I., Passino C., Angeloni D. 'Human pathophysiological adaptations to the space environment'. *Frontiers in Physiology*. 2017;**8**:547.

[89]   Koppelmans V., Bloomberg J.J., Mulavara A.P., Seidler R.D. 'Brain structural plasticity with spaceflight'. *npj Microgravity*. 2016;**2**(1):1–8.

[90]   Roberts D.R., Ramsey D., Johnson K., *et al*. 'Cerebral cortex plasticity after 90 days of bed rest: data from TMS and fMRI'. *Aviation, Space, and Environmental Medicine*. 2010;**81**(1):30–40.

[91]   Liao Y., Zhang J., Huang Z., *et al*. 'Altered baseline brain activity with 72 h of simulated microgravity – initial evidence from resting-state fMRI'. *PLoS One*. 2012;**7**(12):e52558.

[92]   Zhou Y., Wang Y., Rao L.-L., *et al*. 'Disrupted resting-state functional architecture of the brain after 45-day simulated microgravity'. *Frontiers in Behavioral Neuroscience*. 2014;**8**(13):200.

[93]   Li K., Guo X., Jin Z., *et al*. 'Effect of simulated microgravity on human brain gray matter and white matter – evidence from MRI'. *Plos One*. 2015;**10**(8):e0135835.

[94]   Yuan P., Koppelmans V., Reuter-Lorenz P.A., *et al*. 'Increased brain activation for dual tasking with 70-days head-down bed rest'. *Frontiers in Systems Neuroscience*. 2016;**10**:71.

[95]   Werner J. 'Regulatory processes of the human body during thermal and work strain'. *Elsevier Ergon Book Series*. 2005;**3**(C):3–9.

[96]   Zhang X. 'Heat-storage and thermo-regulated textiles and clothing'. *Smart Fibres, Fabrics and Clothing*; 2001. pp. 34–57.

[97]   Das A., Alagirusamy R. 'Science in comfort clothing' in Das A., Alagirusamy R. (eds.). *Science in Clothing Comfort*. New York, USA: Woodhead Publishing India PVT Ltd; 2010. pp. 79–105.

[98]   Parsons K.(ed.) *Human Thermal Environments: The Effects of Hot, Moderate, and Cold Environments on Human Health, Comfort, and Performance*. Third Edition. Boca Raton, FL. USA: CRC Press. Taylor and Francis Group; 2002.

[99]   Nasa. *Staying cool on the ISS [online]*. Nasa Science. 2001.

[100]   Nyberg K.L., Diller K.R., Wissler E.H. 'Model of human/liquid cooling garment interaction for space suit automatic thermal control'. *Journal of Biomechanical Engineering*. 2001;**123**(1):114–20.

[101]   Oka T. 'Psychogenic fever: how psychological stress affects body temperature in the clinical population'. *Temperature*. 2015;**2**(3):368–78.

[102]   Nalishiti V. 'Clinical investigator sleep and circadian rhythm during a short space mission'. *The Clinical Instigator*. 1993;**71**:718–24.

[103]   Hancock P.A., Ross J.M., Szalma J.L. 'A meta-analysis of performance response under thermal stressors'. *Human Factors: The Journal of the Human Factors and Ergonomics Society*. 2007;**49**(5):851–77.

[104]   Hancock P.A., Vasmatzidis I. 'Effects of heat stress on cognitive performance: the current state of knowledge'. *International Journal of Hyperthermia*. 2003;**19**(3):355–72.

[105]   Taylor L., Watkins S.L., Marshall H., Dascombe B.J., Foster J. 'The impact of different environmental conditions on cognitive function: a focused review'. *Frontiers in Physiology*. 2016;**6**(JAN):1–12.

[106]   Stahn A.C., Werner A., Opatz O., *et al.* 'Increased core body temperature in astronauts during long-duration space missions'. *Scientific Reports*. 2017;**7**(1):1–8.

[107]   Strollo F., Vassilieva G., Ruscica M., Masini M., Santucci D., Borgia L. *Changes in stress hormones and metabolism during a 105-day simulated mars mission [online]*. Aerospace Medical Association. 2014.

[108]   Russell E., Koren G., Rieder M., Van Uum S.H.M. 'The detection of cortisol in human sweat: implications for implications for measurement of cortisol in hair'. *Therapeutic Drug Monitoring*. 2014;**36**(1):30–4.

[109]   Steckl A.J., Ray P. 'Stress biomarkers in biological fluids and their point-of-use detection'. *ACS Sensors*. 2018;**3**(10):2025–44.

[110]   Hellhammer D.H., Wüst S., Kudielka B.M. 'Salivary cortisol as a biomarker in stress research'. *Psychoneuroendocrinology*. 2009;**34**(2):163–71.

[111]   Agha N.H., Baker F.L., Kunz H.E., *et al.* 'Salivary antimicrobial proteins and stress biomarkers are elevated during a 6-month mission to the International Space Station'. *Journal of Applied Physiology*. 2020;**128**(2):264–75.

[112]   Stein T.P., Schluter M.D. 'Excretion of IL-6 by astronauts during spaceflight'. *American Journal of Physiology-Endocrinology Metabolism*. 1994;**266**(3):E448–52.

[113]   Rai B., Kaur J., Foing B, Stress F.B.H. 'Stress, workload and physiology demand during extravehicular activity: a pilot study'. *North American Journal of Medical Sciences*. 2012;**4**(6):266.

[114]   Smith J.K. 'Il-6 and the dysregulation of immune, bone, muscle, and metabolic homeostasis during spaceflight'. *npj Microgravity*. 2018;**4**(1):1–8.

[115]   Stowe R.P., Pierson D.L., Barrett A.D. 'Elevated stress hormone levels relate to Epstein-Barr virus reactivation in astronauts'. *Psychosomatic Medicine*. 2001;**63**(6):891–5.

[116]   Mehta S.K., Laudenslager M.L., Stowe R.P., Crucian B.E., Sams C.F., Pierson D.L. 'Multiple latent viruses reactivate in astronauts during space shuttle missions'. *Brain, Behavior, Immunity*. 2014;**41**(6):210–17.

[117]   Mehta S.K., Cohrs R.J., Forghani B., Zerbe G., Gilden D.H., Pierson D.L. 'Stress-Induced subclinical reactivation of varicella zoster virus in astronauts'. *Journal of Medical Virology*. 2004;**72**(1):174–9.

[118] Mehta S.K., Stowe R.P., Feiveson A.H., Tyring S.K., Pierson D.L. 'Reactivation and shedding of cytomegalovirus in astronauts during space-flight'. *The Journal of Infectious Diseases*. 2000;**182**(6):1761–4.

[119] Drummer C., Valenti G., Cirillo M., *et al.* 'Vasopressin, hypercalciuria and aquaporin – the key elements for impaired renal water handling in astro-nauts?' *Nephron*. 2002;**92**(3):503–14.

[120] Jones J.A., Jennings R., Pietryzk R., Ciftcioglu N., Stepaniak P. 'Genitourinary issues during spaceflight: a review'. *International Journal of Impotence Research*. 2005;**17 Suppl 1**(1):S64–7.

[121] Jones J.A., Ciftcioglu N., Schmid J.F., Barr Y.R., Griffith D. 'Calcifying na-noparticles (nanobacteria): an additional potential factor for urolithiasis in space flight crews'. *Urology*. 2009;**73**(1):210.e11–210.e13.

[122] Smith S.M., Zwart S.R., Heer M. Human adaptation to spaceflight: the role of nutrition. 151. Available from https://www.nasa.gov/sites/default/files/human-adaptation-to-spaceflight-the-role-of-nutrition.pdf.

[123] Jones J.A., Pietrzyk R.A., Cristea O., Whitson P.A. 'Renal and genitouri-nary concerns' in Barratt M.R., Baker E.S., Pool S.L. (eds.). *Principles of Clinical Medicine for Space Flight [Internet]*. New York, NY: Springer; 2019. pp. 545–79.

[124] Steller J.G., Alberts J.R., Ronca A.E. 'Oxidative stress as cause, conse-quence, or biomarker of altered female reproduction and development in the space environment'. *International Journal of Molecular Sciences*. 2018;**19**(12):3729.

[125] Mishra B., Luderer U. 'Reproductive hazards of space travel in women and men'. *Nature Reviews Endocrinology*. 2019;**15**(12):713–30.

[126] Pastushkova L.K., Kireev K.S., Kononikhin A.S., *et al.* 'Detection of renal tissue and urinary tract proteins in the human urine after space flight'. *PLoS One*. 2013;**8**(8):e71652.

[127] Cirillo M., De Santo N.G., Heer M., *et al.* 'Low urinary albumin excretion in astronauts during space missions'. *Nephron Physiology*. 2003;**93**(4):102–5.

[128] Mena-Bravo A., Luque de Castro M.D. 'Sweat: a sample with limited present applications and promising future in metabolomics'. *Journal of Pharmaceutical and Biomedical Analysis*. 2014;**90**:139–47.

[129] Braun N., Binder S., Grosch H., *et al.* 'Current data on effects of long-term missions on the International Space Station on skin physiological param-eters'. *Skin Pharmacology and Physiology*. 2019;**32**(1):43–51.

[130] Kassanos P., Seichepine F., Keshavarz M., Yang G.-Z. 'Towards a flexible wrist-worn thermotherapy and thermoregulation device'. *2019 IEEE 19th International Conference on Bioinformatics and Bioengineering (BIBE)*; 2019. pp. 644–8.

[131] Song K., Ha U., Lee J., Bong K., Yoo H.-J. 'An 87-mA·min iontophoresis con-troller IC with dual-mode impedance sensor for patch-type transdermal drug delivery system'. *IEEE Journal of Solid-State Circuits*. 2014;**49**(1):167–78.

[132] Kim J., Campbell A.S., de Ávila B.E.-F., Wang J. 'Wearable biosensors for healthcare monitoring'. *Nature Biotechnology*. 2019;**37**(4):389–406.

[133] Bandodkar A.J., Jia W., Yardımcı C., Wang X., Ramirez J., Wang J. 'Tattoo-based noninvasive glucose monitoring: a proof-of-concept study'. *Analytical Chemistry*. 2015;**87**(1):394–8.

[134] Emaminejad S., Gao W., Wu E., *et al.* 'Autonomous sweat extraction and analysis applied to cystic fibrosis and glucose monitoring using a fully integrated wearable platform'. *Proceedings of the National Academy of Sciences*. 2017;**114**(18):4625–30.

[135] Kassanos P., Seichepine F., Wales D., Yang G.-Z. 'Towards a flexible/stretchable multiparametric sensing device for surgical and wearable applications'. *2019 IEEE Biomedical Circuits and Systems Conference (BioCAS)*; 2019. pp. 1–4.

[136] Heikenfeld J. 'Let them see you sweat'. *IEEE Spectrum*. 2014;**51**(11):46–63.

[137] Katchman B.A., Zhu M., Blain Christen J., Anderson K.S. 'Eccrine sweat as a Biofluid for profiling immune biomarkers'. *PROTEOMICS - Clinical Applications*. 2018;**12**(6):1800010.

[138] Bariya M., Nyein H.Y.Y., Javey A. 'Wearable sweat sensors'. *Nature Electronics*. 2018;**1**(3):160–71.

[139] Sonner Z., Wilder E., Heikenfeld J., *et al.* 'The microfluidics of the eccrine sweat gland, including biomarker partitioning, transport, and biosensing implications'. *Biomicrofluidics*. 2015;**9**(3):031301.

[140] Kinnamon D., Ghanta R., Lin K.-C., Muthukumar S., Prasad S. 'Portable biosensor for monitoring cortisol in low-volume perspired human sweat'. *Scientific Reports*. 2017;**7**(1):1–13.

[141] Rummel J.A., Sawin C.F., Michel E.L., Buderer M.C., Thornton W.T. 'Exercise and long duration spaceflight through 84 days'. *Journal of the American Medical Women's Association*. 1975;**30**(4):173–87.

[142] Narici M.V., de Boer M.D. 'Disuse of the musculo-skeletal system in space and on earth'. *European Journal of Applied Physiology*. 2011;**111**(3):403–20.

[143] Winnard A., Scott J., Waters N., Vance M., Caplan N. 'Effect of time on human muscle outcomes during simulated microgravity exposure without countermeasures–systematic review'. *Frontiers in Physiology*. 2019;**10**:1046.

[144] LeBlanc A., Rowe R., Schneider V., Evans H., Hedrick T. 'Regional muscle loss after short duration spaceflight'. *Aviation, Space, and Environmental Medicine*. 1995;**66**(12):1151–4.

[145] Akima H., Kawakami Y., Kubo K., *et al.* 'Effect of short-duration spaceflight on thigh and leg muscle volume'. *Medicine & Science in Sports & Exercise*. 2000;**32**(10):1743–7.

[146] LeBlanc A., Lin C., Shackelford L., *et al.* 'Muscle volume, MRI relaxation times (T2), and body composition after spaceflight'. *Journal of Applied Physiology*. 2000;**86**(6):2158–64.

[147] Gopalakrishnan R., Genc K.O., Rice A.J., *et al.* 'Muscle volume, strength, endurance, and exercise loads during 6-month missions in space'. *Aviation, Space, and Environmental Medicine*. 2010;**81**(2):91–104.

[148]   Lang T., Van Loon J.J.W.A., Bloomfield S., *et al.* 'Towards human exploration of space: the THESEUS review series on muscle and bone research
priorities'. *npj Microgravity.* 2017;**3**(1):1–10.

[149]   Trappe S., Costill D., Gallagher P., *et al.* 'Exercise in space: human skeletal
muscle after 6 months aboard the International Space Station'. *Journal of
Applied Physiology.* 2009;**106**(4):1159–68.

[150]   Zierath J.R., Hawley J.A. 'Skeletal muscle fiber type: influence on contractile and metabolic properties'. *PLoS Biology.* 2004;**2**(10):e348.

[151]   Gollnick P.D., Sjödin B., Karlsson J., Jansson E., Saltin B. 'Human soleus muscle: a comparison of fiber composition and enzyme activities
with other leg muscles'. *Pflugers Archiv European Journal of Physiology.*
1974;**348**(3):247–55.

[152]   Bagley J.R., a M.K., Trappe S.W. 'Microgravity-induced fiber type shift in human skeletal muscle'. *Gravitational and Space Biology.* 2012;**26**(1):34–40.

[153]   Fitts R.H., Trappe S.W., Costill D.L., *et al.* 'Prolonged space flight-induced
alterations in the structure and function of human skeletal muscle fibres'.
The Journal of Physiology. 2010;**588**(18):3567–92.

[154]   Schulze K., Gallagher P., Trappe S. 'Resistance training preserves skeletal
muscle function during unloading in humans'. *Medicine and Science in
Sports and Exercise.* 2002;**34**(2):303–13.

[155]   Antonutto G., Bodem F., Zamparo P., di Prampero P.E. 'Maximal power and
EMG of lower limbs after 21 days spaceflight in one astronaut'. *Journal
of Gravitational Physiology: A journal of the International Society for
Gravitational Physiology.* 1998;**5**(1):P63–6.

[156]   Lambertz D., Pérot C., Kaspranski R., Goubel F. 'Effects of long-term spaceflight on mechanical properties of muscles in humans'. *Journal of Applied
Physiology.* 2001;**90**(1):179–88.

[157]   Clark B.C., Fernhall B., Ploutz-Snyder L.L. 'Adaptations in human neu-
romuscular function following prolonged unweighting: I. skeletal muscle
contractile properties and applied ischemia efficacy'. *Journal of Applied
Physiology.* 2006;**101**(1):256–63.

[158]   Morseth B., Emaus N., Jørgensen L. 'Physical activity and bone: the importance of the various mechanical stimuli for bone mineral density. A review'.
Norsk Epidemiologi. 2011;**20**(2).

[159]   Vico L., Hargens A. 'Skeletal changes during and after spaceflight'. *Nature
Reviews Rheumatology.* 2018;**14**(4):229–45.

[160]   Smith S.M., McCoy T., Gazda D., Morgan J.L.L., Heer M., Zwart S.R.
'Space flight calcium: implications for astronaut health, spacecraft operations, and earth'. *Nutrients.* 2012;**4**(12):2047–68.

[161]   Dadwal U.C., Maupin K.A., Zamarioli A., *et al.* 'The effects of spaceflight and fracture healing on distant skeletal sites'. *Scientific Reports.*
2019;**9**(1):1–11.

[162]   Rittweger J. 'Maintaining crew bone health' in Seedhouse E., Shayler D.J.
(eds.). *Handbook of Life Support Systems for Spacecraft and Extraterrestrial
Habitats.* Cham: Springer International Publishing; 2019. pp. 1–15.

[163] LeBlanc A., Schneider V., Shackelford L., *et al.* 'Bone mineral and lean tissue loss after long duration space flight'. *Journal of Musculoskeletal & Neuronal Interactions.* 2000;**1**(2):157–60.

[164] Keyak J.H., Koyama A.K., LeBlanc A., Lu Y., Lang T.F. 'Reduction in proximal femoral strength due to long-duration spaceflight'. *Bone.* 2009;**44**(3):449–53.

[165] Coulombe J.C., Senwar B., Ferguson V.L. 'Spaceflight-Induced bone tissue changes that affect bone quality and increase fracture risk'. *Current Osteoporosis Reports.* 2020;**18**(1):1–12.

[166] Tanaka K., Nishimura N., Kawai Y. 'Adaptation to microgravity, deconditioning, and countermeasures'. *The Journal of Physiological Sciences.* 2017;**67**(2):271–81.

[167] McCarthy I.D. 'Fluid shifts due to microgravity and their effects on bone: a review of current knowledge'. *Annals of Biomedical Engineering.* 2005;**33**(1):95–103.

[168] Amin S., Factors M. 'Mechanical factors and bone health: effects of weightlessness and neurologic injury'. *Current Rheumatology Reports.* 2010;**12**(3):170–6.

[169] Kwon R.Y., Meays D.R., Tang W.J., Frangos J.A. 'Microfluidic enhancement of intramedullary pressure increases interstitial fluid flow and inhibits bone loss in hindlimb suspended mice'. *Journal of Bone and Mineral Research.* 2010;**25**(8):1798–807.

[170] Lam H., Brink P., Qin Y.-X. 'Skeletal nutrient vascular adaptation induced by external oscillatory intramedullary fluid pressure intervention'. *Journal of Orthopaedic Surgery and Research.* 2010;**5**:18.

[171] Smith S.M., Heer M. 'Calcium and bone metabolism during space flight'. *Nutrition.* 2002;**18**(10):849–52.

[172] Sibonga J.D., Evans H.J., Sung H.G., *et al.* 'Recovery of spaceflight-induced bone loss: bone mineral density after long-duration missions as fitted with an exponential function'. *Bone.* 2007;**41**(6):973–8.

[173] LKh P., Goncharova A.G., GYu V., Tagirova S.K., Kashirina D.N., Sayk O.V. 'Search for blood proteome proteins involved in the regulation of bone remodeling in astronauts'. *Human Physiology.* 2019;**45**(5):536–42.

[174] Fitzgerald J. 'Cartilage breakdown in microgravity—a problem for long-term spaceflight?' *npj Regenerative Medicine.* 2017;**2**(1):1–2.

[175] Ramachandran V., Wang R., Ramachandran S.S., Ahmed A.S., Phan K., Antonsen E.L. 'Effects of spaceflight on cartilage: implications on spinal physiology'. *Journal of Spine Surgery.* 2018;**4**(2):433–45.

[176] Schlotman T.E., Lehnhardt K.R., Abercromby A.F., *et al.* 'Bridging the gap between military prolonged field care monitoring and exploration spaceflight: the compensatory reserve'. *npj Microgravity.* 2019;**5**(1):1–11.

[177] Kassanos P., Ip H., Yang G.-Z. 'Ultra-low power application-specific integrated circuits for sensing' in Yang G.-Z. (ed.). *Implantable Sensors and Systems: From Theory to Practice.* Cham: Springer International Publishing; 2018. pp. 281–437.

[178] Roda A., Mirasoli M., Guardigli M., *et al.* 'Advanced biosensors for monitoring astronauts' health during long-duration space missions'. *Biosensors and Bioelectronics.* 2018;**111**:18–26.

[179] Kassanos P., Anastasova S., Chen C.M., Yang G.-Z. 'Sensor embodiment and flexible electronics' in Yang G.-Z. (ed.). *Implantable Sensors and Systems: From Theory to Practice.* Cham: Springer International Publishing; 2018. pp. 197–279.

[180] Komorowski M., Fleming S., Mawkin M., Hinkelbein J. 'Anaesthesia in austere environments: literature review and considerations for future space exploration missions'. *npj Microgravity.* 2018;**4**:5.

[181] Dunn J., Huebner E., Liu S., Landry S., Binsted K. 'Using consumer-grade wearables and novel measures of sleep and activity to analyze changes in behavioral health during an 8-month simulated Mars mission'. *Computers in Industry.* 2017;**92-93**(3):32–42.

[182] Fraser K.S., Greaves D.K., Shoemaker J.K., Blaber A.P., Hughson R.L. 'Heart rate and daily physical activity with long-duration habitation of the International Space Station'. *Aviation, Space, and Environmental Medicine.* 2012;**83**(6):577–84.

[183] Guillodo E., Lemey C., Simonnet M., *et al.* 'Clinical applications of mobile health wearable–based sleep monitoring: systematic review'. *JMIR mHealth uHealth.* 2020;**8**(4):e10733):e10733:.

[184] Gundel A., Drescher J., Spatenko Y.A., Polyakov V.V. 'Heart period and heart period variability during sleep on the miR space station'. *Journal of Sleep Research.* 1999;**8**(1):37–43.

[185] Sampson M. 'Ambulatory electrocardiography: indications and devices'. *British Journal of Cardiac Nursing.* 2019;**14**(3):114–21.

[186] Cooke W.H., Ames J.E., Crossman A.A., *et al.* 'Nine months in space: effects on human autonomic cardiovascular regulation'. *Journal of Applied Physiology.* 2000;**89**(3):1039–45.

[187] Seibert F.S., Bernhard F., Stervbo U., *et al.* 'The effect of microgravity on central aortic blood pressure'. *American Journal of Hypertension.* 2018;**31**(11):1183–9.

[188] Villa-Colín J., Shaw T., Toscano W., Cowings P. 'Evaluation of Astroskin bio-monitor during high intensity physical activities'. *Mem Congr Nac Ing Bioméd.* 2018;**5**(1):262–5.

[189] Toscano W., Cowings P., Sullivan P., Martin A., Roy J., Krihak M. *Wearable biosensor monitor to support autonomous crew health and readiness to perform.* 2017. Available from https://ntrs.nasa.gov/search.jsp?R=20190001996 [Accessed 10 May 2021].

[190] Fei D.-Y., Zhao X., Boanca C., *et al.* 'A biomedical sensor system for real-time monitoring of astronauts' physiological parameters during extra-vehicular activities'. *Computers in Biology and Medicine.* 2010;**40**(7):635–42.

[191] Anon. Bio-analyzer: near-real-time biomedical results from space to Earth. 2017. Available from https://www.asc-csa.gc.ca/eng/iss/bio-analyzer.asp.

[192] Zangheri M., Mirasoli M., Guardigli M., *et al.* 'Chemiluminescence-based biosensor for monitoring astronauts' health status during space missions: results from the International Space Station'. *Biosensors and Bioelectronics.* 2019;**129**:260–8.

[193] Parlak O., Keene S.T., Marais A., Curto V.F., Salleo A. 'Molecularly selective nanoporous membrane-based wearable organic electro-chemical device for noninvasive cortisol sensing'. *Science Advances.* 2018;**4**(7):eaar2904.

[194] Anastasova S., Kassanos P., Yang G.-Z. 'Electrochemical sensor designs for biomedical implants' in Yang G.-Z. (ed.). *Implantable Sensors and Systems: From Theory to Practice.* Cham: Springer International Publishing; 2018. pp. 19–98.

[195] Liu Y., Pharr M., Salvatore G.A. 'Lab-on-skin: a review of flexible and Stretchable electronics for wearable health monitoring'. *ACS Nano.* 2017;**11**(10):9614–35.

[196] Costa J.C., Pouryazdan A., Panidi J., *et al.* 'Flexible IGZO TFTs and their suitability for space applications'. *IEEE Journal of the Electron Devices Society.* 2019;**7**:1182–90.

[197] Rosa B.M.G., Yang G.-Z. 'A flexible wearable device for measurement of cardiac, electrodermal, and motion parameters in mental health-care applications'. *IEEE Journal of Biomedical and Health Informatics.* 2019;**23**(6):2276–85.

[198] Rosa B.G., Anastasova-Ivanova S., Yang G.-Z. 'A low-powered and wearable device for monitoring sleep through electrical, chemical and motion signals recorded over the head'. *BioCAS.* IEEE Biomedical Circuits and Systems Conference; 2019. pp. 1–4.

[199] Rosa B.M.G., Anastasova-Ivanova S., Yang G.-Z. 'NFC-powered flexible chest patch for fast assessment of cardiac, hemodynamic, and endo-crine parameters'. *IEEE Transactions on Biomedical Circuits Systems.* 2019;**13**(6):1603–14.

[200] Kassanos P., Anastasova S., Yang G.-Z. 'A low-cost amperometric glucose sensor based on PCB technology'. *2018 IEEE Sensors*; 2018. pp. 1–4.

[201] Kassanos P., Iles R.K., Bayford R.H., Demosthenous A. 'Towards the development of an electrochemical biosensor for hCGβ detection'. *Physiological Measurement.* 2008;**29**(6):S241–.

[202] Anastasova S., Kassanos P., Yang G.-Z. 'Multi-parametric rigid and flex-ible, low-cost, disposable sensing platforms for biomedical applications'. *Biosensors Bioelectronics.* 2018;**102**:668–75.

[203] Kassanos P., Bayford R.H., Demosthenous A. 'Optimization of bipolar and tetrapolar impedance biosensors'. *Proceedings of 2010 IEEE International Symposium on Circuits and Systems*; 2010. pp. 1512–5.

[204] Kassanos P., Bayford R.H., Demosthenous A. 'Towards an optimized de-sign for tetrapolar affinity-based impedimetric immunosensors for lab-on-a-chip applications'. *2008 IEEE Biomedical Circuits and Systems Conference*; 2008. pp. 141–4.

[205] Kassanos P., Yang G.-Z., Ip HMD. 'A tetrapolar bio-impedance sensing system for gastrointestinal tract monitoring'. *2015 IEEE 12th International Conference on Wearable and Implantable Body Sensor Networks (BSN)*; 2015. pp. 1–6.

[206] Kassanos P., Anastasova S., Yang G.-Z. 'Electrical and physical sensors for biomedical implants' in Yang G.-Z. (ed.). *Implantable Sensors and Systems: From Theory to Practice*. Cham: Springer International Publishing; 2018. pp. 99–195.

[207] Gernhardt M.L., Jones J.A., Scheuring R.A., Abercromby A.F., Tuxhorn J.A., Norcross J.R. 'Risk of compromised EVA performance and crew health due to inadequate EVA suit systems'. *NASA's Human Research Program Evidence Book*; 2008. pp. 334–58.

[208] Compton C. *Fit for space: leveraging a novel skin contact measurement technique toward a more efficient liquid cooled garment*. Minnesota, USA: University of Minnesota; 2016. Available from http://conservancy.umn.edu/handle/11299/183307.

[209] Nunneley S.A. 'Water cooled garments: a review'. *Space Life Sciences*. 1970;2(3):335–60.

[210] Ferl J., Hewes L., Aitchison L., Koscheyev V., Leon G., Hodgson E. 'Trade study of an exploration cooling garment'. *SAE Technical Papers*. 2008;**724**:776–90.

[211] Koscheyev V.S., Leon G.R., Treviño R.C. 'An advanced physiological based shortened liquid cooling/warming garment for comfort management in routine and emergency EVA'. *SAE Technical Papers*. 2002;**724**.

[212] Yang K., Jiao Ming-Li., Chen Yi-Song., Li J., Zhang Wei-Yuan. 'Study on heat transfer of liquid cooling garment based on a novel thermal manikin'. *International Journal of Clothing Science and Technology*. 2008;**20**(5):289–98.

[213] Koscheyev V.S., Coca A., Leon G.R. 'Overview of physiological principles to support thermal balance and comfort of astronauts in open space and on planetary surfaces'. *Acta Astronautica*. 2007;**60**(4-7 SPEC. ISS):479–87.

[214] Daniels K. Thermal Performance Analysis of the Liquid Cooling and Ventilation Garment (LCVG) with Respect to Tubing Geometry. *Thesis Diss.* 2019. Available from https://commons.und.edu/theses/2451.

[215] Cao H., Branson D.H., Peksoz S., Nam J., Farr C.A. 'Fabric selection for a liquid cooling garment'. *Textile Research Journal*. 2006;**76**(7):587–95.

[216] Koscheyev V.S., Leon G.R., Dancisak M.J. *Multi-zone cooling/warming garment [online]*. US7089995B2. 2006. Available from https://patents.google.com/patent/US7089995B2/en [Accessed 10/05/2021].

[217] Koscheyev V.S., Leon G.R., Coca A., Ferl J., Graziosi D. 'Comparison of shortened and standard liquid cooling garments to provide physiological and subjective comfort during EVA'. *SAE Technical Papers*. 2004;**724**.

[218] Trevino L.A., Bue G., Orndoff E., Kesterson M., Connell J.W., Smith J.G. 'Flexible fabrics with high thermal conductivity for advanced spacesuits'. *SAE Technical Papers*. 2006;**724**.

[219]   Tao X. 'Smart fibres, fabrics and clothing. smart fibres, fabrics and clothing'. 2001.

[220]   Tabor J., Chatterjee K., Ghosh T.K. 'Smart textile-based personal thermal comfort systems: current status and potential solutions'. *Advanced Materials Technologies*. 2020;**1901155**:1–40.

[221]   Bartkowiak G., Dąbrowska A., Włodarczyk B. 'Construction of a garment for an integrated liquid cooling system'. *Textile Research Journal*. 2015;**85**(17):1809–16.

[222]   Hong S., Gu Y., Seo J.K., *et al.* 'Wearable thermoelectrics for personalized thermoregulation'. *Science Advances*. 2019;**5**(5):eaaw0536.

[223]   Yang G.-Z. *Body Sensor Networks*. 2nd ed. London: Springer-Verlag; 2014.

[224]   Yang G.-Z. *Implantable sensors and systems: from theory to practice*. Cham: Springer International Publishing AG; 2018.

Chapter 10

Future of human–robot interaction in space

Stephanie Sze Ting Pau[1], Judith-Irina Buchheim[2], Daniel Freer[1], and Guang-Zhong Yang[1,3]

Space is a unique environment where experts agree that robots should be sent to explore before humans – regardless of the level of autonomy. Space robots are expected to perform 3D (dull, difficult and dangerous) tasks [1], in an environment where humans have not been. Uncrewed missions have always preceded crewed missions to avoid exposing astronauts to unnecessary risks. Through uncrewed missions, hazards and risks to crewed missions are identified in unexplored territories – in terms of environment and mission concepts.

The acceptance of the use of robots in space is very high, however, space robots are still mostly deployed as passive tools in space commanded by humans. The use of autonomous robots has been limited by deployable technologies (e.g. computing, power) and circumstances of operations: unexplored environments and limited maintenance/repair options. Presently, autonomy that space robots have is usually a localized action that is designed for safety – to protect itself and its surroundings from being damaged, e.g. hazard avoidance software onboard Mars rovers. Uncrewed missions are far from being performed by fully autonomous robots. Humans remain an essential part of both crewed and uncrewed missions – especially in planning and in reacting to unexpected circumstances.

However, the landscape of autonomous technologies is changing. As developments in artificial intelligence (AI) advances and matures, autonomous systems attract interests and investments, which in turn increases the pace of advancement. At the same time, the rise of commercial activities to utilize the space environment as a resource drives additional demands for low-cost operations, where the autonomous robots would be the answer. Crew and robot autonomy are increasingly investigated and considered to ensure mission success. As agencies are pushing crewed missions further from Earth, the challenge of intermittent and delayed

[1]The Hamlyn Centre, Imperial College London, London SW7 2AZ, United Kingdom
[2]Laboratory of Translational Research "Stress and Immunity," Department of Anesthesiology, Hospital of the University of Munich, Ludwig-Maximilians-University (LMU), Marchioninistr. 15, 81377, Munich, Germany
[3]The Institute of Medical Robotics, Shanghai Jiao Tong University, Shanghai 200240, China

communications make teleoperation challenging. Unlike rovers that are presently operating on Mars mostly in teleoperations mode, humans cannot simply switch into an energy-saving mode and wait for an answer to be computed on and transmitted from Earth once per day and overstay their mission by 60 times like the Mars rover Opportunity did. Beyond commercial factors, the need for autonomous systems for the future of space is, paradoxically, driven by crewed missions.

The rise of autonomous space systems drives the need to expand and enrich the field of human–robot interaction (HRI). As opposed to becoming obsolete in the face of autonomous systems, HRI research becomes more important as autonomous systems are far from being a self-planning and self-actuating agent – on Earth or in Space. New tools and approaches are necessary for the successful adoption and integration of autonomous space robots in new human–robot teaming paradigms and in the shift from teleoperations to shared autonomy.

This chapter investigates the existing challenges, trends in capabilities, future opportunities and motivations for developments of HRI in space.

10.1   The challenge of human–robot interaction in space

A human–robot system consists of the agents (humans and robots) and the inter-agent interfaces (human–human, human–robot, robot–robot). A shift in the agents will require a corresponding adjustment in the relevant interface. With the rise of autonomous robotics systems in space, it is crucial to revisit the human–robot interface.

A human–robot interface is one that enables the human and robot to communicate with each other and to work productively as a system and/or a team. The quality of that communication depends on the capability of the human, the capability of the robot and, just as importantly, the capability of the human–robot interface [2], all of which are influenced by the operating environment each entity is situated in (Figure 10.1).

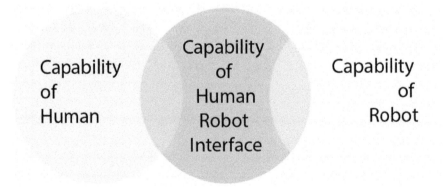

Figure 10.1   *The interface as a bridge between the capabilities of the human and robot*

The DIKW pyramid

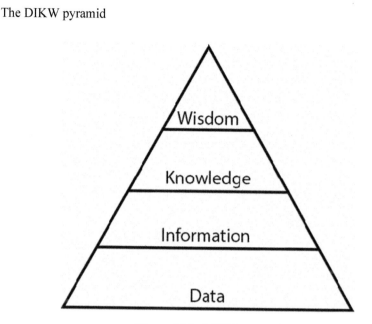

Figure 10.2 DIKW pyramid
The DIKW hierarchy is widely used in many domains, including by information architects. In DIKW, each level depends on and advances from the processing of the levels below it. It is also referred to as the wisdom hierarchy, knowledge hierarchy, or information hierarchy depending on the disciplinary perspectives [3] and we will refer to it as the DIKW pyramid to be neutral.

The meaning of the capability of a human–robot interface can be derived by taking apart the following definition:

A human–robot interface brokers the exchange of insights and decisions between the agents situated in their operating environments, in a productive way.

Exchange of Insights: Insight is a deliberate choice of word. What do we mean by a robot's insights? It depends on the intelligence the robot has. We propose that the robot's insight can be described by the DIKW (data, information, knowledge and wisdom) pyramid (Figure 10.2).

We propose that the least intelligent robot presents data as insights, with rising intelligence the robot presents insights in forms that are higher up in the DIKW pyramid; the fully intelligent robot presents wisdom. Physically, the robot's insights can be presented to any senses of the human agent – most commonly in the form of visual representation on a screen, or more recently as haptics or other more immersive visual representation such as augmented and virtual reality (VR).

As exchange goes both ways, how does a human present an insight to a robot? In terms of the DIKW pyramid, at present it is most likely data or commands (as a result of information, knowledge and wisdom) – where commands can be implicit or explicit. Explicit data and commands are widely practiced. For example, when a human enters data or executes some code via the interface or using joysticks and buttons, game controllers, tablets and exoskeletons with/without haptics feedback to command the space robots. Implicit data/commands can be generated by physiological sensors or inferred information (such as intention and stress) from human physiological signals. Interfaces that use implicit human–computer interaction (HCI) consist of human physiological sensors, e.g. sensors for sweat content, ECG, skin conductance, brain–computer interfacing – measuring EEG, fNIRS and EOG, etc. and algorithms that analyse such data into higher-order knowledge.

Decision: Insights are exchanged between the agents in order to support decision-making – unless the decision has been made by an external system, e.g. a simulation outside of the human–robot team.

In teleoperation mode, decisions are mostly made by human agent(s) – in consultation with other human agents. It is important to note this aspect of consultation with other human agents, as it is more common than not for space operations. With the rise of autonomous space robots, the decision-making process is set to change, disruptively, to shared autonomy modes with highly autonomous robots.

Brokerage: While the exchange is about various forms of insights and decision transmissions, the brokerage is about context awareness to support a successful navigation of the insights and decisions.

A good broker (i) is selective in what is prioritized for exchange depending on the context of all of the agents and (ii) has an appropriate physicality (shape, form, mechanics, graphics design) for the human. After all, the interface is also a connection between the different physical forms and the operating environments of the robot and human agents. Standard interface technologies include visual feedback on-screen and explicit data/command input using buttons and computer interfaces. More novel forms of interface technologies include haptics feedback, mixed-reality with Microsoft HoloLens [4] and brain–computer interface (BCI) [5]. Mixed reality was demonstrated onboard the International Space Station (ISS), despite the experiment was unrelated to robotic control.

Finally, the effects of the environment in which the agents are situated for the operation significantly affect the capability of the agents. To fully understand HRI in space, we must understand the context: the human and robot capabilities, their activities and the implications of the operating environments.

10.1.1   Humans, the complexity of space operations

The operation of using the robotic arm to support the docking with the ISS is one of the most stressful HRI scenarios in space – as the time pressure and the stakes are high. While we have not been able to obtain latest mission transcripts to illustrate this specific event, the complexity of a docking event can be illustrated by a lunar module transposition and docking event in Apollo 14. During this event, the command capsule aims to dock with the lunar module, facilitated by mission control on the ground (capsule communicator is the only nominated spokesperson from

Table 10.1   A short excerpt from the declassified air-to-ground transcripts to illustrate the complexity in robotic operations

Transcript Log

Day-0 03:13:04 Lunar module pilot: And Houston; we're about to dock.
Day-0 03:13:05 Capsule communicator: Roger.
Day-0 03:13:22 Lunar module pilot: We're probably a foot – 18 in to 2 ft out now.
Day-0 03:13:26 Capsule communicator: Roger.
Day-0 03:13:55 Lunar module pilot: And we docked.
Day-0 03:13:57 Capsule communicator: Roger. We could see a slight oscillation.
Day-0 03:14:58 Command module pilot: Okay, Houston. We hit it twice and – Sure looks like we're closing fast enough. I'm going to back out here and try it again.
Day-0 03:15:06 Capsule communicator: Roger.
Day-0 03:16:52 Command module pilot: Man, we'd better back off here and think about this one, Houston.
Day-0 03:16:56 Capsule communicator: Roger.
Day-0 03:16:57 Lunar module pilot: We're unable to get a capture.
Day-0 03:17:00 Capsule communicator: Roger, Ed.
Day-0 03:17:52 Command module pilot: Okay, Houston. We backed out a little bit, and that last time I hit it pretty good and we're just not getting – getting the capture latches in there.
Day-0 03:18:06 Capsule communicator: Roger. We suggest you verify, if you haven't already, the docking probe circuit breakers on panel 8.
Day-0 03:18:12 Command module pilot: That's verified.
Day-0 03:18:15 Capsule communicator: And – stand by 1.
Day-0 03:18:31 Capsule communicator: And, Stu, we suggest you go to EXTEND on the DOCKING PROBE EXTEND/RETRACT switch and check the talkbacks gray.
Day-0 03:18:40 Command module pilot: Okay. We did that when we extended them, but we'll sure do it again.
Day-0 03:18:43 Capsule communicator: And then back to RETRACT.
Day-0 03:18:58 Command module pilot: Okay. We get both gray in the EXTEND position.
Day-0 03:19:02 Capsule communicator: Roger.
Day-0 03:19:06 Command module pilot: And we go RETRACT and both gray.
Day-0 03:19:18 Capsule communicator: 14, Houston. One other suggestion. Go to panel 229 and check the EPS group 4 circuit breakers.
Day-0 03:19:33 Command module pilot: Okay. They're both in.
Day-0 03:19:35 Capsule communicator: Roger.
Day-0 03:20:14 Capsule communicator: 14, Houston. We're about out of ideas here. Suggest you verify you got it – the switch back in RETRACT and then give it another try at docking.

Source: NASA [6]

ground to astronauts). Six attempts were required to successfully achieve capture latch engagement – this is about 90 pages of transcripts (conversations). The script (Table 10.1) is a short excerpt from the declassified air-to-ground transcripts [6] early in the process.

This transcript, covering seven minutes of interaction, has illustrated the complexity in communication in resolving unexpected situations in a docking event between highly trained human agents. It involved one human agent from each of the parts to be docked together (lunar module pilot, command module pilot), one human agent from mission control on Earth (capsule communicator) and the docking probe. The communication from Earth is simplified by channeling any expert suggestions via the capsule communicator, as the only spokesperson. Much has changed from the Apollo missions and the process for docking has been automated. Such manual operation is now only necessary when there is an error that requires manual intervention. Nevertheless, the interaction between human agents in space and on the ground to develop shared situation awareness remains an important aspect of space missions, especially in resolving unexpected situations.

Typically there are two types of human agents in a space mission: operators in space, e.g. astronauts, and operators on Earth, e.g. mission control. As well as their location (thus environmental implication on the capability of the agent), the two types of agents also differ in their role.

Operators in space are highly trained generalists. Operating space robots, such as using the robotic arms attached to the ISS to move cargo, is one of the many tasks they can perform. The astronauts also perform science experiments on behalf of the researchers on Earth, carry out maintenance tasks on the space infrastructure and keep themselves healthy with exercises (2 h per day). All tasks are performed to a carefully prescribed schedule, mostly with step-by-step details. The nature of the role is demanding – time pressure, the ability to memorize and recall a wide range of knowledge and executing experiments and maintenance work with precision.

Most tasks performed by astronauts are unrelated to operating robotics; but when robots must be operated, the location of operation matters as space is not one homogeneous environment. Presently, operations from space take place onboard the ISS in Low Earth Orbit (LEO). In the future, astronauts can be expected to operate robots during spacewalks (outside of the ISS), in the orbit of the moon or Mars, or on planetary surfaces such as the Moon. In orbit (of the Moon, Mars or Earth), microgravity, high-contrast lighting conditions and noise (created by the space station) characterize this space environment. As a result, the human neuro-sensory system, sensory-motor system and motion are altered. In terms of neuro-sensory system, changes in vestibular function, the five senses and proprioception affect the ability to perform a task. Anatomical changes such as "optic disc edema, globe flattening, choroidal folds, hyperopic shifts and raised intracranial pressure" [7, p. 105] result in varying degrees of visual impairment – observed as a transient or persistent effect in about 1 in 5 astronauts. In microgravity, stimuli associated with proprioception are modified and impact spatial orientation suggestive of "degradation in proprioceptive function, or an inaccurate external spatial map, or both" [7, p. 111]. Sensory-motor system coordination, too, is affected but typically locomotion in microgravity poses is quickly learned [7, p. 116]. Motion is affected by Newton's third law of motion: for every action, there is an equal and opposite reaction. In microgravity, astronauts must anchor themselves when they are applying forces – to move a joystick, type on a

Figure 10.3   John Space Center mission control (Source: NASA)

keyboard, etc. Indeed, any impulse on the ISS, such as those introduced by a vehicle docking, thruster firing, or even a crew push-off or crew landing introduces relatively brief and high-magnitude accelerations [8]. Lastly, the impact of the spacesuit on HRI must also be considered. When astronauts perform extravehicular activities (EVAs), the vacuum environment means pressurized spacesuits with life support must be worn. The spacesuit significantly limits the human agent's mobility and dexterity, thus any interface designed to be operated when a human agent is suited must account for mobility and dexterity limitations. At the same time, the spacesuit is an interface linking between the human agent and the space vehicle/station.

Operators on Earth, in contrast to astronauts, are usually specialists, e.g. engineers and scientists, working in mission controls/operations centers (Figure 10.3) and any external support sites (e.g. for IBM Watson). Each specialist is responsible for a different aspect or sub-system of a mission. The commands issued to the space robot, e.g. rover, are based on complex collective decisions from the specialists and the flight director. Only one person from mission control, the capsule communicator, is nominated to directly liaise with the astronauts.

For operators on Earth, the space environments introduce a communication delay with the robot and ground. The further away the space robot is from the operator, the longer it takes for any commands or information to travel that distance. For Mars, the communication delay is between 4 and 24 min [9]. In addition to the delay caused by the distance traveled by the signal, additional communication issues arise due to bandwidth limits and intermittent signal caused by the loss of line of sight between the communicating satellites, mission control and space robots. The issue of bandwidth is notable when large amounts of data are transferred. In Earth orbit, near-constant communication to the ISS is provided by a constellation of communication satellites in geosynchronous orbits called Tracking and Data Relay Satellites (TDRS).

Like the astronauts, controlling space robots constitutes a small percentage of time for the operators on Earth. In mission control centers, the majority of time is spent on planning and organizing (75 percent for Shuttle mission control workers, with 15 percent devoted to training and education) [10].

10.1.2   Space robots, a technological challenge

In a literature review on the frameworks of robot autonomy with respect to HRI, Beer *et al.* have identified two "school of thoughts" to describe levels of autonomy: (i) higher robot autonomy reduces the frequency of interaction and (ii) higher robot autonomy enables more sophisticated interactions [11].

The frequency of interaction is a slightly misplaced concept for discriminating the levels of autonomy, which is evident in the context of space operations. It is possible for space robots to have extremely infrequent interactions and to be operated entirely in teleoperation mode. Flybys and space probes receive commands only when they need to change course or take actions – e.g. Rosetta was transiting in hibernation for 10 years before it arrived at the target asteroid and was then teleoperated by mission control and Voyager has successfully headed to the edge of the solar system without AI. Rovers too – existing Mars rovers receive highly prescriptive commands once per sol (a Mars day) from mission control and only perform actions for a few hours. In both cases, despite the infrequency of control, the robots have low levels of autonomy. As highlighted by the context of space, it is the unpredictability of the operational environment, the number of unexpected events and the limit of physics (i.e. communication latency and delays) that are driving the frequency of interaction – not necessarily robot autonomy.

Levels of robot autonomy defined by the sophistication of the interaction have been described by Beer *et al.* [11] in terms of types of tasks: sense, plan and act. From which the level of autonomy is defined by whether sense, plan and act are performed by human and/or robot. With sense, plan and act all performed by a human in teleoperation, contrasting with sense, plan and act all performed by a robot in full autonomy with seven levels of shared autonomy in between. Mars rovers do sense, plan and act on tasks that have to do with self-preservation, e.g. not falling over. However, existing rovers are still considered to be largely teleoperated, i.e. not autonomous. This illustrated that the level of autonomy of a robot is not defined by whether it is the human's or the robot's task to sense, plan and act. It is the granularity at which this sense, plan, act takes place that defines the autonomy level of a robot. European Corporate for Space Standardization has defined four levels for autonomy onboard spacecraft associated with a different ontology of the sophistication of interaction [12, 13] – where the sophistication of interaction is defined by the types of plans and executions that are generated onboard the spacecraft: mission control from the ground and supported by a limited capability for safety operations onboard (teleoperation), pre-planned/time-based mission loaded onboard for

Level of Autonomy

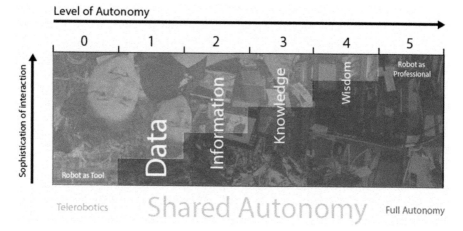

Figure 10.4   Levels of autonomy of space robots (using a picture of astronaut from NASA)

automatic execution, event-based plans for adaptive autonomy and goal-oriented mission re-plans (full autonomy).

To generalize beyond onboard operations for general space robotics (including spacecrafts, rovers and others), we return to the DIKW pyramid, which we have used in the definition of HRI earlier and looked to medical robotics for some inspiration. Yang *et al.* [14] has proposed a model for six levels of autonomy based on tasks types and frequency of control in medical robotics. We took inspiration from the framework, decoupled it from the frequency of control (due to problems discussed above) and amended it with DIKW pyramid to describe levels of autonomy for space robotics.

We proposed six levels of autonomy (Figure 10.4), where the four levels of shared autonomy have the human and robot exchanging in data, information, knowledge and wisdom, in ascending level autonomy and sophistication of interaction. At either end, no insights are exchanged.

- **Level 0:** No autonomy. No negotiation between the human and robot, data and command is the input for the robot. An operator performs the entire task including monitoring, generating performance options, decision-making and execution. The robot performs exactly as the operator commands. Teleoperation shifts more responsibility to the human operator in terms of real-time performance, perception and interpretation, while very little autonomy is left to the robot. The Space Station Robotic Manipulator System (SSRMS), e.g. can be controlled in multiple modes: joint control, end-point control and automatic trajectory control. When using the different control modes, the user must constantly monitor multiple videos of the task space, considering several different frames of reference and coordinate systems [15]. This in itself can be complicated, even before attempting to determine the best way to complete the task at hand. If additionally considering the inclusion of the 14 DoF (Degrees of Freedom) of Dextre, even very simple tasks could turn into non-intuitive and long procedures.

- **Level 1:** Compensatory autonomy. Some negotiation between the human and robot, data is exchanged. Operator maintains control of the system and the task. A robot provides assistance such as automatic safety measures (e.g. collision detection, responding to faults) and motion compensation. Where motion compensation can either refer to the robot automatically moving more quickly when it knows it is safe to do so or compensating for motion artifacts such as vibrations. For example, the automatic trajectory control mode implemented into the SSRMS and Special Purpose Dexterous Manipulator (SPDM) has allowed ground control to more efficiently and intuitively position the robot arms [15]. But even in this more autonomous mode, the control loop is short, so there is higher cognitive demand on all human operators.
- **Level 2:** Task autonomy. Some negotiation between the human and robot, information is exchanged. Operator maintains control of the system. A robot can perform operator-initiated tasks with localized autonomy. For example, the Mars rover Sojourner "has specific behaviours such as a thread-the-needle action that could be triggered in certain situations… It also had contingency capabilities such as reacting to contact bump sensors or not reaching a destination" [16]. However, the autonomous tasks are short in duration and highly localized. Code and command have to be uploaded every sol for the Mars rovers to complete – opportunity was designed for a 90 sol mission to travel 1 000 m and it overstayed its mission by 60 times of the planned duration and traveled 28 miles.
- **Level 3:** Activity autonomy. Some negotiation between the human and robot, knowledge is exchanged. An operator selects and approves a mission plan, and the robot performs the procedure automatically with oversight from the human operator. Here, the distinction between Level 3 and Level 2 is largely the intelligence of the robot. While Level 2 robots would be able to reliably perform a series of tasks that are explicitly defined, Level 3 robots could do so at a higher level of abstraction, e.g. to track and move toward a specific target or to plan a safe path to a specified destination. Such advances are currently being investigated in rovers for future space missions [16]. Adding this intelligence would allow rovers to complete longer and more meaningful tasks, speeding up missions and providing more valuable data.
- **Level 4:** Procedure autonomy. Some negotiation between the human and robot, wisdom is exchanged. A robot can use incoming information to make decisions about a task and perform new actions, without additional input from a qualified

human operator. However, the human operator may override and command the robot to begin a different task if and when necessary.

- **Level 5:** Full automation. No negotiation between the human and robot, wisdom is exercised. No human needs to be in the loop. A robot can perform an entire mission by itself, including real-time decision-making and planning for new procedures and actions.

Ultimately, the robot's capability of achieving autonomy is limited by the hardware and software that enables the capability, which in turn is severely constrained by the space environment.

Gao and Chien [17] have classified space robots into in-orbit robots and planetary robots. Indeed, each of these two distinctive space environments provides their own challenges for robotic operation. In-orbit robots operate under microgravity, which means Newton's third law must be accounted for in motion planning. On planetary surfaces, robots operate with gravity but must deal with many other technical and environmental factors such as latency and natural phenomena like dust storms. For these reasons, the choice of mechanical structure, software and robotics for mobility significantly differs.

We propose to expand the classification further with an additional dimension: shielded and unshielded (from radiation and thermal extremes). For example, crew interactive mobile companion (CIMON®) is a crew-assist robot that operates within a shielded in-orbit space environment, the ISS, and in contrast a space harpoon operates within an unshielded in-orbit space environment. While most existing planetary robots currently operate only in unshielded environments, future command stations built on the Moon, Mars or beyond can additionally make use of shielded robots.

The computational capability of the robot is distinctly dissimilar in shielded and unshielded space environments. To meet the demands of the harsher radiation and thermal conditions in unshielded environments, robots designed to operate in unshielded space environments have severe limitations on computation and data storage capability. Radiation hardened electronics have an architecture that requires more space, thus limiting the computation available on unshielded space robots, which in turn limits the capability of the software. Robots operating beyond LEO and the space between the Van Allen belts are subject to additional radiation hazards, thus increased limitations. To illustrate, in 2011, the clock speed of the computer aboard the Curiosity rover was about 200 MHz – for comparison, in the same year Intel launched quad-core-processors at above 2 GHz. Often, the onboard logic is complemented by commands issued by ground operators, where computation and expertise are more abundant. A small number of autonomous operations (e.g. moving to a new location, taking pictures, or gathering samples) can be commanded by mission control.

Robots working on the outside of the ISS

Most tasks outside of the ISS utilize robotic arms. Multiple robotic arms are attached to different segments of the ISS. The SSRMS, which is also known as the Canadarm2, is 17 m long and 7 DoF and was launched to the ISS in 2001. It is used to perform large-scale tasks from acceptance of cargo to assembling the different station parts [18]. However, a variety of maintenance tasks require more tools and dexterity, so the SPDM, also known as Dextre, was created. This smaller-scale two-armed robot is attached to the end of Canadarm2 for operation in 2008. Each arm of Dextre has an Orbital Replacement Unit/Tool Changeout Mechanism (OTCM) at the end of seven independently controllable joints, providing maximum utility for the task at hand [15, 19]. The OTCM contains several mechanical and electrical interfaces to accommodate repair and ORU exchange, such as a gripper, drive bolt, socket drive and umbilical socket, each of which can interface with standardized fixtures on equipment [19]. The ISS also has a Japanese Experiment Module Remote Manipulator System (JEM-RMS), which consists of a main arm similar in design to the SSRMS (but smaller), and a small fine arm to act as a more dexterous extension, similar to the SPDM [18]. The JEM-RMS is used to support experiments on the Exposed Facility and to perform maintenance tasks on the Japanese module: Kibo. Finally, European Robotic Arm (ERA), due to launch in Summer 2020, is a co-operative development with the Russian Space Agency. It is the only robotic arm designed with an interface for use during a spacewalk.

However, some tasks still require more dexterity than Dextre has to offer. These cases typically require an astronaut to go outside the ISS to perform an EVA. Significant resources have been invested in the development of robots with advanced dexterous manipulation capabilities that could support and/or replace astronauts during EVAs [20, 21]. Robonaut (NASA) and Eurobot (European Space Agency [ESA]) was launched to ISS for preliminary testing in 2011. To operate the robotic arms is already a complex task. For advanced dexterity, anthropomorphic robots and wearable interfaces are explored. The first design of the Robonaut consisted of a base body with two 7 DoF arms, two 12 DoF five-fingered hands, a 6+ DoF "stinger tail," and a mobile stereo camera platform. While this design served as a great research platform, it never made it aboard the ISS. The Robonaut 2, on the other hand, was sent to the ISS in 2012 and was able to successfully power up and complete initial tasks aboard the ISS such as flipping power switches [20, 21]. After an attempted upgrade to add two 7 DoF legs to the Robonaut 2 [22], however, it was left unable to aid crew members with their daily tasks and was returned to Earth for repairs and upgrades [23]. While the goal is for Robonaut to operate in unshielded environments, the current version has only been designed for operation within the ISS.

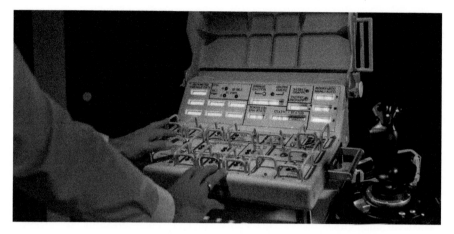

Figure 10.5   *ERA control panel for use in EVAs (Source: ©ESA–SJM Photography)*

In contrast, robots operating in a shielded environment, such as the ISS, can use non-radiation hardened hardware reliably. Indeed, commercial off-the-shelf computing equipment such as laptops and iPads are presently used in the ISS. On the moon and Mars, lava tubes can potentially be classified as a shielded environment. Finally, the size of robots that operate in a shielded environment becomes an important consideration as space for maneuvring is limited. Fortunately, electronics with denser architecture can be used in shielded environments.

10.1.3   Interaction, theory and practice

As outlined in the earlier sections, humans and robots have their corresponding capabilities and constraints in space. A successful human–robot system leverages the capabilities of the agents to overcome limitations. Much of the research and technological advancement in human–robot systems in space is focused on developing capabilities of the robot or the human but rarely, relatively speaking, on the interface between them.

Developments on HRIs are centered around creating better safety, improved performance and trust. Fundamentally, it is the design of the communication between human and robot that underpins these higher objectives. Astronauts are trained in cohorts for a mission so that they learn about each other, as well as the skills they need. They also take language classes to communicate in Russian (necessary for launching in Soyuz, the only human-carrying spacecraft at present, from Star City, Russia). Communication and knowing each other's way of operating and pattern of breaking are part of the critical training for extreme environment operations.

For a human–robot system to operate, there is an equivalent need for communication, which is usually known as interaction, between agents. Donald Norman, in his seminal work, The Design of Everyday Things (formerly called The Psychology of Everyday Things in 1988), has proposed two frameworks for understanding interactions with objects: (i) the interaction between mental models (the designers model, the users model and the system's image) [24, p. 25–32] and (ii) the gulf of execution and evaluation [24, p. 38–40]. Both frameworks have been widely adopted by HCI textbooks.

Mental models, in cognitive psychology, are mental representations created based on the observables. "This model is used to reason about the system, to anticipate system behaviour and to explain why the system reacts as it does." The mental models are a proxy to the true process that is not externally exposed. Norman highlights that there are not two but three mental models involved in an interaction – the user's model (the way the operator reason about the system), the system image (the way the robot presents itself externally) and the designer's conceptual model (the way the designer reason about the system). To paraphrase Norman for our context: for an understandable, useful and possibly error-free interaction, it is the alignment of the mental models of the user and the designer that is important, Therefore, the system image must express and signify the conceptual model of the designer as clearly as possible to the user to facilitate appropriate usage and behavior. To facilitate the development of the mental model, there are several concepts that are useful to consider when designing the interface: affordances, signifiers and mappings. Affordance is a term originated from James J. Gibson [25], a psychologist, to mean the properties of a physical object that conveys information about how people should interact with them. For example, the control panel of ERA (Figure 10.5) and CANADARM affords button pressing in the event of sudden accelerations: the bars around each individual button prevent the finger from straying into the one next to it; and the chunky buttons afford clumsy actions (either due to movement, deteriorating eyesight or because gloves are worn).

Signifiers are the signals a designer built into the object to help the user to perceive the affordances. For example, on CIMON® (Crew Interactive Mobile CompanioN), the friendly face (Figure 10.6) that winks at you is signifying itself as an interactive object that uses modalities familiar to humans – it understands speech.

An interface with appropriate affordances and signifiers should help the user to discover the designer's conceptual model and learn it as their own user model. It could be argued that for HRI in space, since training is so intense, perhaps the designer's conceptual model is already well transferred thus the need to pay attention to affordance and signifiers are less essential. We argue that it should be the other way round – as discussed earlier, whether it is mission control or astronauts, controlling a robot occupies a small percentage of the tasks that have to be performed. Appropriate affordance and signifiers are useful to facilitate the (re-) discovery of the functions. Natural mapping of the interface elements is what helps the memory recall. Mapping is the design of layout of controls and display; **natural mapping** is a mapping that makes the relationship between the controls, actions and intended results easiest to learn and understand. Natural mapping can take inspiration from cultural norms, biology or well-known standards. At the ISS, mappings that take inspiration from cultural norms or sense of direction may not be as natural as they might sound. For example, the mapping of the controls of Canadarm2 (Figure 10.7) (designed in the 1980s) might not look natural, but it is natural for those who have operated the robotic arm on Space Shuttle, the predecessor of the Canadarm2 on ISS, as the interface is very similar. Indeed, the buttons and knob layout are familiar

Figure 10.6   CIMON®, the graphical interface (Source: ESA/NASA/DLR)

to pilots, as all astronauts have a pilot license, this control panel design is of a natural mapping.

But, of course, the mental model is not a static image. The user's mental model is updated as the interaction provides **feedback**. With good feedback, the user mental model iterates efficiently closer to the designer's conceptual model of operation by re-appropriating meanings to the observed affordances, signifiers and mappings after taking an action.

In critical operations like space missions, multiple feedback mechanisms are employed to mitigate risks and provide resilience. The feedback mechanisms for Canadarm2 are three screens, each providing a narrow and specific view of the

Figure 10.7   Robotic workstations in the cupola (main controls, MSSRMS) and in the laboratory (backup controls, JEMRMS) (Source: NASA)

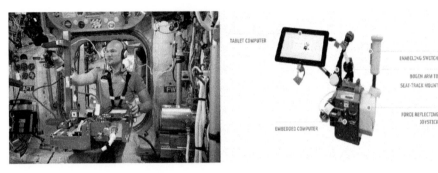

Figure 10.8   *(Left) Haptics-1 control interface (source: ©ESA/NASA); (right)*
Haptics-2 controls interface (source: ©ESA)

robotic arm. As well as these camera views, there are also numbers on the screens that can help the operators to orient. This is important as spatial awareness is tricky with limited camera view and lack of alignment cues [26] – if you can only see black and a few stars behind the arm segment, which way is up? Finally, the interface for Canadarm2 is situated near the window of the cupola, a direct and more holistic view of the robotic arm, moving relative to the frame of the windows and the target object provides a more intuitive (natural), despite not accurate, feedback. Novel forms of multimodal interface and feedback mechanisms (Haptics-1 and Haptics-2, see Figure 10.8) are explored under the METERON project to interact with new rover concepts. These new rovers are designed to be operated by astronauts in orbit close to the surface the rovers are on and are distinctly different from the existing rovers designs: dexterous robotics manipulators, e.g. Justin, a table-mounted humanoid upper body system and two 7-DoF robotic LWR III arms with gripper, are mounted on mobile platforms, forming rollin' Justin and Interact Centaur correspondingly.

Ultimately, the user learns about the interface in order to take an action. **The gulf of execution and the gulf of evaluation** are concepts that describe addressable gaps between the goals (internal objectives) and the world (external actions). The gulf of execution is where the user tries to figure out how things operate and the gulf of evaluation is where the user tries to figure out what happened. Roughly this corresponds to the design (in terms of affordances, signifiers and mappings) of the control mechanisms and the feedback mechanisms, although not exclusively. The larger the gulfs, the more effort is required by the user.

Under the environmental conditions of space, the gulf of execution and evaluation is expanded. The longer the distance between the space robot and the human agent, the longer it takes for the signal to travel between them. The vast distance in space means the latency becomes noticeable, creating a delay in the exchange of insights and feedback. Direct real-time teleoperation is typically only possible with very low latency (0–25 ms), while the latency of transmission from Earth to the ISS is usually several hundred milliseconds, and the latency from Earth to Mars is usually tens of minutes, depending on the distance between Mars and Earth [15]. This makes teleoperation – the predominant mode of HRI in space, at best, very inefficient, at worse, dangerous. One strategy is to create a positive correlation between autonomous functions and the distance between a robot and an operator. This means the further the robot is from the human operator, the more autonomous the robot needs to be. The rovers on Mars must, therefore, contain a large amount of autonomy, or suffer a longer cycle for execution and evaluation, thus lengthening the operation.

In addition to the latency, intermittent communication windows for data transfer [27] introduces additional considerations for a rover mission. At present, this is mostly a problem with scheduling. For Mars rover missions, the communication windows to satellites can occur as few as one or two times per day with lengths depending on visibility, while relaying information via a lander limits the ability of the rover to explore environments beyond the views of the lander.

The gulf of execution and evaluation can be further broken down into three stages of action each. Stages in the process of execution are planning, specifying and performing. Stages in the process of evaluation involve perceiving, interpreting and comparing [24]. In terms of a Mars rover operation, execution of a task can be divided into three distinct phases: first, the engineers at mission control plan the actions for the time period between the next available communication window and the following one; next, the specification of the task is issued to the robot precisely during the communication window; and finally, the performance of the task: in which the rover autonomously carries out a series of subtasks defined in the planning phase. In terms of evaluation, the rovers already contain a certain amount of low-level perception and interpretation which allow them to autonomously react to their environment without direct intervention from a human operator. However, comparisons of intra-sol activities can only be performed by humans using other systems in Earth.

These concepts provide constructs and context to frame the investigation and assessment of HRI research from a human's perspective (user-centered design). For example, haptic feedback can be evaluated by the improvement it has on minimizing the gulf of evaluation. More specifically, did it bridge the comparison, interpretation or perception stage? In what ways? Physiological sensors and BCIs may provide some objective metrics in addition to subjective reports from the user regarding the appropriate design of signifiers and natural mappings as well as the gaps in the gulfs.

10.2 Future of interaction with autonomous robotics in space

10.2.1 Motivations for shared autonomy

The new momentum for autonomous robots in space is driven by commercial activity and technology push.

Space mining (for in situ resource usage) and space debris management are key activities targeted by commercial operators. From well-established companies, such as GMV, to new start-ups, such as Off World, companies are looking to expand their operations on autonomous mining robots developed for Earth to space. Governments, in the US and Luxemburg, have made legislative changes that open the doors to space mining companies. Small satellite manufacturers and operators, such as Astroscale, ClearSpace, SSTL, are racing to develop solutions for space debris collection. Space debris issue was raised by NASA in the 1970s and has made a return to the agenda of the space agencies in 2019 as ESA commissioned the world's first space debris removal mission. The end of NASA support on the ISS in 2024 prompted the rare opportunity for commercialization plans for the ISS and on-orbit servicing and assembly services. Many companies, including SpaceLogistics, Airbus, Space Application Services and Orbital ATK, are heralding the development. Commercial operations, unlike science exploration missions carried out by space agencies, are invested in making viable business. Autonomous robots are seen as one of the ways to keep the cost low in these new types of commercial missions. Meanwhile, NASA and ESA are working toward a crewed orbiter in the Moon orbit (Lunar Gateway), a long-term Moon base with the use of lunar surface resources in the near- to mid-term future, pushing crewed missions to Mars further back. Although SpaceX, a commercial company, has been making claims regarding a more aggressive timeline for crewed missions to Mars. While Mars is far away to incur significant delays in communications and drives a need for more autonomy in robots can improve the efficiency of science missions, the need for autonomous robots on the moon orbit or surface for science missions is less motivated.

The technology push for autonomous robots in space is motivated by the advancement in machine learning software and improvement in computation capabilities. Robots that operate in sheltered environments (cf. Space Robots, a Technology Challenge) or have access to near-constant communication correlates to having better computational power and storage, thus lending itself to the hosting of AI or similar software to support autonomy. The LEO and lunar missions are well-positioned for deploying sophisticated AI onboard the robot. Robots that are operating in the LEO environment can benefit from near-constant communication to the computational power and storage on Earth and act as a thin client of AI. However, with plans like having a high-performance computing (HPC) cluster in-orbit [28] and research on implementing AI with low computational power [29, 30], it is also possible to envision autonomous robots in locations that are not sheltered and are further away from Earth (which will remain the host for most computation and storage resources in the foreseeable future). Improvement on the modes and efficient of power would further support the development of AI in space.

The rise of autonomous robots in space, however, is unlikely to shift all operations from teleoperation (level 0, space robots as tools) to full autonomy (level 5, space robots as independent operators). Autonomous robots are far from creating or setting missions for themselves. Humans will remain in the loop, one way or another, for even the most advanced level of autonomous robots in space. These are the motivations to design for shared autonomy in HRI-AI. We must stress that it is not the case that shared autonomy is the future of HRI in space simply because autonomous technologies are not mature enough. It is because space is an exploration frontier, the idea that an investigation of human futures planned and carried out by fully autonomous robots with zero human agent input and decision requires a non-human centric culture. Culture changes at a much slower pace than technology. Until then, shared autonomy will take a dominant stage.

10.2.2 Capabilities for the future of interaction

The motivations to have increased autonomy in space robotics in turns drive the projected future of HRI in space. The intersection between AI and HCI is not a new topic. HCI is concerned with the interaction between human and computer, or more generally machines in human–machine interaction (HMI), with a strong focus on usability – from the perspective of the human agent. AI is concerned with the evolution of machine intelligence, from the perspective of technology push. The focus is on creating the machine's capability. Schneiderman and Maes debate whether AI should be a metaphor in human interface to computers, others have observed alternating cycles of focus between AI and HCI [31]. These debates implied only one of the philosophies of HCI or AI can be right. However, as alluded to earlier, shared autonomy is the future unless human culture shifts to being non-human centric, HCI and AI, more specifically, embodied HCI and AI, i.e. HRI and autonomous robots need to be advanced in closer collaboration and co-evolve.

The research opportunities lie in evolving the capabilities (models, technologies and autonomy) of human, robot and interface, the interaction paradigm and in the methods and capability for transdisciplinary research. Examples of these can be found on NASA's ISS experiment database. A search for HRI shows 1 000+ results. Of which, excluding the invalid results (not related to human or robot), the majority of the results are related to supporting human health, e.g. ISS Medical Monitoring, and understanding human performance and biology followed by results related to the development of robot capabilities in relation to new mission concepts, e.g. free-flyer platforms [32–34], self-assembling robots and robotic refuel missions [35]. Experiments for human performance has a long history in spaceflight operations, more recently includes the study of neuro-ergonomics.

Beyond the study of human performance and development of robot capabilities, there are experiments conducted to study interface or interaction paradigm. Interface development is mainly focused on the study of natural mappings and feedback mechanisms, e.g. haptics controller (Haptics-1, Haptics-2) and wearable interfaces (Haptics-1, X-Arm-2).

Beyond studying the capabilities of human, robot and interface, experiments were also conducted to explore interaction paradigms, i.e. different configurations of human–robot teaming. For example, an experiment on end-to-end control of robots operating on surfaces with wearable robots on the operator (METRON: SUPVIS-E), one human to multiple robot interactions (SATS-Interact) and fully autonomous operations (Avatar Explore). Last but not least research to understand how to work with the main communication issue in space: working with a time delay (KONTUR).

In the following, we expand on the research opportunities identified, relating to interaction design terminologies that we have introduced in earlier sections – except for autonomous robot capability as it is well-covered in the rest of this book.

10.2.2.1   Research on signifiers from human agent – sensors and neuro-ergonomics

We have discussed how it is important for the human agent to perceive and interact with the robot agent. The signifiers are largely discussed from the perspective of human perceiving robot actions and affordances. With advanced autonomy, research opportunities exist in both directions – not just how humans interact with robots, but how robots interact with humans. Models, tools and metrics need to be developed to understand (i) when and what decisions and actions are best made by human or robot and (ii) when and how information can be easily transferred and consumed between human and robot. New paradigms do not happen if we continue to understand new things using old tools for measurement.

To support the efforts in helping autonomous robots to understand how best to cooperate with human development in BCIs [36], low power physiological sensors (e.g. fast assessment of cardiac, hemodynamic, endocrine parameters and hydration levels [37, 38]) can be used to conceptualize and implement signifiers from the human agent for the autonomous robot agent to work with. This is also known as implicit interaction at ESA – implicit in terms of communication. When the human is considered the controlling agent, this is called intention detection. The development of models for neuro-ergonomics essentially provides the robot or the interface with an ability to extract salient features and interpret decisions and actions and the appropriate response.

10.2.2.2   Research on natural mapping and feedback mechanisms – embodied interaction/humanoids

Humanoids can be considered as attempts to provide a more natural mapping and feedback for mechanical tasks. By basing the function and the form factor of the robots on known human function and form, human operators are expected to have a more natural mapping into understanding the robots' operation, current status and any feedback that the robot may provide. Robonaut 2 and Valkyrie are two humanoids currently controlled using similar techniques to the other robots. These techniques include using custom-developed software to define a sequence of goal poses of the hands, feet or body, with the option to add additional constraints [39]. However, these humanoid robots have the additional benefit of having the human operators compare the robotic movements to movements that a human would perform. In this respect, research into controlling the high DoF of the anthropomorphic hands has considered concepts such as "synergy grasping" [40], which divides different objects into distinct grasp types based on a grasp taxonomy [41]. Different objects can be labeled by human users based on their understanding of how a human would grasp it, and thus can be mapped to a predefined grasping strategy. Real-time grasp type labeling can also be achieved autonomously, by using computer vision to determine the grasp type necessary to interact with a given object [42]. However, as most state-of-the-art strategies for object labeling utilize deep learning and therefore higher computational power, the translation of this technology to unshielded environments must be investigated further.

The search for a more natural embodied interaction continues to push research opportunities in kinematics – sensing, actuation mechanism, control logics. Movement-based interaction has thus far been implemented using exoskeletons, other rigid external devices and computer vision [15] with camera setups, development in soft robotics create a capability for improving the sense of immersion in movement-based interaction, e.g. tendon-based sensing [43]. Additional challenges created by these new research drive the need for high dimensionality and haptic and tactile feedback. Without high-quality feedback to the user, such dexterous and close-contact operation could lead to failed tasks and damage to robotic components.

10.2.2.3   Research to support human capabilities – crew autonomy

As discussed in earlier sections, astronauts are generalists. They are highly trained, but the training was done, in the best case, 6 months before their mission. To carry out experiments in space, the few astronauts in space are supported by hundreds of flight controllers in mission control, each specialized in their own experiments and are available to answer any questions that may arise as the astronaut performs the experiments. Indeed, mission control is not just a support function, they handle most of the running of the ISS station and schedules the astronauts' activities onboard so that astronauts can focus on experiments onboard and maintenance that must be carried out in space. For crew autonomy, astronauts have to be in some way a flight controller.

Research to support crew autonomy is absolutely essential for future deep space missions. Depending on the position of the planets, the round-trip communication between Earth and Mars is 8–48 min. But crew autonomy is not just important for deep space missions, it is also important for missions closer to Earth as astronauts have more situational awareness than on the ground. Crew autonomy would allow better use of crew time, which costs about 10 000 USD per hour.

The support of crew autonomy calls for research on human–AI interaction and development in the interface technologies for shared situation awareness. For example, CIMON® is a project that is aimed at supporting crew autonomy.

10.2.2.4   Research of different interaction paradigms of human–robot teaming

Beyond developments in new interfacing technologies, robots and augmentations of human capabilities, a shift in interaction paradigm can open up research opportunities.

* **Human–Swarm Interaction** Swarms of small robots was raised under NASA's Innovative Advanced Concepts (NIAC) for systematic exploration of small worlds (i.e. Mars's moons Phobos and Deimos) [44] in 2012. HRI between human agents and a robot swarm creates a situation with different cognitive and communication complexity.
* **Human–Robot–Human Interaction** Interaction between human agents mediated by robots is a possibility for teleoperation in discovery missions. Unlike here on Earth, most things are not named or not intuitively named: to pick up

Figure 10.9   NASA's OnSight software created by JPL in collaboration with
Microsoft (Hololens) facilitate scientists and engineers to virtually
walk on Mars (Source: NASA)

a specific rock requires significant orchestration. Naming objects, identifying
targets and actions between human agents are extremely complex tasks in environments where the landscape is seemingly repetitive and lack features – such
as the surface of Mars from the views of a rover. Establishing common grounds
[45–47] is a communication challenge between human and human, mediated
by machines. One possibility is to use object detection by computer vision and
eye gaze tracking to coordinate such actions between the many human agents
on the ground can be a solution. Another possibility is to use mixed reality to
visualize the environment and minimize miscommunication and differences in
data interpretation (Figure 10.9). There is an untapped opportunity for human–
AI interaction to facilitate a shared situation awareness and common ground
between human agents in distributed locations.

New interaction paradigms open-up research development opportunities in
neuro-ergonomics, open-up communication models and HRI theories and provide
opportunities as the extreme context for the need of attention scheduling and establishing common ground [46, 47].

10.2.2.5   Research to simulate operation realism and pressure –
working with time delay

The challenge does not lie only in the creation of technologies but also in finding ways
to simulate operation realism and pressure. It's very easy to get good user feedback in a
lab user test, but operational realism and pressure create a different set of behavior and
stress that cannot easily be observed in lab tests. Space agencies have created many analogue environments, e.g. NEEMO (NASA Extreme Environment Mission Operations).
However, there are more opportunities for researchers to develop smaller scale, low

cost and more localized analogues to simulate the condition of space to ensure research applied to space is relevant to the condition. Delay in communication is an area that is easy to simulate and could easily be more incorporated into research to make the technology more relevant. The time delay to Mars has obvious implications for conversational communication, but the delay to the Moon or ISS is not negligible for robots that operate as thin clients where the servers for AI are on the ground.

10.3 Case study: a future crew assistant

A successful interplanetary mission will critically depend on individual adaptation and performance as well as successful joint crew interaction and appropriate psychological and physiological prevention or countermeasures [48]. Robotic autonomy can be a strategy to reduce workload and therefore reduce stress and its negative attributes by effectively supporting crew. Nicogoassian *et al.* [49] stated in 2001 that "long-duration missions necessitate further technological breakthroughs in tele-operations and autonomous technology". Nearly 20 years later, the demand for a wider range of onboard health care and the need to make non-invasive or minimally invasive stress and health-monitoring tools available to crew is still there. These devices should be capable of monitoring and preventing disease by providing the user with personalized recommendations based on comprehensive, high-quality data. To realize this, existing technologies that were shown to be feasible need to be improved but also innovations are currently being under development. One of these combinations of technology and medicine is the cooperation of human and machine. In science-fiction movies [50, 51] and literature, more or less intelligent robots usually play a significant role. They often support humans in their daily work when exploring space. Throughout the history of space travel, there have been repeated attempts to deploy autonomous robots to assist humans in or outside of spacecraft with limited success, due to the technological feasibility or the performance of hardware and software.

10.3.1 CIMON® – the intelligent astronaut assistant

Today's technology and the growing importance of artificial intelligence make autonomous assistance systems possible for the first time. An idea from the company Airbus in Friedrichshafen for such an autonomous assistance system, intended for use on the ISS, was presented at the Humans in Space Conference in July 2015 in Prague. For the then upcoming horizons mission of German ESA astronaut Alexander Gerst, German Aerospace Center (DLR) Mission Management decided to implement such a technology experiment using artificial intelligence (AI) for the first time.

10.3.1.1 Hypothesis

Apart from numerous scientific experiments that are carried out onboard, maintenance and servicing work is an integral part of the daily routine [52]. An AI that supports the crew in their daily routine, complex experiments and the training of new motor skills could facilitate the crew's everyday work. In the ideal case, an intelligent assistance system will make even better and more efficient use of crew time available in orbit, thereby increasing scientific output. The same can be assumed for training and planning processes if assistance systems in these areas find user acceptance and trust. As mentioned before, groupthink in which group members adapt to the expected, albeit incorrect, group opinion might be prevented by a neutral, non-emotionally involved team member. Crew acceptance, professional handling, autonomous mobility and voice control are critical factors for success. An autonomous robotic sphere was developed equipped with an AI based on "Watson" by IBM. The CIMON® has a size of 32 cm in diameter and 5 kg weight [34] and has access to the AI "Watson" from IBM, which is located on earth. A video screen displays his face (Figure 10.6) that can transport different emotions. He can talk, answer questions and show videos and instructions. In this way, CIMON® creates its own, unmistakable personality. CIMON® took up its function during the horizons mission for the first time in a technology demonstration.

10.3.1.2   Implementation as fast track experiment for the horizons mission

The realization of such a technology was under tight time and budget constraints. Under normal conditions, experiments of this complexity have an average realization time of 5–7 years. However, only 2 years and about half of the required budget from DLR's national program were available for the preparation. After intensive discussions with the industrial partners and the necessary content adjustments, the project was started in July 2016. Design of science experiments and recruitment of volunteers is coordinated by the "Laboratory for Stress and Immunity" at the Department for Anaesthesiology of the University Hospital in Munich and carried out with the help of the ESA. This project posed a particular challenge for all those involved, not only because of its technical complexity, but also because of the necessary fast track implementation, as a whole series of precedents are available. For the first time in the history of the ISS, an autonomous free-flyer is being realized, which must be able to operate safely in all conceivable situations. Due to the large volume of data and because there is no server capability on the ISS, artificial intelligence currently has to be provided from the ground via data down- and uplink. This can result in dialogues with delays of up to 6 s. Additional challenges were posed by data security due to the communication of the free-flying unit with ground, as well as the interface with ESA's Multi-Purpose Communications Computer (MPCC) hub, which is mandatory for all communication of experiments within the COLUMBUS module the European laboratory of the ISS. Furthermore, no suitable flight-qualified European batteries were initially available for CIMON®, which could guarantee more than 1 h of operation. In the course of the discussions with the ESA Safety Panel, an exchange of information with the MIT Astro-Bee project [53], which had to contend with similar technical challenges, was established. The CIMON® project was enabled to use the existing, already qualified lithium-ion batteries onboard the ISS on a loan basis. With the US batteries, an operating time of 2 h or more can be achieved, depending on power consumption or mission profile. In return, the relatively low-noise propellers of the CIMON® air jet propulsion system were adopted for the Astro-Bee project. In addition to the technical challenges, this project also presented safety challenges and safety divisions were needed to accompany all processes in a timely manner and thus ensuring on-schedule delivery. CIMON® was launched to the ISS with SpaceX-15 on June 28, 2018.

10.3.1.3 Functionalities of CIMON® onboard

Successful commissioning and successful proof of safety with the support of ESA astronaut Alexander Gerst (Figure 10.10) took place on the ISS in November 2018.

First experiences with the use as a free-flying, autonomous assistance system have been gained. Initially, technical functions of CIMON® were included in the test portfolio. On the one hand, the focus was on the guidance navigation and control (GNC) system allowing CIMON® to orient himself inside the ISS COLUMBUS module [54]. On the other hand, the feasibility of speech-based communication between human and machine was tested as high background noise levels were possible interference factors. Furthermore, the guarantee of communication via downlink to Earth and the coupling to the AI enabling a fluid conversation played a central role, which completes the function of all human–machine interfaces under real conditions. In addition, the robustness, flexibility and user-friendliness of IBM-Watson is particularly important for the possible further development of autonomous assistance systems. Within the scope of a pilot study, CIMON® shall be evaluated in the coming years in terms of (i) efficiency in assisting routine work, (ii) user satisfaction, (iii) quality as a mobile video documentation system (compared to an otherwise common static camera) and (iv) assistance in learning technical skills. CIMON® will be used exclusively within the COLUMBUS module of the ISS; here, however, it can move autonomously. Voice control allows hands-free work for astronauts and verbal instruction of multi-step procedures. Access to the AI can currently

Figure 10.10   *ESA astronaut Alexander Gerst interacting with CIMON®*
(Source: ESA/NASA/DLR)

only occur when a data connection to ground is available which enables connection to the ground-based AI. The user feedback given in the form of a questionnaire will help to further develop CIMON®'s capabilities and personality and to adapt it even better to the wishes and needs of astronauts. This study will help to gain first insights into whether CIMON® and its AI can be a valuable, efficient and sympathetic team support and could even be considered as a crew member. In the course of the study, its characteristics shall be continuously developed and improved by entrusting CIMON® also with entirely autonomous functions with the scope to reduce daily stress and a monotonous working atmosphere. Over the course of time, CIMON® is expected to evolve, not only his character and personality through crew interaction but also in terms of tasks assigned and add-on functionalities.

10.3.2   The case for crew assistance robot – for space and earth

The return to the lunar surface with the construction of the Lunar Orbital Platform – Gateway (LOP-G), a new space station closer to the Earth's Moon, is a joint goal for the space agencies of the USA (NASA), Europe (ESA), Russia (Roscosmos), Canada (CSA) and Japan (JAXA) [55, 56]. Despite being healthy and very well-trained professionals, astronauts are affected by long-term space missions due to the multitude of stressors that characterize the special living environment "space." By moving further away from Earth, it is likely that the strain for humans working and living on the Gateway will increase even more compared to the current exposure at the ISS. Typical known stressors are isolation, confinement, noise, circadian rhythm misalignment, microgravity, sleep disturbances and radiation. These stressors pose a risk to astronauts' health and have triggered the establishment of prevention strategies such as a strict quarantine prior to flight or daily physical training to maintain muscle mass in space. The perception of stress over a longer period of time can lead to psychological impairment which is a major risk factor for manned deep space missions [57]. Moreover, group dynamics such as peer pressure or the so-called groupthink effect can impair objective reasoning and problem-solving capabilities or disrupt communication [58, 59]. The characteristics can be overestimation of one's own capabilities, high group pressure toward uniformity, not questioning the leader's decisions and an overcritical view of outsiders, i.e. people outside their own group. But chronic exposure to stress also triggers the release of classical stress hormones like cortisol driven by the hypothalamic-pituitary-adrenal (HPA) axis or the catecholamines norepinephrine and epinephrine, that in turn can suppress immune responses [60]. Increased susceptibility to infection was reported already during NASA's Apollo missions [61–64] and a strong body of evidence supports the fact that the reactivation of dormant viruses such as Herpes simplex or Epstein Barr virus is common in flight [65–67]. Furthermore, a high incidence of allergic symptoms and skin rashes was reported among space travelers [68, 69] and the ability to adequately control inflammation is compromised [70]. There is only limited knowledge of the mechanisms behind these immune alterations and their impact on the re-adaptation to the conditions on Earth once an astronaut returned from space [71]. Nevertheless, similar effects, e.g. viral reactivation and the shift toward a proinflammatory immune phenotype can also be observed in the so-called spaceflight analogues [72, 73], research platforms that allow studying of spaceflight stressors on Earth. Data from long-term isolation studies (Mars500), hibernation studies in Antarctica or short-term exposure studies during parabolic flights show the complex interaction between stress, body function, cognitive abilities and behavior also separately from the ISS [74–80]. The success of long-term missions and the safe return to Earth will therefore depend not only on an excellently trained crew, but also decisively on the countermeasures available to avoid stress and its negative effects [48].

Why should we be concerned with these observations and diseases, if they seem to affect only a very little percentage of the world's population, e.g. astronauts? Besides microgravity, most of these stressors are not so different from the stressors the general population is exposed to every day due to our "modern" lifestyle. It is known that chronic exposure to stress greatly impairs immune and cognitive function [81]. Sleep quality has become a major public health topic [82–84] since poor sleep quality seems to interfere with many physiological functions of the human body and can trigger diseases with high socioeconomic impact such as type 2 diabetes [85], hypertension [86] or coronary heart disease [87]. A wide variety of smart phone applications or smart devices are now available that can be used to monitor individual sleep duration and quality. Furthermore, the prevalence of allergic diseases has continued to rise worldwide with an estimated sensitization rate of 40–50 percent in school children [88, 89]. An overview of statistics on allergic diseases can be found on the web page of the American Academy of Allergy, Asthma, and Immunology (AAAI) [90]. Interestingly, constant exposure to a stressful environment together with a constantly present immune activation can result in a pathogenic phenotype showing features of premature immune aging [91–93]. This so-called inflammaging is a low-grade, pro-inflammatory state that renders its host at risk for latent viral infections, allergic or autoimmune disease as seen often in elderly patients [94–97]. So it seems a stressful lifestyle and a change in environment can trigger similar diseases in healthy astronauts that we detect in growing numbers here on Earth. The major difference is, however, that we see these symptoms and immune changes not in elderly astronauts but already after 6 months of exposure to spaceflight and are, at least in part, reversible [70, 98]. That is why astronauts and their specific living and working conditions are important to study. Furthermore, we can test the effectiveness of potential countermeasures that have the purpose to prevent stress-related immune alterations. It is likely that missions to the Moon or even Mars with a much longer duration will have more severe implications. Therefore, a very specific countermeasures protocol for deep space exploration was set up by the research community focusing on the immune system in Space [99]. Most notably, stress-relieving strategies, such as breathing exercises are mentioned, showing that novel, non-invasive countermeasure strategies are becoming more popular among astronauts.

10.4 Recommendations and trends

No space technology discussion ends without a transfer of technology and applications to Earth. Space is an extreme environment in which early developments of human autonomous-system interaction will be adopted and demonstrated beyond the laboratory, as alluded to by the case study of CIMON®. The ongoing research efforts in HRI for space can result in new paradigms, mental models and technologies for human autonomous-robot interaction, which can be applied to human-expert autonomous-system interaction, on Earth, e.g. in robotic surgery. Surgeons, like astronauts, are highly selected and trained individuals, performing tasks in a relatively well-controlled environment. Often, to complete the tasks, endurance is required by the astronauts and surgeons. Internal factors such as stress and external factors, such as situation awareness to complete tasks under high pressure and conditions of endurance are critical areas of study. In contrast, research for the future of HRI in space does not apply to other human–AI interaction where the operators and the environment of operation are not comparable. A temptation is to extend the concepts of autonomy from autonomous vehicles to the space robots – after all, most space robots in space are a type of vehicle (free-flyers, rovers, orbiters, flybys), apart from robotic manipulators, such as the robot arms attached to the outside of ISS. Autonomous vehicles must operate in a relatively unconstrained and interruptive environment, while in general their human operators are not required to develop the deep expertise about the subject matter. Therefore, we anticipate that the exchange of HRI knowledge between space professionals and terrestrial medical professionals will benefit both fields with great synergy. CIMON® illustrates the many opportunities for HRI in space – primarily on the challenge of implementing AI, modes and models for exchanging insights between agents, development of robotics technologies and identifying models for transitioning between different levels of autonomy.

At the start of the chapter, we reviewed some key concepts for HRI originating from design studies, e.g. signifiers, natural mapping, feedback and mental models, and illustrated the complexity introduced by the space environment to HRI. To create the future of HRI in space, we anticipate new research opportunities in advancing technologies and sciences in (i) sensing technologies and neuro-ergonomics – for designing better signifiers; (ii) embodied interaction/humanoids – for designing natural mapping and feedback mechanisms and (iii) developing an understanding of the different interaction paradigm of human–robot teaming, with applications in (i) crew autonomy, (ii) robot autonomy and (iii) complex human–human interaction. To support the development of these applications, auxiliary research opportunities in creating a small scale, low-cost environment for simulating operation realism and pressure, such as working with time delay would facilitate the efficient development of knowledge for HRI.

References

[1]   GMV. *Robotics technology control, navigation and manipulation* [online]. Available from https://www.gmv.com/en/Sectors/space/Space_Segment/Autonomy_and_Robotics.html [Accessed 08 Jan 2020].
[2]   JPL. *Robotics: user interfaces* [online]. Available from https://www-robotics.jpl.nasa.gov/applications/applicationArea.cfm?App=11 [Accessed 08 Jan 2020].
[3]   Rowley J. 'The wisdom hierarchy: representations of the DIKW hierarchy'. *Journal of Information Science.* 2007;**33**(2):163–80.
[4]   Norris J. *Sidekick: investigating immersive visualization capabilities* [online]. 2016. Available from https://www.nasa.gov/mission_pages/station/research/experiments/explorer/Investigation.html? [Accessed 01 May 2020].

[5]   de Negueruela C., Broschart M., Menon C., del R. Millán J. 'Brain–computer interfaces for space applications'. *Personal and Ubiquitous Computing.* 2011;**15**(5):527–37.

[6]   Apollo Spacecraft Program Office. *APOLLO 14 TECHNICAL AIR-TO-GROUND VOICE TRANSCRIPTION*; 1971.

[7]   Clément G. *Fundamentals of space medicine space technology.* New York: Springer-Verlag; 2011.

[8]   McPherson K., Hrovat K., Kelly E., Keller J. *A Researcher's Guide to International Space Station Acceleration Environment.* NASA ISS Program Science Office; 2015.

[9]   ESA. *Time delay between Mars and Earth* [online]. 2012. Available from http://blogs.esa.int/mex/2012/08/05/time-delay-between-mars-and-earth/ [Accessed 08 Jan 2020].

[10] NASA. The people behind the astronauts [online]. 2009. Available from https://www.nasa.gov/audience/foreducators/k-4/features/F_People_Behind_the_Astronauts.html [Accessed 12 Dec 2019].

[11] Beer J.M., Fisk A.D., Rogers W.A. 'Toward a framework for levels of robot autonomy in human-robot interaction'. *Journal of Human-Robot Interaction.* 2014;**3**(2):74–99.

[12] ESA Requirements and Standards Division. *ECSS-E-ST-70-11C – Space segment operability.* 2008. Available from https://ecss.nl/standard/ecss-e-st-70-11c-space-segment-operability/ [Accessed 1 Oct 2020].

[13] UK-RAS. *Space Robotics & Autonomous Systems: Widening the horizon of space exploration*; 2018.

[14] Yang G.-Z., Cambias J., Cleary K., *et al.* 'Medical robotics – regulatory, ethical, and legal considerations for increasing levels of autonomy'. *Science Robotics.* 2017;**2**(4):eaam8638.

[15] Mishkin A., Fong T., Akin D.L., Currie N., Rochlis Zumbado J. 'Space telerobotics: unique challenges to human-robot collaboration in space'. *Reviews of Human Factors and Ergonomics.* 2013;**9**(1):6–56.

[16] Bajracharya M., Maimone M.W., Helmick D. 'Autonomy for Mars rovers: past, present, and future'. *Computer.* 2008;**41**(12):44–50.

[17] Gao Y., Chien S. 'Review on space robotics: toward top-level science through space exploration'. *Science Robotics.* 2017;**2**(7):eaan5074.

[18] Visinsky M. *Robotics on the International Space Station (ISS) and Lessons in Progress.* 2017. Available from https://ntrs.nasa.gov/api/citations/20170002575/downloads/20170002575.pdf?attachment=true [Accessed 1 Oct 2019].

[19] Callen P. *Robotic Transfer and Interfaces for External ISS Payloads.* 3rd Annual ISS Research and Development Conference; Chicago, Illinois, USA, July 2014; 2014.

[20] Hoffman R.R., Elliott L.R. 'Using scenario-based envisioning to explore wearable flexible displays'. *Ergonomics in Design: The Quarterly of Human Factors Applications.* 2010;**18**(4):4–8.

[21] Diftler M.A., Ahlstrom T.D., Ambrose R.O. Robonaut 2 – initial activities on-board the ISS. 2012 IEEE Aerospace Conference; Montana, USA, 2012; 2012. pp. 1–12.

[22] Diftler M.A., Badger J., Joyce C., Potter E., Pike L. *Robonaut 2 – building a robot on the International Space Station*. 2015 ISS Research and Development Conference; Boston, USA, July 2015; 2015.

[23] Ackerman E., Guizzo E. 'Robonaut returns for repairs [News]'. *IEEE Spectrum*. 2018;**55**(4):7–8.

[24] Gibson J. *Senses Considered as Perceptual Systems*. Praeger; 1983.

[25] Currie N.J., Peacock B. 'International Space Station robotic systems operations – a human factors perspective'. *Proceedings of the Human Factors and Ergonomics Society Annual Meeting*. 2002;**46**(1):26–30.

[26] Maurette M. 'Mars Rover autonomous navigation'. *Autonomous Robots*. 2003;**14**(2):199–208.

[27] Norman D.A. *The Design of Everyday Things (The MIT Press)*; 2013.

[28] HPE. *Accelerating space exploration with the spaceborne computer* [online]. 2020. Available from https://www.hpe.com/us/en/newsroom/Accelerating-space-exploration-with-the-Spaceborne-Computer.html [Accessed 08 Jan 2020].

[29] Blacker P., Bridges C.P., Hadfield S. *Rapid prototyping of deep learning models on radiation hardened CPUs*. 2019 NASA/ESA Conference on Adaptive Hardware and Systems (AHS); 2019. pp. 25–32.

[30] Mazouz A., Bridges C.P. *Adaptive hardware reconfiguration for performance tradeoffs in CNNs*. 2019 NASA/ESA Conference on Adaptive Hardware and Systems (AHS); 2019. pp. 33–40.

[31] Winograd T. 'Shifting viewpoints: artificial intelligence and human–computer interaction'. *Artificial Intelligence*. 2006;**170**(18):1256–8.

[32] Micire M., Fong T., Morse T. *Smart SPHERES: a Telerobotic Free-Flyer for Intravehicular Activities in Space*. AIAA Space 2013 Conference and Exposition; San Diego, CA, USA, 10-12 September 2013; 2013.

[33] Bualat M.G., Smith T., Smith E.E., Fong T., Wheeler D.W. *Astrobee: A New Tool for ISS Operations*. 2018 SpaceOps Conference; Marseille, France, 28 May - 1 June 2018; 2018.

[34] Eisenberg T., Schulien P., Koessl C., *et al*. 'CIMON – a mobile artificial intelligent Crew mate for the ISS'. *Proceedings of the 69th International Astronautical Congress*. 2018:5806–10.

[35] Ackerman B.E. *How NASA will grapple and refuel a satellite in low earth orbit – IEEE spectrum* [online]. 2019. Available from https://spectrum.ieee.org/tech-talk/aerospace/satellites/how-nasa-will-grapple-and-refuel-a-satellite-in-low-earth-orbit [Accessed 01 May 2020].

[36] Freer D., Guo Y., Deligianni F., Yang G.-Z. 'On-orbit operations simulator for workload measurement during Telerobotic training'. *arXiv [cs.HC]*. 2020.

[37] Yang G.Z., Rosa B.M.G. *A wearable and battery-less device for assessing skin hydration level under direct sunlight exposure with ultraviolet index calculation*. 2018 IEEE 15th International Conference on Wearable and Implantable Body Sensor Networks (BSN); 2018. pp. 201–4.

[38] Rosa B.M.G., Anastasova-Ivanova S., Yang G.Z. 'NFC-powered flexible chest patch for fast assessment of cardiac, hemodynamic, and endocrine parameters'. *IEEE Transactions on Biomedical Circuits and Systems*. 2019;**13**(6):1603–14.

[39] Baker W., Kingston Z., Moll M., Badger J., Kavraki L.E. 'Robonaut 2 and you: specifying and executing complex operations'. *Proceedings of IEEE Workshop on Advanced Robotics and Its Social Impacts*; ARSO; 2017.

[40] Farrell L.C., Dennis T.A., Badger J., O'Malley M.K. 'Simply grasping simple shapes: commanding a humanoid hand with a shape-based synergy'. The 18th International Symposium ISRR; 2017.

[41] Feix T., Romero J., Schmiedmayer H.-B., Dollar A.M., Kragic D. 'The GRASP taxonomy of human GRASP types'. *IEEE Transactions on Human-Machine Systems*. 2016;**46**(1):66–77.

[42] Wang C., Frccr D., Liu J., Yang G.-Z. 'Vision-based automatic control of a 5-fingered assistive robotic manipulator for activities of daily living'. *IEEE*. 2019:627–33.

[43] Varghese R.J., Lo B.P.L., Yang G.-Z. 'Design and prototyping of a bio-inspired kinematic sensing suit for the shoulder joint: precursor to a Multi-DoF shoulder Exosuit'. *IEEE Robotics and Automation Letters*. 2020;**5**(2):540–7.

[44] Pavone M., Castillo-Rogez J.C., Nesnas I.A.D., Hoffman J.A., Strange N.J. *Spacecraft/rover hybrids for the exploration of small solar system bodies*; 2013. pp. 1–11.

[45] Stubbs K., Hinds P.J., Wettergreen D. 'Autonomy and common ground in human-robot interaction: a field study'. *IEEE Intelligent Systems*. 2007;**22**(2):42–50.

[46] Kiesler S. 'Fostering common ground in human-robot interaction'. *IEEE*. 2005:729–34.

[47] Chai J.Y., She L., Fang R. 'Collaborative effort towards common ground in situated human-robot dialogue'. *IEEE Computer Society*. 2014:33–40.

[48] Manzey D. 'Human missions to Mars: new psychological challenges and research issues'. *Acta Astronautica*. 2004;**55**(3–9):781–90.

[49] Nicogossian A.E., Pober D.F., Roy S.A. 'Evolution of telemedicine in the space program and earth applications'. *Telemedicine Journal and e-Health*. 2001;**7**(1):1–15.

[50] Manthey D. 'Ch.9'. *Die Science-Fiction-Filme*. Hamburg: Zweiter Kino Verlag; 1983. pp. 170–81.

[51] ORMS. [online]. Available from https://www.orms.co.uk/insights/tars-a-study-of-movement-and-design/ [Accessed 13 Nov 2019].

[52] Selye H. 'What is stress?' *Metabolism: Clinical and Experimental*. 1956;**5**(5):525–30.

[53] NASA. *What is astrobee?* [online]. Available from https://www.nasa.gov/astrobee [Accessed 2 Dec 2019].

[54] Schröder V., Regele R., Sommer J., Eisenberg T., Karrasch C. *Gnc system design for the crew interactive mobile companion (CIMON)*. Bremen, Germany; 2018.

[55]   ESA. Angelic halo orbit chosen for humankind's first lunar outpost [online]. Available from https://www.esa.int/Enabling_Support/Operations/Angelic_halo_orbit_chosen_for_humankind_s_first_lunar_outpost [Accessed 13 Nov 2019].

[56]   NASA. *In lunar orbit* [online]. Available from https://www.nasa.gov/topics/moon-to-mars/lunar-gateway [Accessed 13 Nov 2019].

[57]   Kraft N.O., Lyons T.J., Binder H. 'Intercultural crew issues in long-duration spaceflight'. *Aviation, Space, and Environmental Medicine*. 2003;**74**(5):575–8.

[58]   Turner M.E., Pratkanis A.R. 'Twenty-five years of groupthink theory and research: lessons from the evaluation of a theory'. *Organizational Behavior and Human Decision Processes*. 1998;**73**(2/3):105–15.

[59]   Dion K.L. 'Interpersonal and group processes in long-term spaceflight crews: perspectives from social and organizational psychology'. *Aviation, Space, and Environmental Medicine*. 2004;**75**(7 Suppl):C36–43.

[60]   Sorrells S.F., Sapolsky R.M. 'An inflammatory review of glucocorticoid actions in the CNS'. *Brain, Behavior, and Immunity*. 2007;**21**(3):259–72.

[61]   Taylor G.R., Janney R.P. 'In vivo testing confirms a blunting of the human cell-mediated immune mechanism during space flight'. *Journal of Leukocyte Biology*. 1992;**51**(2):129–32.

[62]   Sonnenfeld G., Mandel A.D., Konstantinova I.V., *et al*. 'Spaceflight alters immune cell function and distribution'. *Journal of Applied Physiology*. 1992;**73**(2):S191–5.

[63]   Nefedov Y.G., Shilov V.M., Konstantinova I.V., Zaloguyev S.N. 'Microbiological and immunological aspects of extended manned space flights'. *Life Sciences and Space Research*. 1971;**9**:11–16.

[64]   Ginsberg H.S. 'Immune states in long-term space flights'. *Life Sciences and Space Research*. 1971;**9**:1–9.

[65]   Mehta S.K., Laudenslager M.L., Stowe R.P., Crucian B.E., Sams C.F., Pierson D.L. 'Multiple latent viruses reactivate in astronauts during space shuttle missions'. *Brain, Behavior, and Immunity*. 2014;**41**:210–7.

[66]   Crucian B., Stowe R., Mehta S., *et al*. 'Immune system dysregulation occurs during short duration spaceflight on board the space shuttle'. *Journal of Clinical Immunology*. 2013;**33**(2):456–65.

[67]   Payne D.A., Mehta S.K., Tyring S.K., Stowe R.P., Pierson D.L. 'Incidence of Epstein-Barr virus in astronaut saliva during spaceflight'. *Aviation, Space, and Environmental Medicine*. 1999;**70**(12):1211–3.

[68]   Crucian B., Babiak-Vazquez A., Johnston S., Pierson D., Ott C.M., Sams C. 'Incidence of clinical symptoms during long-duration orbital spaceflight'. *International Journal of General Medicine*. 2016;**9**:383–91.

[69]   Crucian B., Johnston S., Mehta S., *et al*. 'A case of persistent skin rash and rhinitis with immune system dysregulation onboard the International Space Station'. *The Journal of Allergy and Clinical Immunology: In Practice*. 2016;**4**(4):759–62.

[70]   Buchheim J.-I., Matzel S., Rykova M., *et al*. 'Stress related shift toward inflammaging in cosmonauts after long-duration space flight'. *Frontiers in Physiology*. 2019;**10**:85.

[71] Crucian B., Stowe R.P., Mehta S., Quiriarte H., Pierson D., Sams C. 'Alterations in adaptive immunity persist during long-duration spaceflight'. *npj Microgravity*. 2015;**1**:15013.

[72] Mehta S.K., Pierson D.L., Cooley H., Dubow R., Lugg D. 'Epstein-Barr virus reactivation associated with diminished cell-mediated immunity in Antarctic expeditioners'. *Journal of Medical Virology*. 2000;**61**(2):235–40.

[73] Feuerecker M., Crucian B.E., Quintens R., *et al*. 'Immune sensitization during 1 year in the Antarctic high-altitude concordia environment'. *Allergy*. 2019;**74**(1):64–77.

[74] Jacubowski A., Abeln V., Vogt T., *et al*. 'The impact of long-term confinement and exercise on central and peripheral stress markers'. *Physiology & Behavior*. 2015;**152**(Pt A):106–11.

[75] Yi B., Matzel S., Feuerecker M., *et al*. 'The impact of chronic stress burden of 520-d isolation and confinement on the physiological response to subsequent acute stress challenge'. *Behavioural Brain Research*. 2015;**281**:111–15.

[76] Yi B., Rykova M., Feuerecker M., *et al*. '520-d isolation and confinement simulating a flight to Mars reveals heightened immune responses and alterations of leukocyte phenotype'. *Brain, Behavior, and Immunity*. 2014;**40**:203–10.

[77] Schneider S., Abeln V., Popova J., *et al*. 'The influence of exercise on prefrontal cortex activity and cognitive performance during a simulated space flight to Mars (MARS500)'. *Behavioural Brain Research*. 2013;**236**(1):1–7.

[78] Abeln V., MacDonald-Nethercott E., Piacentini M.F., *et al*. 'Exercise in isolation – a countermeasure for electrocortical, mental and cognitive impairments'. *Plos One*. 2015;**10**(5):e0126356.

[79] Wollseiffen P., Vogt T., Abeln V., Strüder H.K., Askew C.D., Schneider S. 'Neuro-cognitive performance is enhanced during short periods of microgravity'. *Physiology & Behavior*. 2016;**155**:9–16.

[80] Pagel J.I., Choukèr A. 'Effects of isolation and confinement on humans-implications for manned space explorations'. *Journal of Applied Physiology*. 2016;**120**(12):1449–57.

[81] Sapolsky R.M. 'Why stress is bad for your brain'. *Science*. 1996;**273**(5276):749–50.

[82] Lallukka T., Kronholm E. 'The contribution of sleep quality and quantity to public health and work ability'. *The European Journal of Public Health*. 2016;**26**(4):532.

[83] Bin Y.S. 'Is sleep quality more important than sleep duration for public health?' *Sleep*. 2016;**39**(9):1629–30.

[84] Chattu V.K., Manzar M.D., Kumary S., Burman D., Spence D.W., Pandi-Perumal S.R. 'The global problem of insufficient sleep and its serious public health implications'. *Healthcare*. 2018;**7**(1):1.

[85] Karthikeyan R., Spence D.W., Pandi-Perumal S.R. 'The contribution of modern 24-hour Society to the development of type 2 diabetes mellitus: the role of insufficient sleep'. *Sleep Science*. 2019;**12**(3):227–31.

[86] Mansukhani M.P., Covassin N., Somers V.K. 'Apneic sleep, insufficient sleep, and hypertension'. *Hypertension*. 2019;**73**(4):744–56.

[87]   Kwok C.S., Kontopantelis E., Kuligowski G., *et al.* 'Self-reported sleep duration and quality and cardiovascular disease and mortality: a dose-response meta-analysis'. *Journal of the American Heart Association.* 2018;**7**(15):e008552.

[88]   Pawankar R. 'Allergic diseases and asthma: a global public health concern and a call to action'. *World Allergy Organization Journal.* 2014;**7**(1):12.

[89]   Lockey RF S.T., Blaiss M.PR.C.G.S.T.H. *'The WAO White Book on allergy – executive summary'.* 2013. Available from https://www.worldallergy.org/wao-white-book-on-allergy [Accessed 10 Oct 2019].

[90]   AAAAI. *American Academy of Allergy Asthma and Immunology.* Available from https://www.aaaai.org/about-aaaai/newsroom/allergy-statistics   [Accessed 10 Oct 2019].

[91]   Franceschi C., Bonafè M., Valensin S., *et al.* 'Inflammaging: an evolutionary perspective on immunosenescence'. *Annals of the New York Academy of Sciences.* 2000;**908**(1):244–54.

[92]   Fagiolo U., Cossarizza A., Santacaterina S., *et al.* 'Increased cytokine production by peripheral blood mononuclear cells from healthy elderly people'. *Annals of the New York Academy of Sciences.* 1992;**663**:490–3.

[93]   De Martinis M., Franceschi C., Monti D., Ginaldi L. 'Inflammageing and lifelong antigenic load as major determinants of ageing rate and longevity'. *FEBS Letters.* 2005;**579**(10):2035–9.

[94]   Pawelec G., Gouttefangeas C. 'T-cell dysregulation caused by chronic antigenic stress: the role of CMV in immunosenescence?' *Aging Clinical and Experimental Research.* 2006;**18**(2):171–3.

[95]   Bauer M.E., Fuente M.Dla. 'The role of oxidative and inflammatory stress and persistent viral infections in immunosenescence'. *Mechanisms of Ageing and Development.* 2016;**158**:27–37.

[96]   Ravaglia G., Forti P., Maioli F., *et al.* 'Increased prevalence of coeliac disease in autoimmune thyroiditis is restricted to aged patients'. *Experimental Gerontology.* 2003;**38**(5):589–95.

[97]   Sanada F., Taniyama Y., Muratsu J., *et al.* 'Source of chronic inflammation in aging'. *Frontiers in Cardiovascular Medicine.* 2018;**5**:12.

[98]   Crucian B.E., Choukèr A., Simpson R.J., *et al.* 'Immune system dysregulation during spaceflight: potential countermeasures for deep space exploration missions'. *Frontiers in Immunology.* 2018;**9**:1437.

[99]   Makedonas G., Mehta S., Choukèr A., *et al.* 'Specific immunologic countermeasure protocol for deep-space exploration missions'. *Frontiers in Immunology.* 2019;**10**:2407.

Part IV

System engineering

Chapter 11
Verification for space robotics

Rafael C. Cardoso[1], Marie Farrell[2], Georgios Kourtis[1], Matt Webster[3], Louise A. Dennis[1], Clare Dixon[1], Michael Fisher[1], and Alexei Lisitsa[4]

Verification and validation are required to help assure the safety, reliability, functional correctness and trustworthiness of systems. Verification demonstrates that the system conforms to its requirements and validation that it meets the needs of the stakeholders. Formal verification involves a mathematical analysis of all behaviours of a system often using logics and tools such as theorem provers or model checkers. Model checkers [1–3] are based on an automated, algorithmic method to show whether a property holds on all runs of the system. As input, model checkers require a model of the system and a property relating to the requirements in some (temporal) logical language. Commonly used logics to express change over time are propositional linear time temporal logic (LTL) [4, 5] and computational tree logic (CTL) [6]. Theorem proving, see, e.g., [7–10] for calculi and tools for temporal resolution, involves specifying both the system and the property in some logical language and using mathematical proof to show that the property is a logical conclusion from the formulae specifying the system. A number of tools and techniques have been developed to carry out theorem proving and model checking. While formal verification has the advantages of being exhaustive (considering all states in the system), precise details of the system often have to be represented in an abstract form to minimise the amount of time and memory required during the formal verification process.

Non-formal techniques can also be used for verification, e.g., simulation-based testing [11], or end-user experiments [12]. While not exhaustive, i.e., not every path through the system will be tested and not every scenario can be examined, the system being tested is more realistic than the abstracted model above.

[1]Department of Computer Science, University of Manchester, Manchester, United Kingdom
[2]Department of Computer Science, Maynooth University, Maynooth, Ireland
[3]School of Computer Science and Mathematics, Liverpool John Moores University, Liverpool, United Kingdom
[4]Department of Computer Science, University of Liverpool, Liverpool, United Kingdom

Different verification methods might be used depending on the system under consideration. A combination of different types of verification using both formal and non-formal techniques can help to improve the confidence in the system [13].

In this chapter, we discuss a range of tools and techniques applicable to the verification and validation of autonomous space systems. In particular we describe:

- verification techniques for robotics and autonomous systems;
- theorem proving for space robotics using modal and temporal logics;
- verifiable space robot architectures;
- simulation and verification of the Mars Curiosity rover;
- verification of astronaut–rover teamwork as modelled by the Brahms agent modelling system originally developed at NASA;
- modelling and verification of multi-objects systems (such as swarms of satellites or sensor network protocols).

11.1   Formal specification and verification techniques

In this section, we provide an overview of formal specification and verification techniques that have been applied to robotics and autonomous systems that can be found in the literature. Additionally, we discuss some particular formal verification techniques and how they might be applied to autonomous space systems. We believe that many of the issues for verifying systems for space are similar to those for robotics and autonomous systems in other extreme and dangerous environments.

11.1.1   Formal specification and verification for autonomous robotic systems

The work summarised in this section was originally published in [14, 15]. In particular, [14] contains a comprehensive survey of the literature in relation to formal specification and verification of autonomous robotic systems. This work identified a number of distinct challenges for formal specification and verification of robotic systems and some of these are described in the position paper which advocates the use of integrated formal methods in this domain [15].

11.1.1.1   Methodology

In this work, we began by identifying the following three research questions:

RQ1: What are the challenges when formally specifying and verifying the behaviour of (autonomous) robotic systems?

RQ2: What are the current formalisms, tools and approaches used when addressing the answer to **RQ1**?

RQ3: What are the current limitations of the answers to **RQ2** and are there developing solutions aiming to address them?

We investigated these questions by carrying out a systematic literature survey on *formal modelling of (autonomous) robotic systems, formal specification of (autonomous) robotic systems* and *formal verification of (autonomous) robotic systems*. This search applied to papers published from 2007 to 2018, inclusive. A summary of the results can be found in [16].

11.1.1.2 Answering RQ1: challenges

By analysing the literature, we were able to identify a number of challenges for formal specification and verification. We partitioned the challenges into those that are *external* and those that are *internal* to the robotic system. The challenges that were deemed as *external* to the robotic system were modelling the physical environment and providing evidence for certification and trust. The challenges that were deemed to be *internal* to the robotic system were agent-based systems, multi-robot systems and self-adaptive and reconfigurable systems. These challenges and current efforts for solving them are summarised in detail in [14, §3–4].

Interestingly, tackling the internal challenges may help to minimise the effects of the external challenges. For example, reconfigurability can help an autonomous robotic system to handle an uncertain and dynamic environment. Similarly, *rational agents*, which can provide reasons for their choices, can help with evidence for certification and public trust.

11.1.1.3 Answering RQ2: formalisms, tools and approaches

To answer RQ2, we quantified and described the formalisms, tools and approaches used in the literature [14, §5–6]. We have summarised these findings briefly in Tables 11.1 and 11.2. In particular, Table 11.1 reveals that most used a state-transition formalism to specify or build a model of their system. Whereas, most used a logic, normally temporal logic, to specify the properties of the system to be verified.

This could be a result of model checkers being the most favourable approach as indicated by Table 11.2, which generally take a state-transition system as input to verify against logical properties, often temporal logic. This choice of model checking may be due to the fact that it is generally easy to explain to stakeholders as 'exhaustive testing', a concept that most are familiar with. The lack of popularity of

Table 11.1 Summary of the types of formalisms for specifying the system and the properties to be checked [14, table 2]

Formalism	System	Property
Set-based	5	0
State-transition	33	0
Logic	6	32
Process algebra	3	1
Ontology	4	0
Other	5	8

Table 11.2   Summary of the verification approaches used throughout the literature [14, table 4]

Approach	Total
Model checking	32
Theorem proving	3
Runtime monitoring	3
Integrated formal methods	8
Formal software frameworks/architectures	10

theorem provers is likely due to their usability issues since they are generally difficult to operate for non-expert users.

11.1.1.4   Answering RQ3: limitations

We address this question in detail in [14, §7]. The most obvious limitation is the lack of adoption of formal methods by robotic software developers. It is often the case that these developers view formal methods as difficult to use and as a complicated additional step in the development process. Further, a lack of appropriate tools often impedes the application of formal methods [16].

As indicated by [14, Table 3], there is a huge variety of tools that have been developed for the same formalism. This indicates a lack of interoperability between different formalisms and tools. The development of a common framework for translating between, relating or integrating different formalisms would be useful in this domain and is an open problem in this domain [15].

Furthermore, formalising the *last link* between specification and implementable code is another limitation not only in this area but also in software engineering, in general. In particular, ensuring that the software implementation matches the associated formal specification requires a formalised translation.

11.1.1.5   Application to space robotics

Of course, this survey [14], the associated position paper [15] and summary [16] are quite broad and do not specifically target space systems, although some of the surveyed materials do. Since space is an appropriate domain where reliable autonomous robotic systems are required, it is important to understand the current state-of-the-art approaches to formal specification and verification of autonomous robotic systems in general and this survey provides the relevant details [14–16].

11.2   Theorem proving for space robotics using modal and temporal logics

We are developing tools and techniques for verification that can be applied to robotics and autonomous systems. In particular, we are interested in the development of tools and techniques for different dimensions such as temporal logics that consider how systems change over time [3], modal logics [17] that consider possible worlds where the relationships between worlds may represent necessity/possibility, belief or knowledge [18], temporal logics that incorporate probabilities such as Probabilistic Computation Tree Logic (PCTL) [19] or specific time bounds like Metric Temporal Logics (MTLs) [20].

We discuss some deductive methods that have been developed, including calculi and associated theorem-proving tools for different temporal and modal logics. These are resolution-based methods that follow the overall approach developed for LTL in [7] and extended to the branching-time temporal logic CTL in [9]. Using this approach, we can encode the system as a logical formula (S) and a property we want to prove of this system (P) in some logic. If we want to show $S \rightarrow P$ is valid, we negate $S \rightarrow P$ obtaining $S \wedge \neg P$ and show that the negated formula is unsatisfiable. The general approach is to translate the original formula φ into another equi-satisfiable formula φ of a particular form (termed a normal form). Following this, a number of proof rules are applied that generate new normal form formulae (often called clauses). The process stops when no new clauses can be derived or a contradiction can be obtained. With the temporal and modal logics we discuss here, the key thing is to make sure that formulae relate to the same world so that the proof rules can be applied.

11.2.1   The multi-modal logic K

There has been interest in modal logics (see, e.g., [21]) in relation to theoretical results, the development of practical tools such as theorem provers and their application to systems. A calculus [22] and theorem prover [23] have been developed for the multi-modal logic K. This modal logic has two modal operators: $\square$ denoted *necessity* and $\lozenge$ *possibility*. These operators can be indexed for different agents, e.g., below $\square_a$ ($\lozenge_a$) denotes that it is necessary (possible) for the astronaut, whereas $\square_r$ ($\lozenge_r$) denotes that it is necessary (possible) for the rover. The example below considers two agents, an astronaut and a rover, where the astronaut can go outside a lunar habitat to survey the moon surface and the rover must accompany the astronaut during dangerous situations.

$\square_a(out \rightarrow danger)$	For the astronaut it is necessarily the case that if they are out of the habitat then they are in danger
$\square_a(survey \rightarrow out)$	For the astronaut it is necessarily the case that if they are surveying then they are out of the habitat
$\square_a(danger \rightarrow \square_r accompany)$	For the astronaut it is necessarily the case that if they are in danger then the rover necessarily accompanies them
$\lozenge_a survey$	It is possible that the astronaut does a survey

From this, we can prove that it is possible that the astronaut is necessarily accompanied by the rover ($\Diamond_a\Box_r accompany$).

A resolution calculus has been developed for this logic that only allows the deduction rules to be applied to the same modal depth of formula [22]. This makes the prover for this calculus [23] perform well on formulae with a high level of nesting of modal formulae compared to other modal theorem provers for this logic.

11.2.2   Metric temporal logic

MTLs have models that are timed sequences of states. In MTL, the temporal operators such as $\Box$ ('now and at every future moment'), $\Diamond$ ('at sometime in the future') and $\bigcirc$ ('in the next moment in time') include an interval that provides temporal constraints about when formulae should hold. For example, $\Box_{[0,4]}\varphi$ denotes that φ holds at all states that occur between zero and four time points from now, and $\Box_{[3,\infty]}\varphi$ denotes that φ holds at all states that occur at least three time points onwards. In the statements below, we provide some MTL formulae describing some formulae for the astronaut–rover scenario.

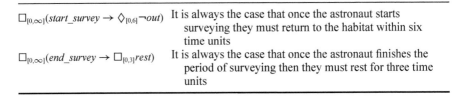

$\Box_{[0,\infty]}(start\_survey \to \Diamond_{[0,6]}\neg out)$	It is always the case that once the astronaut starts surveying they must return to the habitat within six time units
$\Box_{[0,\infty]}(end\_survey \to \Box_{[0,3]}rest)$	It is always the case that once the astronaut finishes the period of surveying then they must rest for three time units

If we consider a natural numbers model where states are mapped to the natural numbers, we can translate such formulae into formulae of LTL. Then we can apply provers that have been developed for LTL to obtain a route to theorem proving for MTL formulae. Two different translations have been developed and applied to two different versions of the semantics. An experimental analysis has been applied translating these alternatives to input to a range of LTL provers to investigate their behaviour [24, 25]. This approach is useful as it allows the re-use of a range of provers for temporal logics that can be applied to problems for MTL over the natural numbers. A related approach is taken in [26], where translations from a similar logic (Mission-Time LTL) into model checkers for LTL are provided.

11.3   Verifiable space robot architectures

Robotic systems combine many hardware and software components, usually represented as node-based architectures. Each node in a robotic system may require different verification techniques, ranging from software testing to formal methods. In fact, *integrating* (formal and non-formal) verification techniques is crucial for the robotics domain [15]. Verification should be carried out using the most suitable technique or formalism for each node. However, linking heterogeneous verification results of individual nodes is difficult and the current state-of-the-art for robotic

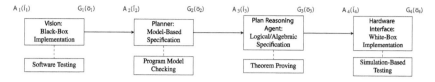

Figure 11.1   *We specify the Assume-Guarantee contracts for each node (denoted by $\mathcal{A}(\bar{i})$ and $\mathcal{G}(\bar{o})$, respectively). These are then used to guide the verification approach applied to each node, denoted by dashed lines, such as software testing for a black-box implementation of the Vision node. The solid arrows represent data flow between nodes and that the assumptions of the next node should follow from the guarantee of the previous node.*

software development does not provide an easy way of achieving this. In this section, we summarise ongoing work that was originally presented in [27].

In Figure 11.1, we consider a simple space robotic system: a planetary rover undertaking a remote inspection task. Here, we have nodes representing the Vision system, a Planner that returns a set of potential plans between the current location and the next point to inspect, an autonomous Plan Reasoning Agent that selects a plan and a Hardware Interface that sends commands to the rover's actuators.

As illustrated in Figure 11.1, we could use logical specifications (e.g., temporal logic), model-based specifications (e.g., Event-B [28] or Z [29]) or algebraic specifications (e.g., Communicating Sequential Processes (CSP) [30] or Common Algebraic Specification Language (CASL) [31]) among others to specify the nodes in a robotic system. Each of these formalisms offers its own range of benefits, and each tends to suit the verification of particular types of behaviour. However, in some cases we may only have access to the black-box or white-box implementation of a node and, so, we must use (simulation-based) testing techniques for verification.

Our approach facilitates the use of heterogeneous verification techniques for the nodes in a robotic system. We achieve this by specifying contracts as properties in First-Order Logic (FOL), as high-level node specifications, and we employ temporal logic for reasoning about the combination of these FOL specifications. Thus, we attach the assumptions ($\mathcal{A}(\bar{i})$) and guarantees ($\mathcal{G}(\bar{o})$) to individual nodes (shown in Figure 11.1). This abstract specification can be seen as a logical prototype for individual nodes and thus the entire robotic system.

11.3.1 FOL contract specifications

For each node, N, we specify $\mathcal{A}_N(\bar{i}_N)$ and $\mathcal{G}_N(\bar{o}_N)$, where $\bar{i}_N$ is a variable representing the input to the node, $\bar{o}_N$ is a variable representing the output from the node and $\mathcal{A}_N(\bar{i}_N)$ and $\mathcal{G}_N(\bar{o}_N)$ are FOL formulae describing the assumptions and guarantees, respectively, of this node.

Each individual node, N, obeys the following implication

$$\forall \bar{i}_N, \bar{o}_N \cdot \mathcal{A}_N(\bar{i}_N) \implies \Diamond \mathcal{G}_N(\bar{o}_N)$$

where '$\Diamond$' is LTL's [4] 'eventually' operator. So, this implication means that if the assumptions, $\mathcal{A}_N(\bar{i}_N)$, hold then *eventually* the guarantee, $\mathcal{G}_N(\bar{o}_N)$, will hold. Note that our use of temporal operators here is motivated by the temporal nature of robotic systems and will be of use in later extensions of this work.

Consider the autonomous Plan Reasoning Agent in Figure 11.1; we can specify the following simple assumption $\mathcal{A}_3(\bar{i}_3)$:

$$\mathcal{A}_3(\bar{i}_3) = \forall p \cdot p \in PlanSet \Rightarrow goal \in p$$

which ensures that every plan that is returned by the Planner contains the *goal* location. Then, we might specify the guarantee that the agent chooses the shortest *plan* as follows:

$$\mathcal{G}_3(\bar{o}_3) = (plan \in PlanSet) \wedge (\forall p \cdot p \in PlanSet) \wedge (p \neq plan) \Rightarrow (length(plan) \leq length(p))$$

Once the FOL assumption and guarantee are specified, then we use these high-level specifications as properties to be verified of the individual nodes. For the autonomous Plan Reasoning Agent, we can use a number of techniques for verifying that it meets its associated FOL specification. For example, we can specify the node using the Gwendolen agent programming language and then use the Agent Java PathFinder (AJPF) model checker to verify that it behaves as specified [32].

Nodes in a modular robotic architecture are linked together and transmit data between them so long as their types/requirements match. Similarly, we can compose the contract specifications of individual nodes in a number of ways and we are working towards a calculus of inference rules that capture this behaviour. To this end, we are developing rules for sequentially composing, joining, branching and looping between nodes.

11.3.2   *Measuring confidence in verification*

A key question is how using these different verification techniques affects our confidence in the verification of the whole system. One might think that a formal proof of correctness corresponds to a higher level of confidence than simple testing methods (especially over unbounded environments). However, formal verification is usually only feasible on an abstraction of the system whereas testing can be carried out on the implemented code. Therefore, it is our view that we achieve higher levels of confidence in verification when multiple verification methods have been employed for each node in the system [13].

We have broadly partitioned current verification techniques into three categories: testing, simulation-based testing and formal methods. We have determined which of these techniques might be employed for each node in our simple example as shown in Table 11.3. We then provide a score for our level of confidence in the verification of the whole system as 9/12, resulting in a confidence measure of 75 percent. Examining how this metric can be calculated for more complex systems with loops is a future direction for this work.

Table 11.3   Verification techniques applied to each node

	Testing	Simulation-based testing	Formal methods
Vision	✓	✗	✗
Planner	✓	✓	✓
Plan reasoning agent	✓	✓	✓
Hardware interface	✓	✓	✗

When verifying complex robotic systems, it is clear that no single verification technique is suitable for every node in the system [15] and so a logical framework that allows us to integrate the results from distinct verification techniques is needed. We have outlined an initial approach to specifying assumptions and guarantees using FOL for individual nodes in robotic systems and we have used a simple, illustrative example of a planetary rover to convey our approach. Once the FOL specifications have been constructed, they are then used to guide the more detailed verification of each node. Furthermore, we introduce the notion of confidence in verification techniques and provide a broad categorisation.

Our current work involves developing a calculus for reasoning about and combining the contract specifications of individual nodes. In the future, we plan to provide tool support for this and to evaluate it using a set of more complex robotic space missions. We also intend to further investigate the suitability of the confidence levels that we have proposed.

11.3.3   Related work

Our approach draws inspiration from Broy's approach to systems engineering [33], which uses logical predicates in the form of assertions, with relationships defined between them that extend to assume/commitment contracts. The treatment of these contracts is purely logical, and we present a similar technique that is specialised to the software engineering of robotic systems – a domain which has not received much attention in this branch of the literature before.

In terms of compositional verification, related work includes CoCoSpec [34], which allows users to specify contracts for reactive systems in terms of assumptions and guarantees. This work is specialised for synchronous communications and thus it differs from the event-based communications that we target here. Further, their contract semantics is more restrictive than ours. It is also not clear how their support for compositional verification can be extended to support heterogeneous components such as those in our example. Other related approaches include OCRA [35] and AGREE [36], although neither explicitly incorporates heterogeneous verification techniques.

11.4   Case study 1: Simulation and verification of the Mars Curiosity rover

Autonomous robots are especially relevant in scenarios with communication bottle-necks. For example, in planetary space exploration it can take a long time for human operators to send commands from Earth to the robot, and then the same amount of time to receive any feedback data from the command that was sent. The Curiosity rover[1] is one of the most complex rovers successfully deployed in a planetary explo-ration mission to date. It was sent by NASA to explore the surface of Mars. Its main objectives include determining signs of life, characterising climate and geology and preparing for human exploration.

However, one of the biggest challenges faced by the Curiosity is the long communication delay between Earth and Mars. Depending on the orbital position of both planets, it can take anywhere from 4 to 24 minutes for a message to be transmitted between Earth and Mars. Thus, if the Curiosity could be controlled autonomously it would be able to perform its activities much faster. One of the major challenges preventing the use of autonomy in such scenarios is the lack of assurance that the autonomous behaviour will work as expected. To this end, it is important to take a corroborative approach [13] when trying to provide assurances about autonomy.

In [37], system scenario tests are described for the validation of the Mars Curiosity rover surface operations. These tests were performed before the launch, over the period of one year, to test high-priority objectives in typical missions that would take place for the duration of a Martian day. Testing-based approaches are essential for validating a system; however, they are not exhaustive and can often miss edge cases, particularly when testing autonomous systems.

The presence of autonomy (without any input from Earth) in the original mission was restricted to the AEGIS (Autonomous Exploration for Gathering Increased Science) component [38]. This component provides the autonomous targeting of surfaces to be processed by the remote geochemical spectrometer. It has been validated through comprehensive testing both in simulation before the launch and on Mars after the launch, but no concrete formal verification was made public. While we do not have the AEGIS in our (much simpler) simulation, we use a rational agent to perform autonomous operations that would usually be delegated to Earth operators and use formal techniques to verify the behaviour of the agent.

In this section, we present a simulation of the Mars Curiosity rover controlled via an autonomous agent. Then, we discuss the formal verification of this agent through the use of model checking. Finally, we verify it at runtime by deploying runtime monitors. This combination of simulation-based testing, and the use of two

[1]https://mars.nasa.gov/msl/

(a)   (b)

Figure 11.2   *The Mars Curiosity rover simulation in Gazebo. (a) The Mars*
Curiosity model in Gazebo. (b) The Mars world in Gazebo.

distinct formal methods, gives us a basis for providing assurances about the use of autonomy in extreme environments that could be transferred and applied to other similar case studies. We refer the reader to [39] for a more in-depth discussion about the use of different verification techniques applied to a similar case study to the one presented in the section.

11.4.1   Simulation

Modular architectures are typically employed to speed up and make the development of robotic systems easier. The Robot Operating System (ROS) [40] is an example of a popular middleware that can be used to develop a modular robotic system. In ROS, nodes are used to effectively capture robotic software in terms of a graph that describes the communication between distinct nodes. Some of the advantages of decoupling the system in this way include more precise failure handling and recovery mechanisms, since failures can be traced to individual nodes and the complexity of the code is reduced when compared to monolithic systems, making it easier to add, replace or remove functionality (i.e., nodes).

Even though the software deployed with the real Curiosity was not ROS-based, a ROS version has been developed by the ROS teaching website The Construct[2] using official data and 3D models of Curiosity and Mars terrain that NASA made public. The simulation uses ROS and runs in Gazebo, a 3D simulator with several high-performance physics engines. The 3D model of the Curiosity running in Gazebo is shown in Figure 11.2a, and the Mars world used in the simulation is shown in Figure 11.2b.

RVIZ is a 3D visualiser tool that displays state information about the virtual model of the robot and live sensor data such as camera feeds, infrared measurements and more. Most of the Curiosity's effectors are included in the simulation, as shown in Figure 11.3. It has all six wheels, along with the suspension system, the complete chassis of the rover, a 7-foot retractable arm with four joints and a retractable mast

---

[2]https://bitbucket.org/theconstructcore/curiosity_mars_rover/src/master/

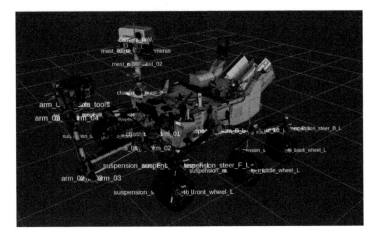

Figure 11.3   RVIZ view with all of the effectors in the simulated Mars Curiosity rover

with two joints and a camera (Mastcam) on top. Some of the sensors are missing (such as chemical and weather sensors), as these would require the sensor data to be simulated in some way.

The standard control of the Curiosity rover in the original simulation was implemented using ROS services and needs to be teleoperated by a human. Services can be provided by ROS nodes and are defined by a pair of request and reply messages. This interaction is similar to a remote procedure call. Action libraries follow a client-server model that is similar to ROS services, both can receive a request to perform some task and then generate a reply. The difference in using action libraries is that the user can cancel the action, as well as receive feedback about the task execution. Thus, action libraries are more suited when autonomy is used, allowing more fine-grained autonomous control of the robot.

We implemented three action libraries: wheels, arm and mast. The client of the wheels can receive high-level action commands to move forward, backward, turn left and turn right, which are then passed over to the server. The server has control over each of the six wheels and publishes speed commands to the appropriate wheels depending on the direction requested in the action. After any movement action, the server always calls a stop action that sets all wheels speed to zero. The arm and mast action libraries are responsible for controlling the joints of the arm and the mast, respectively.

Agent-based control allows a system to dynamically adapt to changes in the environment through the use of modularity, decentralisation, autonomy, scalability and reusability [41]. We use the Gwendolen agent language to program the high-level control and autonomous behaviour of the Mars Curiosity rover. Agent programming languages abstract the environment and other external sources, focusing on programming autonomous control at high-level, resulting in smaller and more modular code than other languages. Furthermore, due to the agent's reasoning cycle and mental attitudes, an execution trace can clearly show how the agent came to a decision, thus providing us with explainability. Finally, using Gwendolen, properties of the agent's reasoning can be formally verified, allowing us to safeguard critical behaviours.

Gwendolen [42] agents follow the Belief-Desire-Intention (BDI) model [43]. These mental attitudes represent, respectively, the information, motivational and deliberative states of the agent. The *belief revision function* is used to process incoming perceptions from the environment (e.g., obtained through sensors) and triggers the update of the *belief* (what the agent believes to be true about its environment and other agents) base. The *option generation function* uses the belief base and the intention base to generate more options and update the *desire* (the desired states that the agent hopes to achieve) base. The *filter* is responsible for updating the *intention* (a sequence of actions that an agent wants to carry out to achieve a desired state) base, taking into account its previous intentions and the belief and desire bases. Finally, the *action selection function* outputs an action from the intention that was chosen to be executed.

It is not possible to integrate Gwendolen, which is implemented in Java, directly with ROS, which is implemented in Python/C++. There are two possible solutions to this problem: use the *rosjava* library or use the *rosbridge* library. The former is a third-party library that is not available in all recent versions of ROS; since it re-implements parts of ROS in Java it can be an arduous process to update it for newer versions. The latter is a compact library that is less dependent on ROS version, as it does not change anything in the core, but rather uses WebSocket communication to interface ROS with external languages through JavaScript Object Notation (JSON) messages.

We developed a Gwendolen environment[3] that can communicate with ROS through the *rosbridge* library. This environment is not domain-specific, i.e., it could be used in other robots as long as they are running ROS. It allows Gwendolen agents to publish and subscribe to ROS topics. Generally, when the agent executes an action in the environment, the action is processed and published to the appropriate ROS topic associated with that action. The environment can also create subscribers that keep listening to predetermined topics, and then when a message is received it is processed and, if it is the case, perceptions are created and sent to the agent.

The Gwendolen agent has access to three high-level actions. The action *control_wheels* has three parameters: direction of movement (forward, backward, left, or right), speed (an integer, if moving backwards should be negative, otherwise it should be positive) and distance (in seconds). This action is defined in the Gwendolen environment, as shown in Listing 11.1, which basically receives the parameters set by the agent, converts it into a *Move3* message (a message type defined in ROS) and then publishes the message to the appropriate ROS topic. The remaining actions

---

[3]Source code is available at: https://github.com/autonomy-and-verification-uol/gwendolen-rosbridge

are *control_arm* and *control_mast*, both require one parameter with possible values
either *open* or *close*.

```
public void control_wheels(String modereq, float speedreq, int distancereq) {
        Publisher wheels = new Publisher("/gwen_wheels", "curiosity/Move3", bridge);
        wheels_control.publish(new Move3(modereq, speedreq, distancereq));
}
```

Listing 11.1 Environment code for the control wheels action

Our simulation[4] contains an inspection mission, where the Curiosity patrols
between four different waypoints (A, B, C and D) that are spread across Mars terrain. The agent has previous knowledge about the terrain, with map coordinates to
each of the waypoints.

The simulation starts with the deployment of the Curiosity and a start-up period
where it initialises all three control modules (wheels, arms and mast). After the
agent receives confirmation that all modules are ready, it autonomously controls the
Curiosity to move to the waypoints. Movement through the waypoints is done in
order and loops back when arriving at the last one (A → B → C → D → A). When
moving to waypoint A and D, the agent ensures that the arm is retracted and that
the mast is extended upwards, since in both of these waypoints, the mission of the
Curiosity is to take images of the terrain. Otherwise, when moving to waypoint B
and C, the arm is fully extended to manipulate soil and rock samples, and the mast
is retracted due to extreme weather conditions, to preserve power and the condition
of the mast.

The plan of the Gwendolen agent that shows the start of the movement from
waypoint A to B is shown in Listing 11.2. The head of the plan (movement_com-
plete) is a trigger event that is activated when the belief with the same name is added
to the agent's belief base. The guard of the plan, enclosed by curly brackets, determines the precondition of the plan (what has to be true for the plan to be selected
for execution), which in this case is the belief that the agent is patrolling waypoint
A and that it is time to turn to go to the next waypoint. The body of the plan begins
after the left arrow sign, with each operation (e.g., the removal of the belief patrol
turn) or action (e.g., closing the mast) separated by a comma and a semicolon after
the last element in the body of the plan.

```
+movement_completed :
    { B patrol("A"), B patrol("turn")   }
    <-
    -movement_completed, -patrol("turn"), print("Turning to go to Waypoint B"),
    control_wheels("right",5,10.0), control_mast("close"), control_arm("open");
```

Listing 11.2 Plan for turning to move to waypoint B

---

[4]Source code is available at: https://github.com/autonomy-and-verification-uol/gwendolen-ros-curiosity

11.4.2   Model checking

Model checking [1] is the process of exhaustively performing a state space search to check if some desired property holds. This can be done either with a formal model of the system, encoded in some specification language, or directly within the program, called program model checking. The property that we want to verify also has to be specified using some language, usually logic-based. For example, we may want to verify a property that states that the Curiosity will not move its arm while it is collecting soil and rock data, to prevent damaging the collection.

AJPF [32] is an extension of Java PathFinder [44], a model checker that works directly on Java program code instead of on a mathematical model of the program's execution. This extension allows for formal verification of BDI-based agent programming languages by providing a property specification language based on LTL that supports the description of terms usually found in BDI agents.

Some of the properties that we verified of the implementation of our agent were:

$$\square(A_{\text{rover}}move\_waypoint(A) \rightarrow \lozenge B_{\text{rover}}(patrol(A)))$$
$$\square(A_{\text{rover}}move\_waypoint(B) \rightarrow \lozenge B_{\text{rover}}(patrol(B)))$$
$$\square(A_{\text{rover}}move\_waypoint(C) \rightarrow \lozenge B_{\text{rover}}(patrol(C))$$
$$\square(A_{\text{rover}}move\_waypoint(D) \rightarrow \lozenge B_{\text{rover}}(patrol(D))$$

These properties state that it is always the case ($\square$) that if the *rover* agent executes the action *move_waypoint* (to either A, B, C or D), then eventually ($\lozenge$) the *rover* agent will believe that it is currently patrolling that waypoint.

11.4.3   Runtime verification

Runtime verification (RV) [45] is a more lightweight approach that is usually more suitable for examining 'black box' software components. RV focuses on analysing only what the system produces while it is being executed and, because of this, it can only conclude the satisfaction/violation of properties regarding the current observed execution.

ROSMonitoring[5] is a framework for runtime monitoring of ROS topics. It creates monitors that are placed between ROS nodes to intercept messages on relevant topics and check the events generated by these messages against formally specified properties in an oracle. We applied this framework to the Curiosity case study using the filter action to intercept external messages sources from the agent that violate our property.

As an example of the filter action, consider an action library in ROS that controls the wheels of the rover. The content of the message includes the parameters discussed previously in the control wheels action: the *direction* for the rover to

[5]https://github.com/autonomy-and-verification-uol/ROSMonitoring

move, the *speed* of the wheels and the *distance* that it should move. The configuration file in ROSMonitoring for this example is shown in Listing 11.3.

```
monitors:
  - monitor:
    id: monitor_0
    log: ./log_0.txt
    oracle:
      port: 8080
      url: 127.0.0.1
    topics:
      - name: wheels_control
        type:
        curiosity_mars_rover_description.msg.Move3
        action: filter
        warning: True
        side: subscriber
          - node: wheels_client
            path: /curiosity/launch/wheels.launch
```

Listing 11.3 Configuration file for the first Curiosity example

Due to the gravity and rocky/difficult terrain in Mars, the Curiosity has to be careful with its speed. Thus, when we intercept a message in the *wheels_control* topic, the message is sent to the oracle to verify the following property:

```
left_speed matches {topic:'wheels_control', direction:'left',
    speed:val} with val <= 10;
right_speed matches {topic:'wheels_control', direction:'right',
    speed:val} with val <= 10;
forward_speed matches {topic:'wheels_control', direction:'forward',
    speed:val} with val <= 15;
backward_speed matches {topic:'wheels_control', direction:'backward',
    speed:val} with val <= 15;
Main = (left_speed \/ right_speed \/ forward_speed \/ backward_speed)*;
```

That is, if the direction is left or right (i.e., a turn action) then the speed cannot be greater than 10, and if the direction is forward or backward then the speed cannot be greater than 15. These are arbitrary numbers that were defined based on testing to prevent the Curiosity from suffering any accidents. After the error is intercepted by the oracle, the agent could use this information to adapt its plan. For instance, using the agent reconfigurability approach introduced in [46], the agent could detect that a failure happened and better understand why it happened to reconfigure itself accordingly.

11.5   Case study 2: Verification of astronaut–rover teams

Sections 11.1, 11.3 and 11.4 used the example of a planetary surface rover performing autonomous inspection and surveying tasks on nearby planetary bodies such as the Moon or Mars, e.g., the Curiosity rover used on the Mars Science Laboratory mission since 2012. Such planetary rovers have been used in several missions starting with the Soviet Union's Lunokhod 1 in 1970, and a number of further rover-based missions are being planned by various space agencies. Early rovers were limited in autonomous operations and were primarily remotely operated. However, autonomous systems are increasingly used to increase the reliability and efficiency of mission activities [47]. The increased use of autonomy also enables the proposed use of astronaut–rover teams, in which astronauts are assisted during missions by autonomous rovers [48–50]. For example, astronaut–rover teams have been evaluated for use in planetary outpost assembly [51], may employ multiagent planning systems to distribute tasks between team members [52] and can be assessed at the mission conception stage using simulators [53] and terrestrial field tests with robot prototypes [54].

As described in the previous sections, it is possible to use formal methods in the form of model checking to formally verify the rover's behaviour in the astronaut–rover team. In Section 11.4, the Gwendolen agent programming language was used to specify the behaviour of a decision-making agent in control of the rover. In this case, however, a different approach was taken due to the need to model the behaviour of one or more astronauts in the astronaut–rover team. This new approach used a multiagent workflow specification language called Brahms [55, 56] to model the behaviour of the astronaut–rover team. Brahms has been used before for modelling human–robot teamwork as part of the Mobile Agents Architecture at NASA Ames Research Center [57]. It has also been used to implement systems for mission control for the International Space Station [58] and health monitoring of astronauts [59].

Brahms consists of a modelling language and an integrated development environment (IDE). The modelling language allows multiagent systems to be specified in terms of interacting agents. Each agent has a set of beliefs that resemble common programming language variables types such as Booleans and integers. Agents can be given a location within a topological 'geography' within the Brahms model. Agent can perform activities or movements that can take a period of time. Each agent's behaviour is specified through workframes and thoughtframes. The former allows the agent to perform activities, move and update beliefs, whereas the latter only allows instantaneous belief updates, known as inferences. Workframes can also be interrupted by the receipt of new information from another agent. Communication between agents is modelled using a primitive 'communicate' construct. The design of the Brahms language allows for detailed models of multiagent systems to be developed in an intuitive way. Once a model has been developed, it is possible to run simulations of the model using the Brahms IDE, known as the 'Composer'. The results of these simulations can be displayed in the form of a timeline showing agent locations, workframes and activities as horizontal bars, and with communications shown as vertical lines (see Figure 11.4).

A Brahms model of an astronaut–rover team was developed. The model was based on a scenario similar to those used during NASA field tests and demonstrations of astronaut–rover teamwork [48, 50, 57], in which astronauts and a rover work together to achieve mission goals at a (simulated) outpost on the Moon or Mars. During the scenario, the rover assists the astronaut and performs autonomous behaviours when its assistance is not needed. For example, when the astronaut is performing construction or geological surveying tasks, the rover will assist the astronaut, e.g., by following the astronaut and providing a mobile platform for tools

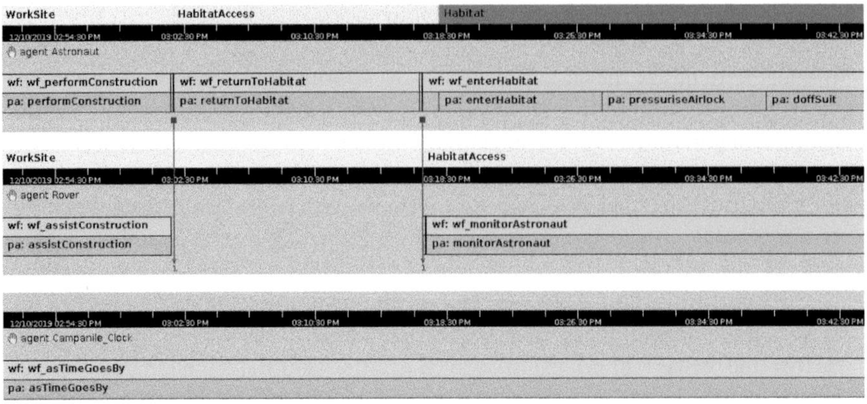

Figure 11.4   Simulation of the astronaut–rover scenario using the Brahms Composer IDE

and materials. During the video extra-vehicular activity (EVA) the rover assists by turning on the camera stream and filming the astronaut so that they can be monitored by the ground team. A similar behaviour is performed when the astronaut enters or leaves the habitat as these operations are especially hazardous. If the astronaut chooses to perform a miscellaneous activity then the rover will recognise that its assistance is not needed and will perform a solo geological survey. Also, whenever the astronaut is in the habitat, the rover performs a habitat-monitoring activity to ensure that the integrity of the habitat's life-support systems is maintained.

The Brahms model of the scenario used three agents: one to model an astronaut's behaviour, the second to model the autonomous rover's behaviour and the third, called the campanile clock, to assist in measuring the passage of time. The model examines the events over the course of a typical work day. The astronaut begins the day in the habitat and at some point decides to leave the habitat and start work. Leaving the habitat involves donning a spacesuit, depressurising the airlock and moving outside through the external door. Once the astronaut is outside, they can choose from a number of different behaviours: construction, geological surveying, video-recorded EVA or miscellaneous activities. At the end of the work day the astronaut returns to and enters the habitat. After entering the astronaut repressurises the airlock and doffs the spacesuit. The astronaut then remains in the habitat for the rest of the work day. The astronaut's choice between different behaviours is modelled as a non-deterministic choice between workframes in the astronaut agent.

As mentioned earlier, the Brahms Composer IDE can be used to display the results of simulations using a Brahms model. An excerpt of a simulation of the astronaut–rover model is shown in Figure 11.4. The behaviour of the three agents (astronaut, rover and campanile clock) are shown as horizontal bands. Time progresses from the left to right. The locations of the agents are shown at the top of the bands. For example, at the start of this excerpt the astronaut agent is at the work site, is in the 'perform construction' workframe and is performing the 'perform construction' activity. While this is happening the rover is in the 'assist construction' workframe and is performing the 'assist construction' activity. After completing the construction task the astronaut decides to return to the habitat. The rover stops working while the astronaut moves to the habitat access point. When the astronaut enters the habitat the rover begins monitoring the astronaut by entering the 'monitor astronaut' workframe. During the entire simulation the campanile clock agent is monitoring the time and announcing it to the other agents. The campanile clock is used as a means of synchronising agent behaviours. For example, at the end of the work day the campanile clock informs the other agents that the work day has ended. When this happens the astronaut agent will stop work and return to the habitat.

It is possible to simulate this scenario many times using Brahms to determine whether the behaviour of the autonomous rover satisfies mission requirements. However, it is difficult to tell using simulation whether these requirements are satisfied in all cases. Therefore it may be useful to perform an exhaustive analysis using model checking to determine whether requirements hold for the agent-based decision-making system. This can be done for the Brahms model of the astronaut–rover scenario by translating the Brahms model code into the input language for a model checker. This was done automatically using the BrahmsToPromela software, which translates Brahms model code into Promela, the input language for the Spin model checker [2]. Once translation is complete, Spin can be used to exhaustively analyse the Promela model to determine whether it satisfies requirements encoded formally as *properties* in LTL. The results of the model checking will also apply to the Brahms model as long as we have validated the automatic translation performed by BrahmsToPromela. Validation was achieved by developing BrahmsToPromela with respect to a formal semantics of Brahms [60] and through extensive applications in other human–robot team scenarios [61, 62].

To demonstrate the approach, an initial set of four mission- or safety-critical requirements were examined:

1. The rover should perform a solo geological survey whenever the astronaut is performing a miscellaneous activity. [Mission-critical.]
2. The rover should assist the astronaut during construction tasks. [Mission-critical.]
3. The rover should monitor the astronaut when they are leaving the habitat. [Safety-critical.]
4. The rover should monitor the habitat whenever the astronaut is located inside the habitat. [Safety-critical.]

These requirements were formalised as four LTL properties. These are shown in Table 11.4 LTL allows the formalisation of concepts relating to time, e.g., 'now and at all points in the future' (via the $\square$ operator), 'now or at some point in the future' ($\lozenge$) and 'in the next state' ($\bigcirc$) [63]. This enables formalisation of safety requirements (something bad never happens, $\square\neg$bad), liveness properties (e.g., something good eventually happens, $\lozenge$good) and fairness properties (e.g., if one thing occurs infinitely often so does another, e.g., $\square\lozenge$send $\implies \square\lozenge$receive). Using BrahmsToPromela extends Spin's property specification language with a *belief*

Table 11.4   *Properties verified for the Brahms model of the astronaut–rover scenario*

Req.	Property	Description
1	$\square \begin{bmatrix} B_{Astro}(\text{goalPerformMisc}) \Longrightarrow \\ \lozenge\ B_{Rover}(\text{goalSoloGeoSurvey}) \end{bmatrix}$	It is always the case that if the astronaut agent believes that it is performing a miscellaneous activity (i.e., an activity that does not require the assistance of the rover), then the rover will perform a solo geological survey after a period of time.
2	$\square \begin{bmatrix} B_{Astro}(\text{goalPerformConstr}) \Longrightarrow \\ \lozenge\ B_{Rover}(\text{goalAssistConstr}) \end{bmatrix}$	It is always the case that if the astronaut agent believes that it has a goal to start construction, then the rover agent will form a corresponding goal to assist in the construction task after a period of time.
3	$\square \begin{bmatrix} B_{Astro}(\text{goalLeaveHabitat}) \Longrightarrow \\ \lozenge\ B_{Rover}(\text{cameraStream}) \end{bmatrix}$	It is always the case that if the astronaut decides to leave the habitat, the rover will start monitoring by setting the camera stream variable to true, indicating that a video stream is being sent back to the habitat (and then back to the ground station for monitoring, if needed).
4	$\square \begin{bmatrix} \text{Astro.location} = \text{Habitat} \Longrightarrow \\ \lozenge\ B_{Rover}(\text{goalSoloMonitorHab}) \end{bmatrix}$	It is always the case that if the astronaut is in the habitat, then the rover will form a goal to autonomously monitor the habitat.

operator, 'B'. This allows us to specify that an agent has a belief, e.g., $B_{Rover}x$ means that the Rover agent believes x is true.

No errors were found by the model checker and therefore all properties held for the Promela model, meaning that the autonomous behaviour of the robot, was correct with respect to the requirements. Using this approach we were able to determine that it is possible to use formal methods, in particular, model checking, to formally verify the behaviour of an autonomous rover within a realistic astronaut–rover scenario.

11.6   Modelling and verification of multi-objects systems

11.6.1   *Motivation*

Sections 11.1, 11.3 and 11.4 presented various ways to model and verify the behaviour of a single autonomous planetary rover operating on nearby planetary bodies such as the Moon or Mars. Section 11.5 provided a methodology to model and verify the rover's behaviour in an astronaut–rover team. In this section, we consider generalisations of the above scenarios with two or more *identical* rovers working in cooperation. Such scenarios present many challenges with regard to coordination and resource management, hence it is important to address these challenges at the appropriate level of abstraction.

To motivate our presentation, let us discuss a simple generalisation of the scenario in Section 1.5. In our version of the scenario, we have a team of an astronaut working with k ($k > 1$) autonomous planetary rovers $r_1, \ldots, r_k$ to perform an action. Our mission- and safety-critical requirements will be as follows:

1. Each rover should perform a solo geological survey whenever the astronaut is performing a miscellaneous activity. [Mission-critical.]
2. One rover should assist the astronaut during construction tasks. [Mission-critical.]
3. One rover should monitor the astronaut when they are leaving the habitat. [Safety-critical.]
4. One rover should monitor the habitat whenever the astronaut is located inside the habitat. [Safety-critical.]

Given the above requirements, we would like to state analogues of the properties in Table 11.4. To simplify matters, let us forget about the belief operators in Section 5.

If the number k is fixed, the most obvious way to encode the above requirements is to use LTL. For the astronaut, we can introduce four propositional variables 'astrGoalPerformMisc', 'astrGoalPerformConstruction', 'astrGoalLeaveHabitat' and 'astrInHabitat', to be viewed as stating, respectively, that 'the astronaut has a goal to perform miscellaneous activity', 'the astronaut has a goal to start construction', 'the astronaut decides to leave the habitat' and 'the astronaut is in the habitat'. For the rover requirements, we can introduce for each rover r_i ($1 \leq i \leq k$) four propositional variables 'rvGoalSoloSurvey$_i$', 'rvGoalAssistConstruction', 'rvCameraStream$_i$' and 'rvGoalSoloMonitorHab$_i$', to be viewed as stating, respectively, that 'rover r_i will perform a solo geological survey', 'rover r_i will assist the astronaut's construction task', 'rover r_i will send a camera stream back to the ground station' and 'rover r_i will autonomously monitor the habitat'. Now, requirements 1–4 can be stated as follows:

1. $\Box[\text{astrGoalPerformMisc} \implies \bigwedge_{1 \leq i \leq k} \Diamond \text{rvGoalSoloSurvey}_i]$
2. $\Box[\text{astrGoalPerformConstruction} \implies \bigvee_{1 \leq i \leq k} \Diamond \text{rvGoalAssistConstruction}_i]$
3. $\Box[\text{astrGoalLeaveHabitat} \implies \bigvee_{1 \leq i \leq k} \Diamond \text{rvCameraStream}_i]$
4. $\Box[\text{astrInHabitat} \implies \bigvee_{1 \leq i \leq k} \Diamond \text{rvGoalSoloMonitorHab}_i]$

The above specification of our requirements has two main disadvantages. First, it depends on the number k being fixed. Thus, it can only answer the question 'given a value of k (e.g., 5), does the above system of the astronaut and the rovers $r_1, \ldots, r_k$ have a given property $\mathcal{P}$?', whereas it would be ideal to answer the more general question 'does the system have the property $\mathcal{P}$ *for all* values of k?'. Second, the above specification is not succinct, since it requires each rover to be mentioned individually in every formula that refers to the totality of rovers. This also makes it difficult to keep track of messages in the system in more complex scenarios that involve communication. It is natural then to seek more expressive languages than LTL that allow each individual in a system like the above to be referred to in a more abstract manner.

11.6.2   Logics for parameterised systems

Recognising the need for better abstractions and formal languages in the verification of parameterised systems, i.e., systems comprising arbitrary numbers of identical components (such as the above system of rovers), various approaches have been proposed in recent decades. Two of the most popular are model checking for parameterised and infinite state-systems [64, 65] and constraint-based verification using counting abstractions [66–68]. The model-checking approach has been applied to several scenarios verifying safety properties and some liveness properties, but is in general incomplete. Constraint-based approaches [67] do provide complete procedures for checking safety properties, but these procedures have non-primitive recursive upper bounds, and thus do not scale well for large instances. In addition, they usually lead to undecidability when applied to liveness properties.

Another approach is *first-order temporal logic* (FOTL), which can be viewed as a first-order generalisation of LTL. Although this logic is incomplete (not finitely axiomatisable) [69] and generally undecidable [70], it is valuable from a practical standpoint because a certain syntactic restriction to it, referred to as *monodic* FOTL, is *finitely axiomatisable* [71], in many cases *decidable* [70] and can naturally model systems of identical, communicating finite-state machines arising frequently in the verification of distributed systems and protocols [72, 73]. In the ensuing part, we briefly present the syntax and semantics of FOTL, as well as the syntactic restriction of monodicity and show how it can be used in the specification and verification of practical systems.

The symbols used in FOTL are predicate symbols $P_0, P_1, \ldots$, each of fixed arity (0-ary or nullary predicate symbols are allowed and correspond to propositions); variables $x_0, x_1, \ldots$, constants $c_0, c_1, \ldots$; the propositional constants $\top$ (true) and $\bot$ (false); the usual Boolean connectives ($\neg, \vee, \wedge, \Rightarrow, \Leftrightarrow$); the quantifiers $\forall$ (for all) and $\exists$ (exists); and the temporal operators $\square$ (always in the future), $\Diamond$ (sometime in the future), $\bigcirc$ (at the next moment), $\mathcal{U}$ (until) and **start** (at the first moment). Neither equality nor function symbols are allowed. The syntax of FOTL is as follows [70, 74]:

- $\top$ and $\bot$ are atomic FOTL-formulae;
- if P is an n-ary predicate symbol and t_i, $1 \leq i \leq n$, are variables or constants, then $P(t_1, \ldots, t_n)$ is an atomic FOTL-formula;
- if ϕ and ψ are FOTL-formulae, so are $\neg\phi$, $\phi \wedge \psi$, $\phi \vee \psi$, $\phi \Rightarrow \psi$, and $\phi \Leftrightarrow \psi$;
- if ϕ is an FOTL-formula and x is a variable, then $\forall x \phi$ and $\exists x \phi$ are FOTL-formulae;
- if ϕ and ψ are FOTL-formulae, then so are $\square\phi$, $\Diamond\phi$, $\bigcirc\phi$, $\phi \, \mathcal{U} \, \psi$, and **start**.

FOTL-formulae are interpreted in *first-order temporal structures*, i.e., sequences $\mathfrak{M} = \mathfrak{A}_0, \mathfrak{A}_1, \ldots$ of first-order structures over a common universe A. In more detail, if A is a non-empty set, each $\mathfrak{A}_n$ ($n \in \mathbb{N}$) is a pair $\langle A, I_n \rangle$, where I_n is an interpretation of predicate and constant symbols over A, assigning to each predicate symbol P a predicate $P^{\mathfrak{A}_n}$ on A of the same arity as P (if P is a nullary predicate, $P^{\mathfrak{A}_n}$ is simply one of the propositional constants $\top$ or $\bot$), and to each constant symbol c an element

$$
\begin{array}{lll}
\mathfrak{A}_n \models^\mathfrak{a} \top & \text{iff} & \mathfrak{A}_n \not\models^\mathfrak{a} \bot \\
\mathfrak{A}_n \models^\mathfrak{a} \textbf{start} & \text{iff} & n = 0 \\
\mathfrak{A}_n \models^\mathfrak{a} P(t_1, \ldots, t_m) & \text{iff} & \langle I_n^\mathfrak{a}(t_1), \ldots I_n^\mathfrak{a}(t_m) \rangle \in I_n(P), \text{ where } I_n^\mathfrak{a}(t_i) = I_n(t_i) \\
& & \text{if } t_i \text{ is a constant, and } I_n^\mathfrak{a}(t_i) = \mathfrak{a}(t_i) \text{ if } t_i \text{ is a variable} \\
\mathfrak{A}_n \models^\mathfrak{a} \neg \phi & \text{iff} & \mathfrak{A}_n \not\models^\mathfrak{a} \phi \\
\mathfrak{A}_n \models^\mathfrak{a} \phi \wedge \psi & \text{iff} & \mathfrak{A}_n \models^\mathfrak{a} \phi \text{ and } \mathfrak{A}_n \models^\mathfrak{a} \psi \\
\mathfrak{A}_n \models^\mathfrak{a} \phi \vee \psi & \text{iff} & \mathfrak{A}_n \models^\mathfrak{a} \phi \text{ or } \mathfrak{A}_n \models^\mathfrak{a} \psi \\
\mathfrak{A}_n \models^\mathfrak{a} \phi \Rightarrow \psi & \text{iff} & \mathfrak{A}_n \models^\mathfrak{a} \neg \phi \vee \psi \\
\mathfrak{A}_n \models^\mathfrak{a} \phi \Leftrightarrow \psi & \text{iff} & \mathfrak{A}_n \models^\mathfrak{a} (\phi \Rightarrow \psi) \wedge (\psi \Rightarrow \phi) \\
\mathfrak{A}_n \models^\mathfrak{a} \forall x \phi & \text{iff} & \mathfrak{A}_n \models^\mathfrak{b} \phi \text{ for every assignment } \mathfrak{b} \text{ that may differ} \\
& & \text{from } \mathfrak{a} \text{ only in } x \text{ and such that } \mathfrak{b}(x) \in A \\
\mathfrak{A}_n \models^\mathfrak{a} \exists x \phi & \text{iff} & \mathfrak{A}_n \models^\mathfrak{b} \phi \text{ for some assignment } \mathfrak{b} \text{ that may differ} \\
& & \text{from } \mathfrak{a} \text{ only in } x \text{ and such that } \mathfrak{b}(x) \in A \\
\mathfrak{A}_n \models^\mathfrak{a} \bigcirc \phi & \text{iff} & \mathfrak{A}_{n+1} \models^\mathfrak{a} \phi; \\
\mathfrak{A}_n \models^\mathfrak{a} \Diamond \phi & \text{iff} & \text{there exists } m \geq n \text{ such that } \mathfrak{A}_m \models^\mathfrak{a} \phi; \\
\mathfrak{A}_n \models^\mathfrak{a} \Box \phi & \text{iff} & \text{for all } m \geq n, \mathfrak{A}_m \models^\mathfrak{a} \phi \\
\mathfrak{A}_n \models^\mathfrak{a} \phi \, \mathcal{U} \, \psi & \text{iff} & \text{there exists } m \geq n, \text{ such that } \mathfrak{A}_m \models^\mathfrak{a} \psi \text{ and,} \\
& & \text{for all } i \in \mathbb{N}, n \leq i < m \text{ implies } \mathfrak{A}_i \models^\mathfrak{a} \phi.
\end{array}
$$

Figure 11.5   Semantics of FOTL

$c^{\mathfrak{A}_n}$ of A. We require that the interpretation of constants be *rigid*, i.e., $I_n(c) = I_m(c)$, for all $n, m \in \mathbb{N}$. Intuitively, each $\mathfrak{A}_n$ $(n \in \mathbb{N})$ represents the state of the world at time n and truth values in different worlds are associated via temporal operators. An *assignment* $\mathfrak{a}$ in A is a function from the set of variables $\{x_0, x_1, \ldots\}$ to A. The *truth* relation $(\mathfrak{M}, \mathfrak{A}_n) \models^\mathfrak{a} \phi$ (or simply $\mathfrak{A}_n \models^\mathfrak{a} \phi$) in the model $\mathfrak{M}$ is defined in a manner analogous to LTL. See Figure 11.5 for details. We say that $\mathfrak{M}$ is a *model* for a formula ϕ or that ϕ is *true* in $\mathfrak{M}$ if there exists an assignment $\mathfrak{a}$ such that $\mathfrak{A}_0 \models^\mathfrak{a} \phi$. A formula is *satisfiable* if it has a model and *valid* if it is true in any temporal structure under any assignment.

As discussed earlier, FOTL is incomplete [69] and generally undecidable [70]. We now define a subset of all possible FOTL-formulae, the *monodic FOTL-formulae*, that allow us to regain finite axiomatisability and in many cases decidability. An FOTL-formula is called *monodic* if any subformula of ϕ of the form $\bigcirc \psi$, $\Diamond \psi$, $\Box \psi$, or $\psi_1 \, \mathcal{U} \, \psi_2$ has *at most one* free variable. For example, the formulae $\Box \forall x (p(x) \Rightarrow \bigcirc q(x))$ and $\forall x \Diamond \exists y \, p(x, y)$ are monodic, whereas the formula $\forall x \forall y (p(x, y) \Rightarrow \bigcirc q(x, y))$ is *not* monodic. The set of all monodic FOTL-formulae form the *monodic fragment*, abbreviated MFOTL, of FOTL. MFOTL is *finitely axiomatisable* [71] and *decidable* (in 2-NExpTime) if, roughly speaking, one restricts the pure classical (first-order) part of monodic formulae to any decidable fragment of first-order logic [70].

For a larger example, let us revisit the requirements for the autonomous planetary rovers considered earlier. We shall keep the propositional variables (nullary predicates) 'astrGoalPerformMisc', 'astrGoalPerformConstruction', 'astrGoalLeaveHabitat' and 'astrInHabitat' and their intended meaning. But instead of the propositional variables corresponding to each individual rover, we now use four unary predicates 'rvGoalSoloSurvey(x)', 'rvGoalAssistConstruction(x)', 'rvCameraStream(x)' and 'rvGoalSoloMonitorHab(x)' referring to an arbitrary rover x, to be viewed as stating, respectively, that 'rover x will perform a solo geological survey', 'rover x will assist the astronaut's construction task', 'rover x will send a camera stream back to the ground station' and 'rover x will autonomously monitor the habitat'. Using MFOTL, requirements 1–4 stated previously, can now be stated as follows:

1. $\Box$[astrGoalPerformMisc $\implies \forall x \Diamond$ rvGoalSoloSurvey(x)]
2. $\Box$[astrGoalPerformConstruction $\implies \exists x \Diamond$ rvGoalAssistConstruction(x)]
3. $\Box$[astrGoalLeaveHabitat $\implies \exists x \Diamond$ rvCameraStream(x)]
4. $\Box$[astrInHabitat $\implies \exists x \Diamond$ rvGoalSoloMonitorHab(x)]

Notice that at no point is a reference made to an individual rover or the number of rovers in the system. If our specification is proved correct, its correctness applies to systems of rovers of *any* size.

From a practical standpoint, two theorem provers are available for FOTL: TeMP [75] and TSPASS [76, 77]. TeMP has been successfully applied to problems from several domains [78], in particular, to examples specified in the temporal logics of knowledge (the fusion of propositional LTL with multi-modal S5) [79–81]. TSPASS has been used (among other things) to reason about contract violations [82] and accountability [83, 84] in distributed protocols as well as the behaviour of robots and robot swarms [85]. We remark that the above provers implement FOTL with so-called *expanding domain* semantics. This allows them to use a simplified clausal resolution calculus [86]. In contrast, in our presentation of FOTL we used so-called *constant domain* semantics. This detail does not affect the reader: satisfiability with constant domain semantics and satisfiability with expanding domain semantics can be reduced to each other with only a polynomial increase in the size of the formulae [87].

11.6.3   *Translating broadcast protocols to MFOTL*

Writing specifications in MFOTL can seem difficult at first to people unfamiliar with logic. Since a lot of verification tasks involve distributed protocols, it is natural to ask whether some commonly used language for the specification of distributed protocols can automatically be translated to MFOTL. Indeed, distributed protocols are often represented as collections of finite-state machines exchanging messages. We now give a brief overview of a distributed system model comprising an arbitrary number of identical finite-state machines communicating by broadcasting messages. This model is quite expressive, capturing many interesting and useful systems, and distributed protocols described in its terms can be *automatically* be translated in MFOTL [72, 73]. In particular, it is rich enough to describe (possibly with small extensions) such diverse systems as cache coherence protocols [67] or distributed atomic commitment protocols including the two- and three-phase commit protocols [88, 89] and their modifications [90, 91].

In the aforementioned distributed system model, we have a collection of k ($k > 1$) *identical* finite-state machines in a network environment. The transitions of these finite-state machines correspond to three types of actions: (a) broadcast a message μ (denoted μ); (b) receive a message μ (denoted $\overline{\mu}$) and (c) local (i.e., an action not related to the network). The delivery of messages in the network is guaranteed. At each moment of time, each machine in the network performs an action depending on its local state at that time or is *idle*, i.e., performs no action at all. (The latter is useful for modelling asynchrony.) See [72, 73] for more technical details.

Now, given such a distributed system, say D, we can construct a MFOTL-formula $\mathcal{T}_D$ such that to each valid execution (run) of D corresponds a temporal model of $\mathcal{T}_D$ and vice versa. In other words, $\mathcal{T}_D$ completely captures the operation of D. (Again, for more details, see [72, 73].) Thus, to check whether the operation of D has a property $\mathcal{P}$, we can (assuming that $\mathcal{P}$ is expressible in MFOTL) check whether the MFOTL-formula $\mathcal{T}_D \Rightarrow \mathcal{P}$ is valid. This obviates the need to specify the operation of D in MFOTL: the specification in MFOTL can be obtained *automatically* from the state machine description of the system. $\mathcal{T}_D$ is written over the signature Σ_D containing a unary predicate symbol $P_q(x)$ for each state q of the machines, a unary predicate $A_\tau(x)$ for each transition τ of the machines (corresponding to actions (a), (b) or (c) above) and a nullary predicate (proposition) μ for each message μ that can be broadcast. Intuitively, $P_q(x)$ is to be viewed as stating that 'machine x is in state q', $A_\tau(x)$ as stating that 'machine x performs action τ' and μ as stating that 'the message μ is in transition'.

To clarify the above, let us consider an example relevant to the system of planetary rovers described earlier. Suppose that all rovers in the system are away from the astronaut, each performing a solo geological survey as described in requirement 1. Suppose, then, the astronaut requests the assistance of a rover for a construction task. Adhering to requirement 2, one of the rovers must move towards the astronaut for assistance. Suppose, further, that, for energy conservation, we would like exactly one of the rovers to go to the astronaut. (This is something that we can include in our requirements.) In distributed systems terminology, this is a scenario in which the rovers must achieve *consensus* [92]. That is, the rovers must agree on which of them will go to the astronaut. Suppose, now, there is a proposal that rover r_i ($1 \le i \le k$), e.g., go to the astronaut, and the rovers vote on whether to reject (0) or accept (1) the proposal.

Let the rovers use the following simplified (asynchronous) variant of the *FloodSet algorithm* [92, p. 105] modelled in FOTL in [73]. Each rover has a preset default bit d (0 for reject and 1 for accept). Each rover has a result bit r, which will eventually contain the result of each rover's decision (0 for reject and 1 for accept). The goal of the algorithm is for all rovers to reach a consensus, i.e., to eventually produce the same result bit. It is also required that if all rovers have been initialised

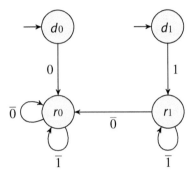

Figure 11.6   Each rover's voting behaviour

with the same default bit, that bit should be produced as a result. The protocol proceeds in the following rounds:

- At the first round, every rover broadcasts (the value of) its preset default bit.
- At every round the result bit is set to the minimum value ever received so far.

Thus, each rover's behaviour is described by the finite-state machine in Figure 11.6, where the states d_0, d_1, r_0 and r_1 denote, respectively, that 'the preset default bit is 0', 'the preset default bit is 1', 'the result bit is 0' and 'the result bit is 1'; and the transitions 0, 1, $\bar{0}$ and $\bar{1}$ correspond, respectively, to 'broadcasting 0', 'broadcasting 1', 'receiving 0' and 'receiving 1'. Now, as discussed earlier, let $\mathcal{T}_{\text{Flood}}$ be the MFOTL-formula that captures the above system, over the signature Σ_{Flood} containing the unary predicates $P_{d_0}(x)$ ('rover x is in state d_0'), $P_{d_1}(x)$ ('rover x is in state d_1'), $P_{r_0}(x)$ ('rover x is in state r_0') and $P_{r_1}(x)$ ('rover x is in state r_1'); $A_0(x)$ ('rover x broadcasts 0'), $A_1(x)$ ('rover x broadcasts 1'), $A_{\bar{0}}(x)$ ('rover x receives 0') and $A_{\bar{1}}(x)$ ('rover x receives 1'); and the nullary predicates (propositions) 0 ('message 0 in transition') and 1 ('message 1 in transition'). Recall that $\mathcal{T}_{\text{Flood}}$ can be obtained automatically from Figure 11.6, so we need not know its content. To prove that the algorithm achieves the informal goal stated above, we check that the following MFOTL-formula is valid:

$$\mathcal{T}_{\text{Flood}} \implies \Diamond(\forall x\, P_{r_0}(x) \lor \forall x\, P_{r_1}(x)).$$

Thus using FOTL provides a route to verifying multi-robot systems with an arbitrary number of identical components. As we have already mentioned, while full FOTL is incomplete and in general undecidable, using monodic FOTL-formulae, we regain finite axiomatisability and in many cases decidability. However, the disadvantage is that this restriction limits what we can express in the logic. Also the theorem-proving tools for FOTL are not as developed in terms of usability as many model checkers.

11.7   Conclusions, recommendations and future trends

Verification techniques such as formal verification, simulation and testing are useful when ensuring systems are safe, trustworthy and meet their stated requirements. They are needed for space robotics as failures in space may be much more critical, costly and harder to resolve. We have discussed several tools and techniques for verification of space robotics with reference to some simple space scenarios.

Recommendations include designing systems for verification using a modular approach separating concerns, embedding verification and validation into engineering process, and using a range of tools and techniques to improve confidence in space systems. Future trends include the greater need for and use of autonomy in space robotics, e.g., to support planetary missions with robots working closely with astronauts, where safety and functional correctness is crucial. Additionally, verification and validation is needed for the New Space sector to conform regulation and standards for applications such as satellite communication, imaging, navigation, space tourism and mining.

References

[1]   Clarke E.M., Grumberg O., Peled D. *Model checking*. MIT Press; 1999.
[2]   Holzmann G.J. *The spin model checker: primer and reference manual*. Addison-Wesley; 2003.
[3]   Fisher M. *An introduction to practical formal methods using temporal logic*. Wiley; 2011.
[4]   Pnueli A. 'The temporal logic of programs'. *Symposium on the Foundations of Computer Science*; IEEE; 1977. pp. 46–57.
[5]   Gabbay D., Pnueli A., Shelah S., *et al*. 'On the temporal analysis of fairness'. *ACM SIGPLAN-SIGACT Symposium on Principles of Programming Languages*; 1980. pp. 163–73.
[6]   Clarke E.M., Emerson E.A. 'Design and synthesis of synchronisation skeletons using branching time temporal logic'. *Workshop on the Logic of Programs. Vol. 131 of LNCS*. Springer; 1981. pp. 52–71.
[7]   Fisher M., Dixon C., Peim M. 'Clausal temporal resolution'. *ACM Transactions on Computational Logic*. 2001;**2**(1):12–56.
[8]   Hustadt U., Konev B. 'TRP++ 2.0: a temporal resolution prover'. *Automated Deduction—CADE-19. Vol. 2741 of LNAI*. Springer; 2003. pp. 274–8.
[9]   Zhang L., Hustadt U., Dixon C. 'A resolution calculus for the branching-time temporal logic CTL'. *ACM Transactions on Computational Logic*. 2014;**15**(1):1–38.
[10]   Zhang L., Hustadt U., Dixon C. 'CTL-RP: a computation tree logic resolution prover'. *AI Communications*. 2010;**23**(2-3):111–36.
[11]   Araiza-Illan D., Western D., Pipe A.G. 'Systematic and realistic testing in simulation of control code for robots in collaborative human-robot interactions'.

Towards Autonomous Robotic Systems. Vol. 9716 of LNCS. Springer; 2016. pp. 20–32.

[12]   Salem M., Lakatos G., Amirabdollahian F. 'Would you trust a (faulty) robot?: effects of error, task type and personality on human-robot cooperation and trust'. *ACM/IEEE International Conference on Human-Robot Interaction*; 2015. pp. 141–8.

[13]   Webster M., Western D., Araiza-Illan D., *et al.* 'A corroborative approach to verification and validation of human–robot teams'. *The International Journal of Robotics Research.* 2020;**39**(1):73–99.

[14]   Luckcuck M., Farrell M., Dennis L.A., *et al.* 'Formal specification and verification of autonomous robotic systems'. *ACM Computing Surveys.* 2019;**52**(5):1–41.

[15]   Farrell M., Luckcuck M., Fisher M. 'Robotics and integrated formal methods: necessity meets opportunity'. *Integrated formal methods. Vol. 11023 of LNCS.* Springer; 2018. pp. 161–71.

[16]   Luckcuck M., Farrell M., Dennis L.A. 'A summary of formal specification and verification of autonomous robotic systems'. *Integrated Formal Methods. Vol. 11918 of LNCS.* Springer; 2019. pp. 538–41.

[17]   Cresswell M.J., Hughes G.E. *A new introduction to modal logic.* Taylor and Francis; 1997.

[18]   Fagin R., Halpern J.Y., Moses Y., *et al. Reasoning about knowledge.* MIT Press; 1995.

[19]   Hansson H., Jonsson B. 'A logic for reasoning about time and reliability'. *Formal Aspects of Computing.* 1994;**6**(5):512–35.

[20]   Koymans R. 'Specifying real-time properties with metric temporal logic'. *Real-Time Systems.* 1990;**2**(4):255–99.

[21]   Blackburn P., van Benthem J., Wolter F. (eds.). *Handbook of modal logic: vol. 3 of studies in logic and practical reasoning.* North-Holland; 2007. Available from https://www.sciencedirect.com/bookseries/studies-in-logic-and-practical-reasoning/vol/3/suppl/C.

[22]   Nalon C., Dixon C., Hustadt U. 'Modal resolution: proofs, layers, and refinements'. *ACM Transactions on Computational Logic.* 2019;**20**(4).

[23]   Nalon C., Hustadt U., Dixon C. 'A resolution-based theorem prover for kn: architecture, refinements, strategies and experiments'. *Journal of Automated Reasoning.* 2020;**64**(3):461–84.

[24]   Hustadt U., Ozaki A., Dixon C. 'Theorem proving for metric temporal logic over the naturals'. *Automated Deduction - CADE 26. vol. 10395 of LNCS.* Springer; 2017. pp. 326–43.

[25]   Hustadt U., Ozaki A., Dixon C. 'Theorem proving for pointwise metric temporal logic over the naturals via translations'. *Journal of Automated Reasoning.* 2020;**64**(8):1553–610.

[26]   Li J., Vardi M.Y., Rozier K.Y. 'Satisfiability checking for mission-time LTL' in Dillig I., Tasiran S. (eds.). *Computer Aided Verification – 31st International Conference, CAV 2019, New York City, NY, USA, July 15–18, 2019, Proceedings, Part II. vol. 11562 of Lecture Notes in Computer Science.* **2019**. Springer; 2019. pp. 3–22.

[27] Farrell M., Cardoso R.C., Dennis L.A., *et al.* 'Modular verification of autonomous space robotics'. *Assuring Autonomy for Space Missions Workshop*; 2019.

[28] Abrial J.R. *Modeling in Event-B: system and software engineering.* Cambridge University Press; 2010.

[29] Spivey J.M. *Understanding Z: a specification language and its formal semantics.* **3**. Cambridge University Press; 1988.

[30] Hoare C.A.R. 'Communicating sequential processes'. *Communications of the ACM.* 1978;**21**(8):666–77.

[31] Astesiano E., Bidoit M., Kirchner H., *et al.* 'CASL: the common algebraic specification language'. *Theoretical Computer Science.* 2002;**286**(2):153–96.

[32] Dennis L.A., Fisher M., Webster M.P., *et al.* 'Model checking agent programming languages'. *Automated Software Engineering.* 2012;**19**(1):5–63.

[33] Broy M. 'A logical approach to systems engineering artifacts: semantic relationships and dependencies beyond traceability—from requirements to functional and architectural views'. *Software & Systems Modeling.* 2018;**17**(2):365–93.

[34] Champion A., Gurfinkel A., Kahsai T. CoCoSpec: A mode-aware contract language for reactive systems. *International Conference on Software Engineering and Formal Methods. vol. 9763 of LNCS*; Springer; 2016. pp. 347–66.

[35] Cimatti A., Dorigatti M., Tonetta S. OCRA: A tool for checking the refinement of temporal contracts. *International Conference on Automated Software Engineering (ASE)*; IEEE; 2013. pp. 702–5.

[36] Cofer D., Gacek A., Miller S., *et al.* 'Compositional verification of architectural models'. *NASA Formal Methods Symposium. vol. 7226 of LNCS*; Springer; 2012. pp. 126–40.

[37] Kornfeld R.P., Prakash R., Devereaux A.S., Greco M.E., Harmon C.C., Kipp D.M. 'Verification and validation of the Mars science laboratory/Curiosity rover entry, descent, and landing system'. *Journal of Spacecraft and Rockets.* 2014;**51**(4):1251–69.

[38] Francis R., Estlin T., Doran G., *et al.* 'AEGIS autonomous targeting for ChemCam on mars science laboratory: deployment and results of initial science team use'. *Science Robotics.* 2017;**2**(7):eaan4582.

[39] Cardoso R.C., Farrell M., Luckcuck M., *et al.* 'Heterogeneous verification of an autonomous curiosity rover'. *NASA Formal Methods.* Cham: Springer International Publishing; 2020. pp. 353–60.

[40] Quigley M., Conley K., Gerkey B., *et al.* 'ROS: an open-source robot operating system'. *Workshop on Open Source Software.* Japan: IEEE; 2009.

[41] Leitão P. 'Agent-based distributed manufacturing control: a state-of-the-art survey'. *Engineering Applications of Artificial Intelligence.* 2009;**22**(7):979–91.

[42] Dennis L.A., Farwer B. 'Gwendolen: a BDI language for verifiable agents'. *Logic and the Simulation of Interaction and Reasoning.* AISB; 2008. pp. 16–23.

[43] Rao A.S., Georgeff M. 'BDI agents: From theory to practice'. *International Conference on Multi-Agent Systems*. AAAI; 1995. pp. 312–9.

[44] Visser W., Havelund K., Brat G., *et al.* 'Model checking programs'. *Automated Software Engineering*. 2002;**10**(2):3–11.

[45] Leucker M., Schallhart C. 'A brief account of runtime verification'. *The Journal of Logic and Algebraic Programming*. 2009;**78**(5):293–303.

[46] Cardoso R.C., Dennis L.A., Fisher M., EMAS. 'Plan library reconfigurability in BDI agents'. *International Workshop on Engineering Multi-Agent Systems*; 2019. pp. 1–16.

[47] Grotzinger J.P., Crisp J., Vasavada A.R., *et al.* 'Mars science laboratory mission and science investigation'. *Space Science Reviews*. 2012;**170**(1):5–56.

[48] Trevino R.C., Kosmo J.J., Ross A., Cabrol N. *'First astronaut-rover interaction field test'*. International Conference On Environmental Systems; 2000.

[49] Landis G.A. 'Robots and humans: synergy in planetary exploration'. *AIP conference proceedings. Vol. 654*. American Institute of Physics; 2003. pp. 853–60.

[50] Pedersen L. *'Field demonstration of surface human-robotic exploration activity'*. 9. AAAI Spring Symposium; 2006.

[51] Medina A., Pradalier C., Paar G., *et al.* 'A servicing rover for planetary outpost assembly'. *Advanced Space Technologies for Robotics and Automation*; 2011.

[52] Ransan M., Atkins E.M. *'A collaborative model for astronaut-rover exploration teams'*. AAAI Spring Symposium; 2006. pp. 52–8.

[53] Heiskanen P., Heikkilä S., Halme A. *'Development of a Dynamic Mobile Robot Simulator for Astronaut Assistance'*. *Workshop on Advanced Space Technologies for Robotics and Automation*; 2008.

[54] Fong T., Nourbakhsh I. 'Peer-to-peer human-robot interaction for space exploration from interfaces to intelligence'. *AAAI Fall Symposium: The Intersection of Cognitive Science and Robotics*; 2004.

[55] Sierhuis M. Modeling and Simulating Work Practice. BRAHMS: a multiagent modeling and simulation language for work system analysis and design [PhD. Thesis]. Social Science and Informatics (SWI), University of Amsterdam, SIKS Dissertation Series No. 2001-10, Amsterdam, The Netherlands; 2001.

[56] Sierhuis M., Clancey W.J. 'Modeling and simulating practices, a work method for work systems design'. *IEEE Intelligent Systems*. 2002;**17**(5):32–41.

[57] Sierhuis M., Clancey W.J., Alena R.L., *et al.* 'NASA's mobile agents architecture: a multi-agent workflow and communication system for planetary exploration'. International Symposium on Artifical Intelligence, Robotics and Automation in Space; 2005.

[58] Sierhuis M., Clancey W.J., Vanhoof R., *et al.* 'ASA's OCA mirroring system: an application of multiagent systems in mission control'. International Conference on Autonomous Agents and Multi-Agent Systems; 2009.

[59] Clancey W.J., van Hoof R. *'The metabolic rate advisor: Using agents to integrate sensors and legacy software'*. NASA; 2013.

[60] Stocker R., Sierhuis M., Dennis L.A. 'A formal semantics for Brahms'. *International workshop on computational logic in multi-agent systems. Vol. 6814 of lncs.* Springer; 2011. pp. 259–74.

[61] Stocker R., Dennis L.A., Dixon C., *et al.* 'Verification of brahms human–robot teamwork models'. *European Conference on Logics in Artificial Intelligence. vol. 7519 of LNCS*; Springer; 2012. pp. 385–97.

[62] Webster M., Dixon C., Fisher M., *et al.* 'Toward reliable autonomous robotic assistants through formal verification: a case study'. *IEEE Transactions on Human-Machine Systems.* 2016;**46**(2):186–96.

[63] Fisher M. *An introduction to practical formal methods using temporal logic.* Wiley; 2011.

[64] Abdulla P.A., Jonsson B., Nilsson M., *et al.* 'Regular model checking for LTL(MSO)'. *International Conference on Computer Aided Verification. vol. 3114 of LNCS*; Springer; 2004. pp. 348–60.

[65] Abdulla P.A., Jonsson B., Rezine A., *et al.* 'Proving liveness by backwards reachability'. *International Conference on Concurrency Theory. vol. 4137 of LNCS*; Springer; 2006. pp. 95–109.

[66] Delzanno G. 'Automatic verification of parameterized cache coherence protocols'. *International Conference on Computer Aided Verification. vol. 1855 of LNCS*; Springer; 2000. pp. 53–68.

[67] Delzanno G. 'Constraint-based verification of parameterized cache coherence protocols'. *Formal Methods in System Design.* 2003;**23**(3):257–301.

[68] Esparza J., Finkel A., Mayr R. 'On the verification of broadcast protocols'. *Symposium on Logic in Computer Science.* IEEE Computer Society Press; 1999. pp. 352–9.

[69] Szalas A., Holenderski L. 'Incompleteness of first-order temporal logic with until'. *Theoretical Computer Science.* 1988;**57**(2–3):317–25.

[70] Hodkinson I., Wolter F., Zakharyaschev M. 'Decidable fragments of first-order temporal logics'. *Annals of Pure and Applied Logic.* 2000;**106**(1–3):85–134.

[71] Wolter F., Zakharyaschev M. 'Axiomatizing the monodic fragment of first-order temporal logic'. *Annals of Pure and Applied Logic.* 2002;**118**(1–2):133–45.

[72] Fisher M., Konev B., Lisitsa A. 'Practical infinite-state verification with temporal reasoning'. *Verification of infinite state systems and security. Vol. 1 of NATO security through science series: information and communication.* IOS Press; 2006. pp. 91–100.

[73] Dixon C., Fisher M., Konev B., *et al.* 'Practical first-order temporal reasoning'. *International Symposium on Temporal Representation and Reasoning. IEEE*; 2008. pp. 156–63.

[74] Degtyarev A., Fisher M., Konev B. 'Monodic temporal resolution'. *ACM Transactions on Computational Logic.* 2006;**7**(1):108–50.

[75] Hustadt U., Konev B., Riazanov A., *et al.* 'TeMP: A temporal monodic prover'. *International Joint Conference on Automated Reasoning. vol. 3097 of LNCS*; Springer; 2004. pp. 326–30.

[76]   Ludwig M., Hustadt U. 'Fair derivations in monodic temporal reasoning'. *International Conference on Automated Deduction*; Springer; 2009. pp. 261–76.

[77]   Ludwig M., Hustadt U. 'Implementing a fair monodic temporal logic prover'. *AI Communications*. 2010;**23**(2–3):69–96.

[78]   Fernández-Gago M.C., Hustadt U., Dixon C., *et al.* 'First-order temporal verification in practice'. *Journal of Automated Reasoning*. 2005;**34**(3):295–321.

[79]   Dixon C., Fisher M., Wooldridge M. 'Resolution for temporal logics of knowledge'. *Journal of Logic and Computation*. 1998;**8**(3):345–72.

[80]   Dixon C. 'Using temporal logics of knowledge for specification and verification—a case study'. *Journal of Applied Logic*. 2006;**4**(1):50–78.

[81]   Dixon C., Fisher M., Konev B. 'Is there a future for deductive temporal verification?' *International Symposium on Temporal Representation and Reasoning. IEEE Computer Society Press*; 2006. pp. 11–18.

[82]   Halle S. 'Causality in message-based contract violations: A temporal logic "Whodunit"'. *International Enterprise Distributed Object Computing Conference*. IEEE; 2011. pp. 171–80.

[83]   Benghabrit W., Grall H., Royer J.C. 'Checking accountability with a prover'. *Computer Software and Applications Conference*. **2**. IEEE; 2015. pp. 83–8.

[84]   Benghabrit W., Grall H., Royer J.C. 'Abstract accountability language: Translation, compliance and application'. *Asia-Pacific Software Engineering Conference. IEEE*; 2015. pp. 214–21.

[85]   Behdenna A., Dixon C., Fisher M. 'Deductive verification of simple foraging robotic behaviours'. *International Journal of Intelligent Computing and Cybernetics*. 2009;**2**(4):604–43.

[86]   Konev B., Degtyarev A., Dixon C., Fisher M., Hustadt U. 'Mechanising first-order temporal resolution'. *Information and Computation*. 2005;**199**(1–2):55–86.

[87]   Wolter F., Zakharyaschev M. 'Decidable fragments of first-order modal logics'. *Journal of Symbolic Logic*. 2001;**66**(3):1415–38.

[88]   Gray J. 'Notes on database operating systems'. *Operating systems: an advanced course*. Springer; 1978. pp. 393–481.

[89]   Skeen D. 'Nonblocking commit protocols'. *SIGMOD International Conference on Management of Data. ACM Press*; 1981. pp. 133–42.

[90]   Chkliaev D., van der Stok P., Hooman J. 'Mechanical verification of a nonblocking atomic commitment protocol'. *Workshop on Distributed System Validation and Verification*. IEEE; 2000. pp. 96–103.

[91]   Chkliaev D., Hooman J., van der Stok P. 'Mechanical verification of transaction processing systems'. *International Conference on Formal Engineering Methods. IEEE*; 2000. pp. 89–97.

[92]   Lynch N.A. *Distributed algorithms*. Elsevier; 1996.

Chapter 12

Cyber security of New Space systems

Carsten Maple[1], Ugur Ilker Atmaca[1], Gregory Epiphaniou[1], Gregory Falco[2], and Hu Yuan[1]

Space systems are an essential part of the national and international infrastructure of data communication, environment monitoring, scientific experiments, global positioning, navigation, timing, and so forth. The space environment is currently experiencing a significant transformation by the adoption of advancing devices with autonomous systems and the entrance of new players. Traditionally such space missions have been a governmental research-based initiative; however, it is now tightly coupled with commercialisation. The term New Space (and indeed Alt Space and NewSpace) has been used to describe this commercialisation of space. Furthermore, the Internet of Space Things (IoST), which facilitates connectivity at low cost, is also gaining momentum. The barriers to entry to space have fallen and the industry is expanding beyond the traditional large players such as state-level space agencies by the advances in such technologies [1]. These IoST systems feature a myriad of components and may be developed by a range of companies from a number of countries. Such satellite systems are launched carrying multiple payloads, each comprising many computing components. These developments are leading to a change in the cyber security threat landscape of space systems. Consequently, there will be a significant number of attack vectors for adversaries to exploit, and previously infeasible threats will now need to be managed. With the vast proliferation of cyber threats, the operation theatre is gradually changing, hindering significant gaps in cyber decision-making capabilities when it comes to defence. Space systems emerge as a new operation theatre in warfare with attacks against them no longer confined to the realm of conventional communication means. This trend is also underpinned by our limited capacity to accurately quantify the imminent security risk of these systems taking into consideration the rapid change and variability in threat landscapes and associated vulnerabilities.

This chapter presents the details of our recent works on the security of such emerging space systems. Specifically, we present a method for identifying the attack surface of New Space systems. We describe a reference architecture (RA) to

[1]Warwick Manufacturing Group, University of Warwick, Coventry, CV4 7AL, United Kingdom
[2]Stanford University, Freeman Spogli Institute, Stanford, CA 94305, United States

provide a visual aid to support the identification of attack surfaces. We then describe approaches for threat modelling that can be used for space system and discuss the requirements for effective threat modelling. We also present the challenges in cyber risk management for space systems and highlight the key characteristics for a thorough risk management approach and framework following works in [2, 3]. Furthermore, we describe a method for assurance of space systems which integrates threat modelling into the formal verification methodology of the system [4].

12.1   A reference architecture for attack surface analysis in space systems

As the New Space systems offer increasing connectivity and autonomy, they may be subject to cyber-physical threats which were previously inapplicable. It is necessary to identify which components of the system are accessible to a threat actor, what vulnerabilities those components have and via which interaction a threat actor can propagate a cyberattack. This is known as the attack surface analysis. One approach to analyse the attack surface of a system is defining the RA of the system and instantiating it with components applicable for a specific use case or scenario. RAs have been used previously for other industries such as smart homes and connected and autonomous vehicles for a better understanding of changes where there are new agile entrants akin to New Space systems. Both the transition and sustainability of these systems require a wider systems design approach. A systems approach is required for investigating the interaction of New Space components with networks and for a holistic view of security from the component to systems-level integration for commercial and military applications.

RAs can abstractly model the system via multiple viewpoints. The viewpoints need to be carefully selected to avoid over-specification of the system for the RA's intended use case. Typical viewpoints of RAs include [5, 6]:

- Functional: how the components work and what tasks they have
- Communication: how the components interact
- Implementation: how the components are implemented
- Enterprise: the relation between organisations and users
- Usage: concerns of the expected use of the system
- Information: the types of information handled by the system
- Physical: the physical objects in the system and their connections.

To analyse the attack surface of the system, the functional and communication viewpoints are vital since a threat actor may use communication channels to interact with the system to alter how it functions [7]. The implementation of the system is not only useful when analysing a low-level aspect of it, but also when performing a high-level analysis to specify how each component is implemented. Such information is usually unavailable. The enterprise and usage viewpoints are not usually used for a cyber security analysis, but they may be useful to investigate human interaction with the system's cyber security. The information viewpoint is important to specify what information is handled by the system, how information flows and how it is transformed. From the attack surface analysis perspective, it is not strictly required to specify the information flow but it is needed to know which functional components have access to information. Since New Space systems must be concerned with cyber-physical threats, it is also needed to define the physical components to understand the impacts of the attacks against them. This is an important aspect, as electronic attacks can now manifest an adverse effect in the physical and natural domain within which space systems usually integrate. Therefore, the functional, communication and physical viewpoints are specified in this study. To reduce the complexity, it is useful to avoid including viewpoints that are less relevant for a cyber security analysis. For example, including an information viewpoint would be helpful if confidentiality is vital for a system. However, for a general RA used for a cyber security analysis, this viewpoint is difficult to define without a priori knowledge of the system.

RAs have been previously used for space systems in a variety of contexts. These include an on-board software (AMASS Reference Tool Architecture [8], ORSA-P [9, 10], SAVOIR OSRA [11]), data systems (RASDS [12]) and mission design [13, 14]. These RAs could be adapted for the use of cyber security analysis of space systems with limited collective security measures in place. However, the additional information on RAs is usually too detailed for a high-level cyber security analysis. Thus we define high-level specifications of space systems in terms of their functionality and the interactions of the components. It provides opportunities to perform the initial cyber security analysis of a space system before focusing on the identified areas of interest. Our reference architecture for attack surface analysis (RASA) focuses on specifying a hybrid functional-interaction viewpoint to simplify the cyber security analysis of existing and emerging space systems including robotic applications in space. The robotic applications include satellite actuators such as robotic grabbing arms to enable the physical interactions with other objects in space and planetary robots. The planetary robots are mainly used for

- Exploration of a planetary surface: In this case, the surface robot or rovers involve the exploration of an unstructured terrain to find and classify a number of predefined targets such as mineral products. Once the target is identified, a required activity would be performed on the target depending on the category of drilling or sample capture and return. These robots are autonomous based on information preloaded and from the on-board sensors.
- Construction on a planetary surface: In this case, the tasks are controlled and monitored by an operator through the planetary surface station. The main difference from the full autonomous scenario is the robots are fully controlled and the tasks are explicitly specified, such as rescuing other robots.

According to [15], a cyberattack can be carried out by compromising either an endpoint instantiation or a communication medium. The endpoint instantiation attacks aim to disable the functioning or controlling of an operator or artificial intelligence whereas the threat actor can eavesdrop into the network or inject malware to manipulate the functioning of the robot by compromising the communication medium.

The attack surface is identified by the interactions that cross the boundary of the system, namely a trust domain, and the path a threat actor can take to compromise further components has been specified by the internal system interactions. We have divided New Space systems into three sub-architecture as satellites, planetary robots and the ground segment, and we have specified the key interactions such as (i) internal/external communication, (ii) internal/external interaction and (iii) environment interaction/sensing. In this work, we design and deploy our RASA to identify the trust boundaries of a New Space system by defining the latter's capabilities and characteristics. We also describe the methodology used using an autonomous debris collection as a use case.

An autonomous debris collection use case has been used to instantiate our RASA. Such systems aim to reduce the potential for collisions with this debris [16]. The European Space Agency (ESA) estimates that there are almost 129 million pieces of debris and approximately 22 000 of them are catalogued in ESA's Database and Information System Characterising Objects in Space [17]. Several techniques have been proposed to collect and remove debris such as capturing with arms or nets [18] or electro-dynamic tethers [19]. We consider an autonomous satellite which relies upon ground-based debris detection and in-orbit sensors [20]. It physically captures and de-orbits debris where there is no human-in-the-loop controlling the satellite. We have described the RAs in Figures 12.1 and 12.2 to model the functional components of autonomous debris collection and analyse the attack surface.

To describe the attack tree, we identified all system functions with relevant components from the RAs and then the corresponding threats for each of these components. The threats have been classified based on STRIDE [21], and the attack surface has been decomposed in three sub-domains: (i) space debris database in the ground segment, (ii) satellite-to-ground segment communication and (iii) satellite as shown in Figure 12.3.

The satellite receives information of debris positions from the database in the ground segment, which could be then stored in their on-board data storage unit, and the debris database is updated accordingly. Based on this communication, a number of attacks could be possible:

- Software updates could be maliciously falsified (tampering) to cause fuel wastage or adverse financial impact on rivals.
- The communication channel could be remotely spoofed by a threat actor.
- The communication channel can be jammed (a type of DoS) to block commands or database updates from being delivered to the satellite.
- Use a programming or debugging interface to read or reprogram a device.
- Replace or bypass hardware or software pieces on the devices or subsystems.
- Extract information from the device by examining buses or individual components.

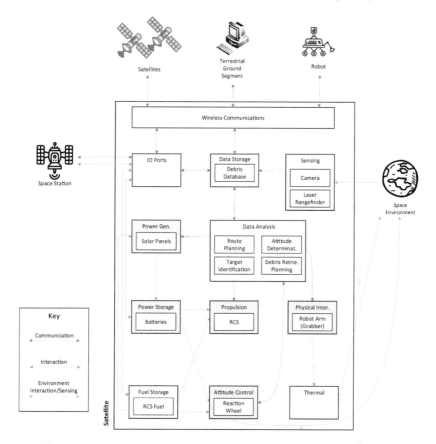

*Figure 12.1   Instantiation of functional viewpoint of satellite reference
architecture for autonomous debris collection [1]*

Such a use case can be also compromised by conducting an attack on the satellite in charge of the collection task. The IO ports and wireless communications can be targeted to exploit an elevation of privilege attack. The on-board satellite sensors can be blinded to exploit a DoS attack. The likelihood of conducting these attacks is highly low for the majority of existing threat actors in Table 12.1. However, the likelihood will also rise by the change of space ecosystem by the reduction of deployment barriers.

12.2   Threat modelling

Threat modelling (TM) is the structured way of identifying a system's vulnerabilities, threat actors, cyber risks and impacts as well as recommending appropriate countermeasures to mitigate the risks and impacts. TM approaches are usually divided into three main classes as attacker-centric, asset-centric and software-centric regarding

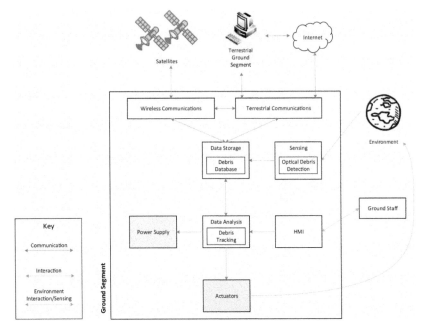

Figure 12.2   Instantiation of functional viewpoint of ground segment reference architecture for autonomous debris collection [1]

their focus point. However, PASTA [22] is considered as a risk-centric approach apart from this classification. The attacker-centric approaches focus on identifying the threat actors, evaluating their goals, motivations and the skill set to make predictions on how they can achieve these goals. The asset-centric approaches identify each asset and find the value of those assets for a system or organisation. They often utilise the information derived from asset classification schemes and attack-tree scenarios to visualise the ways an asset can be attacked. The focus of these approaches is to find the vulnerabilities and deploy mitigation controls prioritised and scaled according to the value of the corresponding assets. The software-centric approaches also referred to as system-centric, design-centric or architecture-centric. It begins by modelling the design of the system. The focus of these approaches is to discover the possible attacks that target the instantiating (i.e., assets or functions) of the system model.

Space systems are a specific type of cyber-physical systems (CPSs) which includes collaborating computation and physical capabilities to enable explore, observe and visit the extraterrestrial environment beyond the earth [23]. These systems may include sort of control, sensor-based, robotic and autonomous subsystems. The instantiations of these subsystems may be distributed and physically or logically accessible by skilful and motivated adversaries. This seems to push the security boundaries between engineering and information systems that often these subsystems access and expand the threat landscapes associated with the coupling

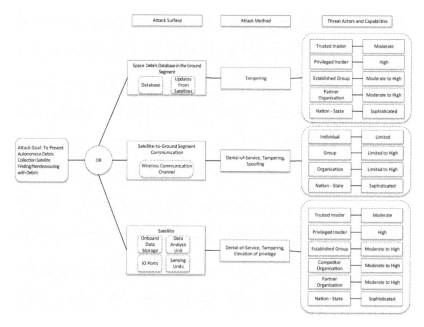

Figure 12.3   *Attack tree for adversary aiming to prevent autonomous debris collection [1]*

of physical objects with computational resources. Besides, CPS are also prone to cyberattacks which target their data management and communication layers. Due to their complex nature, it is increasingly challenging to secure a CPS against a diverse range of possible attacks [24]. It is critical to identify the system vulnerabilities, the attack surface and their potential impact.

To analyse the attack surface of a specific subsystem instantiating, it is necessary to understand the threat actor profiles including their motivations, capabilities, environment and resources. Current changes in the space ecosystem create new avenues such as commercialisation opportunities for third-party suppliers, space colonisation, space tourism and capitalisation of space resources. This will create and enable opportunities for cyber-organised crime rings to derive business models and change their motivation, goals and impact [25]. The abstract threat actors are shown in Table 12.1 which are derived from [26] and the dimensions to describe them are derived from [27]. The focus of the table considers only adversarial threats sources; thus, the accidental, structural and environmental threat sources exist in [26] are excluded. However, such threat sources also have security implications and they will need to be analysed and mitigated by other techniques. For instance, high radiation in space may lead to change the bits in security-critical areas of memory.

Table 12.1   Threat actors with space ecosystem specific examples [1]

	Threat actor	Example	Goals and motivations	Capabilities	Environment	Resources
Individual	Outsider	Hacktivist	Personal satisfaction; passion; ideology. Does not believe in climate change, wants to impact functioning of climate satellite	Limited	Remote access	Minimal
	Insider	Cleaner	Financial gain; discontent	Limited	Permissionless internal access	Internal knowledge
	Trusted insider	Contractor	Financial gain; discontent	Moderate	Internal access with some permissions	Internal knowledge
	Privileged insider	Employee	Financial gain; discontent	High	Internal access with high permissions	Internal knowledge
Group	Ad hoc	A group coming together over a time-critical event (e.g., Brexit, or a collective movement of Extinction Rebellion)	Dependent on group purpose: ideological, financial, political	Limited to moderate	Remote access	Limited knowledge and financial
	Established	A group (e.g., the anonymous group)		Moderate to high	Remote access	Moderate knowledge and financial

(Continues)

Table 12.1 Continued

	Threat actor	Example	Goals and motivations	Capabilities	Environment	Resources
Organisation	Competitor	An organisation about to compete for a tender for services	Corporate espionage; financial gain; reputation damage	Organisation size-related	Remote access	Organisation size-related
	Supplier	A supplier who fears their services are soon to be relinquished	Information gain; financial gain		Remote access; knowledge of internal structure	
	Partner	A partner with whom a relationship is starting to sour or is soon to end	Information gain; financial gain		Limited internal access; knowledge of internal structure	
	Customer	A customer who feels they have had poor or unfair service	Information gain; financial gain		Remote access; knowledge of internal structure	
	Nation state	Geopolitical rival	State rivalry; geopolitics	Sophisticated; coordinated; access to state secrets	Remote and internal access	Extensive knowledge; extensive financial; advanced equipment

12.2.1　Cyber security requirements

Information security and cyber security are usually used as synonymous terms, but the two definitions are not exactly the same [28, 29]. Cyber security is not limited to the security of the information in a computation system. However, information security is considered as a subset of information governance, and it is not solely limited to the computation systems alone [30]. Similar to other CPS, securing space systems entails the protection of both physical assets and information systems these assets access to fulfil their critical mission tasks and objectives. Authors of [31] argue that in CPS the attack prevention can work by obfuscating context from the knowledge and limit access to adversaries. The argument is based on the assumption that access to any kind of resources or data can provide adversaries with enough context to derive information on how successful exploitation can be achieved. Simply put, it covers security in both hardware (i.e., physical instantiating) and software components, in a term often described as converged security [32]. Thus, the CIA triad principles (Confidentiality, Integrity and Availability) of information systems do not cover all of the security concerns regarding such systems but it should be complemented with Authenticity and Safety [2]:

- Confidentiality: It refers to the capability of preventing the disclosure of information by an unauthorised individual or a system [33]. It may be enforced by encrypting the messages during transmission, by limiting the places where they can be accessed such as databases, log files and backups, and by limiting access of individuals with appropriate access control in place. The countermeasures against confidentiality breaches hinder threat actors from inferring from the system by eavesdropping on the communication channels used for the CPS or between multiple cooperating CPSs.
- Integrity: It refers that the information cannot be modified by an unauthorised individual or a system. Achieving integrity for CPS can be defined as preventing and detecting the deception cyberattack on the information on transmission between the instantiations of CPS or multiple cooperating CPSs [34].
- Availability: The services of CPS is aimed to be available and functioning when it is needed. The availability of the services can be stopped or restricted by preventing computing, controlling, communication corruptions due to hardware failures, power outages or denial-of-service cyberattacks by a threat actor [35].
- Authenticity: The transactions and communications between space systems and ground segments must be genuine and distinguished from malicious messages. In New Space systems, authentication relates to all the relevant process such as sensing, communication and actuations [36, 37].
- Safety: Safety is described as the avoidance of hazards to the physical environment due to operations by ISO 60601 for medical electrical equipment and systems. Such definition of safety can be also applied to cyber security of New Space system by expanding the scope of hazard to operational failures due to a cyberattack [36]. As an example of this, a threat actor may attempt to alter

the orbiting path of a satellite to cause a physical damage for other satellites, consequently to prevent a critical operation.

These principles are the main requirements to meet for securing space systems. With these principles in mind, consideration must be made for particular features that enable evaluation metrics that can be used in threat modelling. Below, the proposed features are listed and briefly explained. As a note, the features of ground segments are not included since they are highly similar to the security of enterprises and organisations.

- Ease-of-access by threat actors: Space systems may be attacked by adversaries of varying capabilities from nation states to individuals with limited technical skills. Over time, accessibility to space technology has become easier and cheaper with developing states using private instantiations to launch their own satellites rather than time-sharing existing ones. With these barriers to access and affordability lowered, potential threat actors are finding it easier to gain access to these systems' architectures and, thus, identify vulnerabilities and adapt their attack strategies. Recent works indicate that the number of cyberattacks against Critical National Infrastructure (CNI) is alarmingly increasing due to the integration of legacy and critical mission systems in the cyberspace [38–40].
- Variety of location: The physical instantiations in the space systems encompass a wide variety of physical locations in space [41]. That makes satellite applications rather unique, as proximity and time constraints play a key role in their effectiveness in completing a task. For example, the visibility of a ground station to the satellite in a given time window within which a task must be completed may be prohibited by skilful and resourceful adversaries. Given also the fact that satellites often outnumber ground stations, the competition for access to infrastructures available might be another operation theatre for adversarial actions.
- Link to the CNI: Many space systems provide the functionality for other aspects of the CNI, so there is a need to analyse knock-on effects. The space industry is of increasing importance to nations around the world, and for many states, the space sector and its associated infrastructure are considered to be part of the CNI. According to the UK Centre for the Protection of National Infrastructure, CNI is defined as 'the facilities, systems, sites, information, people, networks and processes, necessary for a country to function and upon which daily life depends' [42]. As such, securing space-related CNI is a key priority for the UK Government.
- Wireless communication: Space systems will typically involve wireless communication, so threat modelling needs to take this into account [43].
- Utilising sensors: The majority of space applications involve some form of sensing (e.g., surveillance of the environment, scientific objectives, planetary objectives) [44]. It is expected that the adoption of autonomy and communication

capabilities of the robotic systems will significantly increase in the future. Thus, it may change the existing ground segment-based control model.

- Extreme environmental conditions: The space systems are exposed to many unique environmental threats which may be degrading for materials and obstruction for the system running. The environmental threats include vacuum, intense ultraviolet radiation, ionising radiation, electrostatic charge, micro-meteoroids, debris impacts and thermal cycling (commonly between −175 and 160 °C) [45].
- Human-in-the-loop: Many space systems are expected to act without a human-in-the-loop, or work with a human operator with respect to supporting the decision-making by recommendations. Such space systems may facilitate human–machine interaction [46].
- Very low fault tolerance: Space missions typically tolerate very low likelihoods of failure due to the high cost of missions. Cyber security threats may or may not be treated in the same way [47].
- Long lifespan: Existing space systems are relatively simple as they typically involve satellites communicating with the ground segment. However, future space systems will encounter much greater complexity with satellite-to-satellite communications, autonomous interactions and further deployments elsewhere in the solar system. The threat modelling approaches will need to be able to model the potential issues of these space systems which are expected to be deployed in the near future and encounter the existing and emerging threat actors.

The selected TM approach should cover the cyber security requirements [48] used in the evaluation metrics to evaluate several common TM approaches from the CPS perspective. We have derived our evaluation metrics from the metrics from this study such as, *maturity, adjustability, coverage of safety-security dependency, coverage of hardware and software threats* and *documentation*. As a conclusion, the TM approach should be *mature*, which refers how well an approach is defined and how often it has been applied to similar use cases; *adjustable*, which refers to how flexible is an approach to be tailored for the specific needs of the system; *cover the safety-security dependency* due to the link to CNI and human-in-the-loop; *cover both the hardware and the software threats* since we have focused on the functions in our RA which includes physical and logical instantiations of the system; and finally *the documentation* of the approach must be sufficiently rich.

12.2.2   Evaluation of threat modelling approaches

The commonly used TM approaches, which can be potentially used for New Space systems, are reviewed and the distinguishing features are highlighted in Table 12.2. Among the selected threat modelling approaches, STRIDE, PASTA, OCTAVE and composite threat modelling provide structured frameworks, which may ease the systematic implementation by time and the growth of the system. STRIDE, PASTA and composite threat modelling frameworks include data flow diagrams. The maturity level of space-domain specific threat assessment approach has been considered as

Table 12.2   Summary of the threat modelling approaches

Threat modelling approach	Feature
Attack trees [7]	• Easy to adopt • Helps to understand the system, and identify the treatments for risk mitigation
Composite threat modelling [49]	• Uses data flow diagrams • Borrows the threat method classification from STRIDE • Tailored for future vehicular systems
OCTAVE [50]	• Helps to identify the treatments for risk mitigation • Directly intended to build risk management • Designed to be scalable • Implementation takes a lot of time, and the documentation is not precisely clear
PASTA [22]	• Helps to identify the treatments for risk mitigation • Directly intended to build risk management. • Has a rich documentation • Requires an extensive labour
SAE J3061-TARA [51]	• Provides a classification of cyberattack impacts as operational, safety, financial and privacy • Uses STRIDE threat classification
Space-Domain Specific Threat Assessment [52, Section 4]	• Uses the generic threat assessment methodologies of NIST SP 800-30 [53] and ISO 27005:2011 [54] • Makes a classification of space missions • Makes a list of generic applicable threats for each mission and assesses the occurrence possibilities
STRIDE [21]	• Helps to identify the treatments for risk mitigation • Has been commonly used in the literature • Has a threat classification used in some other approaches as well

not accomplished. This is due to the fact that it has not considered the evolution of space systems over time while the main focus has been on a generic threat identification process.

Composite threat modelling and SAE J3061-TARA has been specifically tailored for the future transportation systems. OCTAVE has been developed to be an iterative process for organisations for cyber risk mitigation rather than specific systems. Among the remaining approaches PASTA and STRIDE use data flow diagrams in their procedure which may help to analyse threats against CPS. STRIDE has been mainly developed for information systems rather than CPS. STRIDE variants are STRIDE-per-element and STRIDE-per-interaction. In the first case, the threat analysis is focused on how an adversary can interfere with a data flow using certain elements such as external entities, data stores, data flows and processes. The latter enumerates threats by tuples of origin, destination and interaction. It mainly focuses on the main elements and the interaction that each element has and maps the STRIDE threats applicable to each interaction. Attack trees have been considered as an invaluable tool towards the understanding of a system including a physical viewpoint as well. The framework of PASTA also contains generating attack trees.

Unfortunately, none of these approaches fully meet the cyber security requirements of the New Space systems, based on the evaluation in Table 12.3. In this work, we choose the PASTA approach as the fundamental framework of our TM approach and customise the framework by incorporating EVITA impact analysis framework from SAE J3061-TARA. The EVITA framework provides a severity classification scheme for the potential impact assessment of cyberattacks relevant to our use case. It includes safety, operational, financial and privacy aspects of the impact. In the original EVITA framework, the impact aspects are scaled from 0 to 4, and they are accumulated to calculate total impact value. Analysing the overall impact and its constituent parts (safety, financial and operational) is important. However, the potential impact aspects of a cyberattack against the New Space systems must be broken down into direct and indirect impacts as they are linked to the CNI. The chosen approach is a variant of PASTA using our RASA for system model, STRIDE threat classification and a variant of SAE J3061-TARA to classify impact: safety $\otimes$ financial $\otimes$ operational.

12.3   Risk management

The terms threat and risk are often confusingly misused or interchanged by practitioners. ISO/IEC 27005 defined threat as the potential reason for an incident that may cause harm to a system and organisation [54]. However, the risk is defined as the likelihood that a particular threat maliciously impacts a system by exploiting a particular vulnerability by the the Consultative Committee for Space Data Systems (CCSDS) [52]. Risk management strategies generally include mitigating, avoiding, transferring or accepting risk with regard to the likelihood of the cyberattack and impact on the system. Before developing a strategy for risk management to ensure resilience of space systems, it is important to assess the extent of the risk.

It is a challenging task to conduct an accurate risk assessment. In simple terms, the risk is a function of likelihood and impact *Risk = Likelihood × Impact* (*i.e. Harm*), whereas the threat is *Threat = Capability × Intent* [52]. The risk assessment aims to answer these fundamental questions [55]:

- What can go wrong?
- What is the likelihood that it would go wrong?
- What are the consequences?
- Did you validate threats and what lessons have you learnt?

However, it requires an expert knowledge and measurable data to quantitatively answer these questions. Otherwise, the findings of a risk assessment may be misleading, providing a false snapshot for a real situation [56]. Risk assessment

Table 12.3 Summary of the evaluation

Threat modelling approach	Maturity	Adjustability	Coverage of safety-security dependency	Coverage of hardware and software threats	Documentation
Attack trees [7]	✓				✓
Composite threat modelling [49]	✓		✓	✓	✓
OCTAVE [50]	✓			✓	✓
PASTA [22]	✓	✓			✓
SAE J3061-TARA [51]	✓		✓	✓	✓
Space-Domain Specific Threat Assessment [52] Section 4			✓	✓	✓
STRIDE [21]	✓	✓			✓

is challenging for such complex systems because: (i) the likelihood and impact of future cyberattacks are dependent on uncertain variables and therefore difficult to measure; (ii) the quantitative risk assessments are usually not very effective due to the reasons that it is hard to evaluate the actual harm of an incident with a concrete metric and measure the consequence of human interaction and (iii) the threats against cyber systems can swiftly evolve, unlike any other ecosystems. For these reasons, risk values are not calculated in the CCSDS 350.1-G-2 Security Threats Against Space Missions Report [52].

In summary, there is much uncertainty in evaluating the potential impact and securing systems for an attack that has not happened yet. Complicating these matters further, if a risk mitigation countermeasure consumes resources that would have otherwise been deployed on accomplishing a business objective, the asset is also likely to accept the risks [57]. The qualitative risk-assessment methods are usually in use for industrial operations. Although they are not comprehensive but require particular format for situations, they can still give a helpful insight into and the efficient management of risks. The space systems are usually more complex than industrial operations. Thus, it is more challenging to make an effective risk assessment on time [58].

The ISO/IEC 27005 standard presents a generic framework for risk management which has considerably large room for interpretation of implementer organisations. In this regard, Mehari is designed to provide a formally structured approach that is compliant to the requirements of ISO/IEC 27005. It is a de facto methodology in risk management of space systems as it aims to form a formally acceptable risk quantification and to determine which assets or subsystems should be secured against which threat actors [59]. According to the designers, a risk management methodology can be implemented based on three models depending on the needs of the organisation:

- a permanent working model
- a model that works in parallel with other security management practices
- an occasionally working model for regular practices.

Mehari enables the evaluation of the security risks based on qualitative and quantitative analysis of organisational and risk-reduction factors. These factors include the knowledge bases such as manuals and guides to describe different modules (stakes, risks, vulnerabilities) which are used in the tasks and actions of security management [60, 61]. The methodology is composed of three main phases [62]:

1. Risk assessment: Identification, Analysis, Assessment
2. Risk treatment options: Accept, Reduce, Transfer, Avoid
3. Risk management process: Define the plan, Implement the plan, Monitor and review.

NASA has published a handbook for risk management in their organisational and procedural requirements which consists of two stages: (i) Risk-Informed Decision-Making and (ii) Continuous Risk Management. The handbook classifies the performance shortfalls of the missions into safety, technical, cost and schedule. These stages jointly work to achieve effective risk management for the projects and programs of NASA. The first stage aims to address the risk-informed selection of decisions to ensure effectively achieving the objectives. The second stage aims to address the challenges in the implementation of the selected decisions to meet the requirements. The risks are classified as: (i) the scenarios which lead to performance shortfall because of injury, fatality and destruction of key assets; scenarios leading to exceeding mass limits; scenarios leading to cost overruns and scenarios leading to schedule slippage, (ii) the qualitative or quantitative likelihood of these risk scenarios and (iii) the consequence in terms of qualitative or quantitative severity of performance shortfall [63].

Applying an effective risk-assessment framework requires to follow cost analysis and the engagement of senior management. Kapalidis *et al.* [2]. proposed the Cyber Risk Management Process which is drawn by the US National Institute of Standards and Technology (NIST) Cybersecurity Framework and structured around the cyber security principles in subsection 2.1. The steps of the framework as follows:

1. Identification of threat: The profiles of threat actors vary from nation state level to the individual hacktivist as it is explained in detail in Table 12.1. important to understand the motivation, equipment and profiles of the threat actors to be able to make an effective risk assessment [57].
2. Identification of vulnerabilities: Identify vulnerabilities needs a comprehensive mapping of existing components, applications and communication protocols employed in a system. However, system mapping is often not done at the system development stages, which makes it more difficult to assess the risks when the vulnerabilities are later identified.
3. Assessment of risk exposure: The potential incidents should be identified, and the potential impacts should be analysed and compared based on the likelihood of the vulnerabilities being exploited.
4. Development of protection measures: Both physical and software viewpoints should be considered in the development of the appropriate countermeasures to mitigate the likelihood of the risks due to the exploited vulnerabilities.
5. Development of contingency plans: Design an effective contingency plan in which the prioritisation of response actions should be defined beforehand according to the significance of each vulnerability discovered in a space system. The cyber security requirements should also be considered when the response prioritisation is done.
6. Establishing incident response and post-event recovery mechanisms: It is the final step of the framework which is important for having a decisive effect on the impact mitigation of a cyberattack.

To ensure the security of robots, the methodologies, tools and development frameworks used should be secured as well [64]. Robotic Development Frameworks (RDF) are designed to simplify the development of robotic applications by providing code reusability, hardware abstraction and knowledge sharing about the developer communities. RDF also bring cyber security risk to the robotic domains such as code reuse without appropriate mechanisms increase the likelihood of malware injection. However, the major RDF does not include security mechanisms in their design frameworks, and similarly Robotic Operating Systems are designed without integrating sufficient security measures and tools [65]. Robots have performed significant roles in a wide range of applications in the space from mobility systems to robotic arms and manipulators. Besides that future challenging space missions require robots which can employ a higher level of autonomy and sensing on-board. NASA addressed the requirement of a new design standard development for robots and system integration in space missions [66]. Gao *et al.* classified the development need areas of space robotic and autonomous systems with respect to their objectives and challenges as (i) sensing and perception, (ii) mobility, (iii) high-level autonomy for system and subsystems, (iv) astronaut-to-robot and robot-to-robot interaction and (v) system engineering [67]. Such advancements in the space robotics and autonomous systems are also increasing the need for cyber security measures and tools.

12.4   Security-minded verification of space systems

Existing and emerging space systems are expanding in terms of complexity by deploying machine learning-based autonomous systems and IoST. Such a complex progress coupled with the commercialisation of space has resulted in an ecosystem that presents new security challenges. Verification strategies are commonly deployed to ensure and validate functional and safety properties of a system. However, the definition of a precise and systematic way to carry out verification of complex systems still remains an open research challenge [68]. On the other hand, informal methods such as simulation are recognised to have limitations, in particular, their effectiveness to discover corner-case bugs decreases over time [69]. While it is generally challenging to employ formal verification of a system's operations, it becomes even more complicated when taking malicious interaction into account for such systems.

There is an increasing interest in autonomy in the space industry. This is partially attributed to issues related to remote operations, time lags, distance and communications. Recent studies of formal specification and verification techniques [70, 71] have identified that improvement is still required prior to deploying them in such large, complex and autonomous systems. For an extensive review of formal verification in space systems, the reader is advised to see works in [72–77]. Similarly, cyber security of space systems has been studied in [52, 58, 78–81]. Although security and formal verification are important aspects towards the development of resilient space systems in the presence of an adversary, a combined methodology has not yet been formalised for the space sector.

12.4.1   Security-minded verification methodology

In general, formal verification and threat modelling are distinct processes which are usually performed independent of one another. However, both processes may

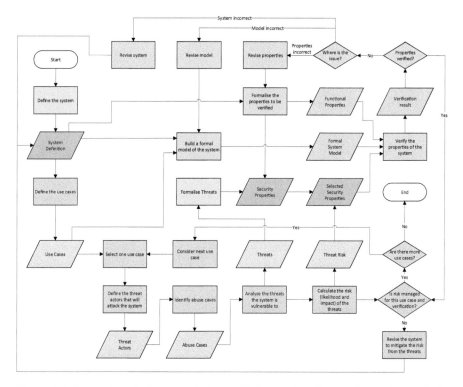

Figure 12.4   *Integrated approach to combining verification and security analysis which allows to use results from the threat modelling [4]*

seriously alter the state of a system while they dictate iterations until the system meets the predefined requirements. Following such an approach may be ineffective since formal verification and threat modelling may encourage teams to make significant changes to existing systems despite the two approaches potentially not converging on similar matters.

The security-minded verification methodology illustrated in Figure 12.4 seeks to integrate formal verification with threat modelling [4]. The fundamental change in the methodology is that the security properties have been formalised and checked as part of the verification process. The proper quantification of cyber risks enables the prioritisation and verification of security properties. By assessing the cyber security risks, it is possible to select which security properties to be prioritised and formally verified. The formal verification methodology usually begins with the system definition derived from the formal model and properties to be verified. Similarly, threat modelling usually begins with the system definition derived from a cyber security perspective of the use case. Such similarity provides a unified starting point for the security-minded verification methodology. The results from the threat model are used to devise the formal security properties at the *Formalise Threats* step in Figure 12.4. At the *Threat Risk* step, it is required to select which of these security

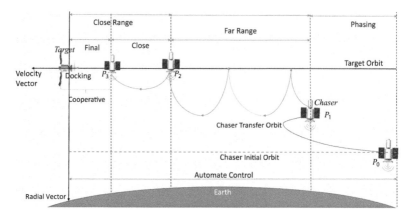

Figure 12.5   *Autonomous on-orbit docking between a chaser and a target is comprised multiple stages including far range and close range rendezvous [4]*

properties need to be prioritised and subsequently formally verified. If the verification is successfully completed, it should then be checked for whether the risk is managed and proceed to examine further use cases. Security-minded verification has been applied to an autonomous docking use case of two satellites in Figure 12.5 [4]. Maple *et al.* introduced the novel definition of an anti-cooperative robot (i.e., satellite in this use case), which is a compromised robot and under the control of a threat actor to complement the definitions of cooperative, non-cooperative and semi-cooperative robots. The target satellite is cooperative and maintains its attitude while the chaser satellite attempts to dock. However, the target may be an anti-cooperative satellite and actively tries to avoid docking to cause a violation of fuel consumption.

The potential threat actors and their motivations are distinguished in Table 12.1 for the use case as privileged insider, competitor and nation state. Two likely scenarios have analysed that a threat actor may attempt to compromise the system either (i) by forcing the chaser satellite to consume excessively or (ii) by forcing the chaser satellite out of its intended region. The following security properties and functional properties are formalised in the formal model based on these threats:

- SEC 1: Maximum x kg of fuel can be consumed to attempt to dock.
- SEC 2: After successfully docking or aborted docking, the position (or orbit) cannot be more than $\pm y\%$ of the desired position (or orbit).
- FUN 1: The distance between the chaser and the target satellites is always decreasing.
- FUN 2: The satellites can successfully dock only if the appropriate conditions are met.

These properties should hold in a successful docking. The reader is advised to see works in [4] for an explanation of the methodology and the use case in detail.

The approach is novel at linking the independent methodologies for threat analysis and formal verification. A more similar approach is the combination of security and verification to extract finite state machines from bank cards using machine learning in EMV (Europay-MasterCard-Visa) protocol [82, 83].

12.5   Discussion

Historically, the legacy space industry prided itself on meticulous safety vetting of systems commissioned for missions, focusing more on scientific discovery and national security, and less on profitability and scale. The New Space market is accelerating at an unprecedented pace, which has been accompanied by untested vendors and developers of space assets racing towards scale and profit. While scale and profitability do not preclude meticulous development, such attention to safety comes with a cost, which is less tolerable for the New Space industry. Security is a recent addition to the scope of safe space asset operations and will undoubtedly add resource and associated cost overhead to mission assets. Security should not stifle innovation and progress in the space industry; yet, it must be addressed at scale across the various assets that will be launched in the coming years. The proposed RA to aid in discovering attack surface for space assets will make security accessible to New Space companies that may have otherwise dismissed the need for security. Like with industrial control systems, the barrier to entry to secure space assets is high. By reducing the barrier to security engagement through providing resources such as the RA and threat assessments that can be leveraged for New Space assets, companies do not have to deploy considerable capital to conduct precautionary security checks on their space assets. Enabling the New Space industry with tools and methods as done in this paper will help to foster growth, while helping to maintain the safety-centric legacy of the space industry.

RASA focuses on the functions of system components and interactions between functions which comprise entities in a space system. The components related to human aspects in space missions is purposely not included to concentrate on the robotic and autonomous systems in space. However, RASA can be extended to comprehend the scenarios including life support systems in orbits or on a planetary surface. RASA is designed to be used for high-level analysis, but not suitable for low-level analysis since the potential threat vectors due to the implementation of systems such as vulnerabilities in software and supply chain and the other implementation.

Our security-minded verification methodology aims to ensure robust security and verification with short iterations and quick convergence on the requirements by combining threat modelling and formal verification. An autonomous docking use case is explored for the use of the methodology yet. However, it is planned to develop detailed formal models to perform verification of the properties for future works.

12.6   Conclusion

In this chapter, we have analysed the current changes in the space industry driven by the commercialisation of the sector and technologies such as the connectivity of space systems and the autonomy of those systems. Many dimensions of the space ecosystem are affected by those changes including the cyber security landscape. We have introduced a reference architecture (RASA) from a functional-interaction viewpoint to analyse paths of cyber security threats. We have demonstrated the utility of the RASA on an autonomous debris collection use case. Then, we developed a cyber security requirement analysis for New Space systems and reviewed the existing threat modelling approaches and challenges of cyber security risk management. Finally, we have proposed a novel security-minded verification methodology that combines the distinct methodologies for threat modelling and formal verification. The methodology is demonstrated in the autonomous docking use case. Our works aimed to analyse the changing cyber security landscape of New Space systems, and propose tools and methods to be used in future projects.

References

[1]   Bradbury M., Maple C., Yuan H., *et al*. Identifying attack surfaces in the evolving space industry using reference architectures,. 2020 IEEE Aerospace Conference; 2020.

[2]   Kapalidis C., Maple C., Bradbury M., *et al*. 'Cyber risk management in satellite systems'. *Living in the Internet of Things*. 2019:1–8.

[3]   Falco G., Eling M., Jablanski D., *et al*. 'Cyber risk research impeded by disciplinary barriers'. *Science*. 2019;**366**(6469):1066–9.

[4]   Maple C., Bradbury M., Yuan H., *et al*. 'Security-minded verification of space systems'. 2020 IEEE Aerospace Conference; 2020.

[5]   Weyrich M., Ebert C. 'Reference architectures for the Internet of things'. *IEEE Software*. 2016;**33**(1):112–6.

[6]   Lin S.W., Miller B., Durand J. *Industrial Internet Reference Architecture. Industrial Internet Consortium (IIC). Tech Rep*; 2015.

[7]   Secrets S.B. *Lies: Digital security in a networked world*. New York, NY, USA: John Wiley & Sons, Inc; 2000.

[8]   Alaña E., Herrero J., Urueña S. 'A reference architecture for space systems'. *Proceedings of the 12th European Conference on Software Architecture: Companion Proceedings. ECSA'18*. New York, NY, USA: ACM; 2018. pp. 11:1–11:2.

[9]   Bos V., Rugina A., Trcka A. 'On-board software reference architecture for payloads'. *DASIA 2016 - Data Systems in Aerospace*. **736**. of ESA Special Publication; 2016.

[10] Panunzio M., Vardanega T. 'On software reference architectures and their application to the space domain,' in Favaro J., Morisio M. (eds.). *Safe and Secure Software Reuse*. Springer Berlin Heidelberg; 2013. pp. 144–59.

[11] European Space Software Repository. *OSRA – Onboard Software Reference Architecture [online]*. 2019. Available from https://essr.esa.int/project/osra-onboard-software-reference-architecture [Accessed 09 sep 2019].

[12] Shames P., Yamada T. *Tools for describing the reference architecture for space data systems. Jet Propulsion Laboratory*. 2004[online]. Available from https://trs.jpl.nasa.gov/bitstream/handle/2014/38436/04-0804.pdf?sequence=1&isAllowed=y.

[13] The ISECG reference architecture for human lunar exploration. 2010ISECG International Architecture Working Group. Available from https://www.lpi.usra.edu/lunar/strategies/ISECGLunarRefArchitectureJuly2010.pdf [Accessed 15 May 2021].

[14] Drake B.G., Hoffman S.J., Beaty D.W. 'Human exploration of Mars, design reference architecture 5.0'. 2010 IEEE Aerospace Conference; 2010. pp. 1–24.

[15] Bonaci T., Herron J., Yusuf T., *et al.* 'To make a robot secure: an experimental analysis of cyber security threats against teleoperated surgical robots'. *arXiv preprint arXiv*. 2015;**150404339**.

[16] Schaub H., Jasper L.E.Z., Anderson P.V., *et al.* 'Cost and risk assessment for spacecraft operation decisions caused by the space debris environment'. *Acta Astronautica*. 2015;**113**(5–6):66–79.

[17] Jehn R., Viñals Larruga S., Klinkrad H. 'Discos – the European space debris database'. *44th Congress of the International Astronautical Federation*; 1993. pp. 93–742.

[18] Forshaw J.L., Aglietti G.S., Navarathinam N., *et al.* 'RemoveDEBRIS: an in-orbit active debris removal demonstration mission'. *Acta Astronautica*. 2016;**127**(1):448–63.

[19] Nishida S.-I., Kawamoto S., Okawa Y., *et al.* 'Space debris removal system using a small satellite'. *Acta Astronautica*. 2009;**65**(1):95–102.

[20] Freiwald D.A., Freiwald J. 'Range-gated laser and ICCD camera system for on-orbit detection of small space debris'. *Proc. SPIE 2214, Space Instrumentation and Dual-Use Technologies*1994. p. 116.

[21] Shostack A. *Experiences Threat Modeling at Microsoft*. International Conference on Model Driven Engineering Languages and Systems (MODELS), Workshop on Modeling Security (MODSEC08); Toulouse, France, 22; 2008.

[22] UcedaVelez T., Morana M.M. *Risk centric threat modeling: process for attack simulation and threat analysis*. John Wiley & Sons; 2015.

[23] Klesh A.T., Cutler J.W., Atkins E.M. 'Cyber-physical challenges for space systems'. *2012 IEEE/ACM Third International Conference on Cyber-Physical Systems. IEEE*; 2012. pp. 45–52.

[24] Alguliyev R., Imamverdiyev Y., Sukhostat L. 'Cyber-physical systems and their security issues'. *Computers in Industry*. 2018;**100**(5):212–23.

[25] Epiphaniou G., Pillai P., Bottarelli M., Al-Khateeb H., Hammoudesh M., Maple C. 'Electronic regulation of data sharing and processing using smart ledger technologies for supply-chain security'. *IEEE Transactions on Engineering Management*. 2020;**67**(4):1059–73.

[26]   Ross R.S. ''Guide for conducting risk assessments'. National Institute of Standards and Technology'. *SP 800-30 Rev.* 2012;**1**.

[27]   Do Q., Martini B., Choo K.-K.R. 'The role of the adversary model in applied security research'. *Computers & Security*. 2019;**81**(8):156–81.

[28]   Bishop M. 'What is computer security?' *IEEE Security & Privacy*. 2003;**1**(1):67–9.

[29]   von Solms R., van Niekerk J. 'From information security to cyber security'. *Computers & Security*. 2013;**38**(6):97–102.

[30]   Kissel R. *NISTIR 7298: 'Glossary of key information security terms, Revision 2'.United States Department of Commerce*. National Institute of Standards and Technology; 2013.

[31]   Suh S.C., Tanik U.J., Carbone J.N. *Applied cyber-physical systems*. Springer Publishing Company, Incorporated; 2013.

[32]   Aleem A., Wakefield A., Button M. 'Addressing the weakest link: implementing converged security'. *Security Journal*. 2013;**26**(3):236–48.

[33]   Pham N., Abdelzaher T., Nath S. 'On bounding data stream privacy in distributed cyber-physical systems'. 2010 IEEE International Conference on Sensor Networks, Ubiquitous, and Trustworthy Computing; 2010. pp. 221–8.

[34]   Madden J., McMillin B., Sinha A. 'Environmental obfuscation of a cyber physical system-vehicle example'. *2010 IEEE 34th A*nnual Computer Software and Applications Conference Workshops. IEEE; 2010. pp. 176–81.

[35]   Work D., Bayen A., Jacobson Q. 'Automotive cyber physical systems in the context of human mobility'. *National Workshop On High-Confidence Automotive Cyber-Physical Systems*; 2008. pp. 3–4.

[36]   Banerjee A., Venkatasubramanian K.K., Mukherjee T., *et al.* 'Ensuring safety, security, and sustainability of mission-critical cyber–physical systems'. *Proceedings of the IEEE*. 2011;**100**(1):283–99.

[37]   CCSDS. *CCSDS guide for secure system interconnection*. The Consultative Committee for Space Data Systems (CCSDS); 2019. CCSDS 350.4-G-2. 2019[online]. Available from https://public.ccsds.org/Pubs/350x4g2.pdf.

[38]   RISI. *Industry attacks growing*. 2013[online]. Available from https://iss-source.com/risi-industry-attacks-growing/.

[39]   Henrie M. 'Cyber security risk management in the SCADA critical infrastructure environment'. *Engineering Management Journal*. 2013;**25**(2):38–45.

[40]   Maybury M. 'Toward the assured cyberspace advantage: air force cyber vision 2025'. *IEEE Security & Privacy*. 2013;**13**(1):49–56.

[41]   Gorman A. 'The archaeology of orbital space'. *In: th NSSA Australian Space Science Conference: 14 to 16 September 2005 Hosted by RMIT University*. **338**. Melbourne, Australia: RMIT University; 2005.

[42]   CPNI. *Critical national infrastructure*. Available from https://www.cpni.gov.uk/critical-national-infrastructure-0.

[43]   Sheynblat L., SnapTrack Inc, 1999. *Satellite positioning system augmentation with wireless communication signals*. U.S. Patent 5,999, 124.

[44] Hunter G., Xu J., Dungan L., *et al.* 'Smart sensor systems for aerospace applications: from sensor development to application testing'. *ECS Transactions.* 2008;**16**(11):333–44.

[45] Yang J C., de Groh K. 'Materials issues in the space environment'. *MRS Bulletin.* 2010;**01**(35):12–19.

[46] Hall L. Human-in-the-loop decision support. NASA. 2016[online]. Available from https://www.nasa.gov/directorates/spacetech/esi/esi2016/Human-in-the-loop_Decision_Support/.

[47] Farooq M., Iqbal M.W., Rana T.A. *'Comparative analysis of fault-tolerance techniques for space applications'.*VFAST Transactions on Software Engineering; 2014. pp. 1–10.

[48] Shevchenko N., Frye B., Woody C. *Threat modelling for cyber-physical system-of-systems:* Methods Evaluation. Software Engineering Institute, Carnegie Mellon University. 2018. Available from https://resources.sei.cmu.edu/library/asset-view.cfm?assetid=526365.

[49] McCarthy C., Harnett K., Carter A., *et al. Characterization of potential security threats in modern automobiles: A composite modeling approach.* United States: National Highway Traffic Safety Administration; 2014.

[50] Alberts C., Dorofee A., Stevens J., *et al. Introduction to the OCTAVE approach.* Carnegie-Mellon University; 2003.

[51] SAE J3061 Vehicle Cybersecurity Systems Engineering Committee. 'Cybersecurity Guidebook for Cyber-Physical Vehicle Systems'. *SAE International*; 2016.

[52] CCSDS. *Security threats against space missions.* The Consultative Committee for Space Data Systems (CCSDS); 2015. CCSDS 350.0-G-3. 2015[online]. Available from https://public.ccsds.org/Pubs/350x1g2.pdf.

[53] NIST S. '800-30 Risk Management Guide for Information Technology Systems'. *National Institute for Standards and Technology*; 2002.

[54] ISO/IEC. *ISO/IEC 27005: 2011 Information technology - Security techniques- Information security risk management*; 2011.

[55] Kaplan S., Garrick B.J. 'On the quantitative definition of risk'. *Risk Analysis.* 1981;**1**(1):11–27.

[56] Cherdantseva Y., Burnap P., Blyth A., *et al.* 'A review of cyber security risk assessment methods for SCADA systems'. *Computers & Security.* 2016;**56**(9):1–27.

[57] Parker D.B. 'Risks of risk-based security'. *Communications of the ACM.* 2007;**50**(3):120.

[58] Livingstone D., Patricia L. 'Space, the final frontier for cybersecurity?' *Chatham House.* 2016.

[59] Fischer D., Spada M., Wallum M. 'Building blocks to support implementation of a secure system engineering lifecycle at ESA'. 2018 SpaceOps Conference; 2018. p. 2479.

[60] Jouas J.P., Roule J.L., Buc D., *et al. MEHARI – overview.* France: CLUSIF; 2019.

[61]   Syalim A., Hori Y., Sakurai K. 'Comparison of risk analysis methods: Mehari, Magerit, NIST800-30 and microsoft's security management guide'. International Conference on Availability, Reliability and Security; 2009. pp. 726–31.

[62]   Roule J.L., Jouas J.P. *MEHARI – Risk analysis and treatment guide*. France: CLUSIF; 2010.

[63]   Dezfuli H., Benjamin A., Everett C. NASA risk management Handbook. version 1.0. NASA/SP-2011-3422. Washington, D.C: National Aeronautics and Space Administration NASA Headquarters; 2011. Available from https://ntrs.nasa.gov/api/citations/20120000033/downloads/20120000033.pdf [Accessed 12 May 2021].

[64]   Cerrudo C., Apa L. Hacking robots before Skynet. IOActive; 2017. pp. 1–17. Available from https://ioactive.com/pdfs/Hacking-Robots-Before-Skynet.pdf [Accessed 12 May 2021].

[65]   Lera F.J.R., Balsa J., Casado F., *et al. Cybersecurity in autonomous systems: evaluating the performance of hardening ROS.* **47**. Málaga, Spain; 2016.

[66]   Dischinger H.C., Mullins J.B. *A Robotics Systems Design Need: A Design Standard to Provide the Systems Focus that Is Required for Longterm Exploration Efforts*. SAE Technical Paper; 2005.

[67]   Gao Y., Jones D., Ward R., *et al. Space Robotics and Autonomous Systems: Widening the Horizon of Space Exploration. UK-RAS White Paper*; 2016.

[68]   Poddey A., Brade T., Stellet J.E., *et al.* 'On the validation of complex systems operating in open contexts'. *arXiv preprint arXiv*. 2019;**190210517**.

[69]   Bhadra J., Abadir M.S., Wang L.-C., Ray S., *et al.* 'A survey of hybrid techniques for functional verification'. *IEEE Design & Test of Computers*. 2007;**24**(02):112–22.

[70]   Luckcuck M., Farrell M., Dennis L.A.,*et al.* 'Formal specification and verification of autonomous robotic systems'. *ACM Computing Surveys*. 2019;**52**(5):1–41.

[71]   Farrell M., Luckcuck M., Fisher M. 'Robotics and ntegrated formal methods: necessity meets opportunity'. *Integrated Formal Methods*Springer; 2018. pp. 161–71.

[72]   Brat G., Drusinsky D., Giannakopoulou D., *et al.* 'Experimental evaluation of verification and validation tools on Martian rover software'. *Formal Methods in System Design*. 2004;**25**(2-3):167–98.

[73]   Brat G., Denney E., Giannakopoulou D. 'Verification of autonomous systems for space applications'. *IEEE Aerospace Conference*. IEEE; 2006.

[74]   Alves M.C.B., Drusinsky D., Michael J.B., *et al.* 'Formal validation and verification of space flight software using statechart-assertions and runtime execution monitoring'. International Conference on System of Systems Engineering; 2011. pp. 155–60.

[75]   Tarasyuk A., Pereverzeva I., Troubitsyna E., *et al.* 'Formal development and assessment of a reconfigurable on-board satellite system'. International Conference on Computer Safety, Reliability, and Security. vol. 7612 of LNCS; 2012. pp. 210–22.

[76] Schumann J., Moosbrugger P., Rozier K.Y. 'R2U2: monitoring and diagnosis of security threats for unmanned aerial systems'. *Runtime Verification. Vol. 9333 of LNCS.* Springer; 2015. pp. 233–49.

[77] Rozier K.Y., Schumann J. 'R2U2 in Space: system and software health management for small satellites. *9th Annual Workshop on Spacecraft Flight Software*2016.

[78] Housen-Couriel D. 'Cybersecurity and anti-satellite capabilities (ASAT) new threats and new legal responses'. *Journal of Law & Cyber Warfare.* 2015;**4**(3):116–49.

[79] Drozhzhin A. *Russian-speaking cyber spies from Turla APT group exploit satellites.* 2015[online]. Available from https://www.kaspersky.co.uk/blog/turla-apt-exploiting-satellites/6210/.

[80] Falco G. 'Cybersecurity principles for space systems'. *Journal of Aerospace Information Systems.* 2019;**16**(2):61–70.

[81] Harrison T., Johnson K., Roberts T.G. *'Space threat assessment 2018'.* Center for Strategic & International Studies; 2018.

[82] Aarts F., De Ruiter J., Poll E. 'Formal models of bank cards for free'. 2013 IEEE Sixth International Conference on Software Testing, Verification and Validation Workshops. IEEE; 2013. pp. 461–8.

[83] De Ruiter J., Poll E. 'Formal analysis of the EMV protocol suite'. *Joint Workshop on Theory of Security and Applications.* Springer; 2011. pp. 113–29.

Index